21世纪软件工程专业规划教材

Java Web开发技术

李雷孝 邢红梅 王 慧 编著

清华大学出版社
北京

内 容 简 介

本书涵盖了 Java Web 开发技术的全部知识点，内容由浅入深，主要包括：Web 编程基础、Servlet 基础、状态管理与作用域对象、JSP 语法基础、过滤器和监听器、JavaBean 组件、MVC 设计模式、EL 表达式、JSP 标签、Java Web 中的中文乱码处理、异常处理等。

全书结构严谨，层次清晰，语言生动，理论论述精准深刻，程序实例丰富实用。本书要求读者具有 Java 语言开发编程基础，通过学习本书，读者可以具备使用 Java Web 技术进行应用开发的能力。

本书不仅适合用作普通高校或者职业培训教材，更是 Java Web 初学者和程序员的首选参考书。

本书封面贴有清华大学出版社防伪标签，无标签者不得销售。

版权所有，侵权必究。举报：010-62782989，beiqinquan@tup.tsinghua.edu.cn。

图书在版编目(CIP)数据

Java Web 开发技术/李雷孝，邢红梅，王慧编著．--北京：清华大学出版社，2015（2024.8重印）
21 世纪软件工程专业规划教材
ISBN 978-7-302-39958-2

Ⅰ．①J… Ⅱ．①李… ②邢… ③王… Ⅲ．①JAVA 语言－程序设计－高等学校－教材
Ⅳ．①TP312

中国版本图书馆 CIP 数据核字(2015)第 085909 号

责任编辑：张　玥　薛　阳
封面设计：常雪影
责任校对：胡伟民
责任印制：沈　露

出版发行：清华大学出版社
　　网　　　址：https://www.tup.com.cn, https://www.wqxuetang.com
　　地　　　址：北京清华大学学研大厦 A 座　　邮　编：100084
　　社　总　机：010-83470000　　邮　购：010-62786544
　　投稿与读者服务：010-62776969, c-service@tup.tsinghua.edu.cn
　　质　量　反　馈：010-62772015, zhiliang@tup.tsinghua.edu.cn
　　课　件　下　载：https://www.tup.com.cn, 010-83470236

印　装　者：北京建宏印刷有限公司
经　　　销：全国新华书店
开　　　本：185mm×260mm　　印　张：23.25　　字　数：580 千字
版　　　次：2015 年 6 月第 1 版　　印　次：2024 年 8 月第 10 次印刷
定　　　价：69.50 元

产品编号：063161-02

Java 语言以其简单易学、适用范围广泛等优点,成为近年来最为流行的编程语言之一。2013 年 8 月,TIOBE 公布了编程语言排行榜,Java 语言位列榜首。近年来,Java 在 TIOBE 公布的排名中,始终保持位列三甲的成绩。尤其在 Web 应用开发方面,Java 更具有得天独厚的优势。随着 Java 语言的推广和应用,各种针对 Web 开发的 Java Web 技术也应运而生。本书从 Java Web 开发的基础技术入手,以实际工程项目为主线,重点讲解了 Java Web 开发技术知识点在实际项目开发中的应用。

本书是一本既培养学生软件开发技术,又培养学生工程实践能力的教材。教材以 IT 企业对开发人员技术能力要求为基础,以工程能力培养为目标,梳理了软件工程对计算机语言要求的知识点,并形成相应知识单元;按照工程需求顺序进行课程内容组织,便于学习和掌握;本书提供一定量的案例,注重实践能力的培养。使用本教材,可以提高学生的工程能力和软件开发能力。本书既可以作为计算机类专业各层次学生教材,还可以作为 Java Web 应用开发者参考用书。

全书共分为 11 章,章节安排以综合项目工程应用为主线展开,内容讲解由浅入深,层次清晰,通俗易懂。第 1 章介绍 Web 编程技术中的相关基础内容,第 2 章介绍 Servlet 的定义与特点、编写与配置、工作流程、基本结构与生命周期、编程接口以及 Servlet 的应用编程,第 3 章介绍状态存储技术 Cookie 与 Session、作用域对象,第 4 章介绍 JSP 基本概念、JSP 元素、JSP 内置对象、JSP 注释以及 Java Web 程序开发中的路径问题,第 5 章介绍过滤器和监听器的基本概念、开发、使用,第 6 章介绍 JavaBean 的概念、使用和作用域,第 7 章介绍 JSP 开发模型和 MVC 设计模式,第 8 章介绍 EL 访问数据、EL 内置对象、EL 运算符以及禁用 EL,第 9 章介绍自定义标签的开发、配置和使用以及常用 JSTL 标签的使用,第 10 章介绍字符编码、Java Web 应用开发中中文乱码产生的原因、解决方法,第 11 章介绍 Java Web 应用中的异常概述、异常处理一般准则、异常处理以及利用 Web 服务器对异常的处理。

本书具有以下特点。

(1) 遵照专业教学指导委员会最新计算机科学与技术和软件工程专业及相关专业的培养目标和培养方案,合理安排 Java Web 开发技术知识体系,结合 Java 开发技术方向的先行课程和后续课程,组织相关知识点与内容。

(2) 注重理论和实践的结合,教材融入面向对象软件开发过程和工程实践背景的项目案例,使得学生在掌握理论知识的同时提高学生在程序设计过程中分析问题和解决问题的实践动手能力,启发学生的创新意识,使学生的理论知识和实践技能得到全面发展。

(3) 每个知识点都包括基础案例、每章都有一个综合案例,知识内容层层推进,使得学生易于接受和掌握相关知识内容。每章综合案例以"大学生成绩管理系统"为基础,以开发过程为主线,将知识点有机地串联在一起,便于学生掌握与理解。

(4) 教材提供配套的课件、例题案例、章节案例和综合案例的源码。

本书由李雷孝、邢红梅和王慧共同编写。其中,李雷孝编写了第 4、5、10 和 11 章并统稿,邢红梅编写了第 6、7、8 和 9 章,王慧编写了第 1、2、3 章。在编写过程中,参阅了甲骨文(Oracle)公司、安博教育集团、青岛软件园、上海杰普软件等公司的教学科研成果,也吸取了国内外教材的精髓,对这些作者的贡献表示由衷的感谢。本书在出版过程中,得到了刘利民教授、马志强副教授和刘建兰老师的支持和帮助;还得到了清华大学出版社的大力支持,在此表示诚挚的感谢。本教材受到全国高等学校计算机教育研究会 2015 年度高等学校计算机教材建设项目资助。

由于作者水平有限,书中难免有不妥和疏漏之处,恳请各位专家、同仁和读者不吝赐教和批评指正,并与笔者讨论,邮箱 llxhappy@126.com。

编者
2014 年 12 月

目 录

第1章 Web 编程基础 ………………………………………………………………………… 1
  1.1 软件开发体系结构 …………………………………………………………………… 1
    1.1.1 C/S 体系结构 …………………………………………………………………… 1
    1.1.2 B/S 体系结构 …………………………………………………………………… 1
    1.1.3 C/S 和 B/S 体系结构的比较 …………………………………………………… 2
  1.2 HTTP ………………………………………………………………………………… 2
  1.3 Web 应用程序工作原理 ……………………………………………………………… 3
  1.4 Web 应用开发技术 …………………………………………………………………… 4
    1.4.1 客户端开发技术 ………………………………………………………………… 4
    1.4.2 服务器端开发技术 ……………………………………………………………… 5
  1.5 Java Web 开发环境的搭建 …………………………………………………………… 6
    1.5.1 Web 服务器 …………………………………………………………………… 6
    1.5.2 Tomcat 的安装与启动 ………………………………………………………… 7
    1.5.3 集成开发工具 MyEclipse 与 Tomcat 的集成 ………………………………… 9
  1.6 Java Web 项目的创建、目录结构及部署 …………………………………………… 12
    1.6.1 Java Web 项目创建 …………………………………………………………… 12
    1.6.2 Java Web 项目目录结构 ……………………………………………………… 13
    1.6.3 Java Web 项目部署 …………………………………………………………… 14
  1.7 Java Web 应用成功案例简介 ………………………………………………………… 17
  1.8 案例 …………………………………………………………………………………… 19
    1.8.1 案例设计 ………………………………………………………………………… 20
    1.8.2 案例演示 ………………………………………………………………………… 25
    1.8.3 代码实现 ………………………………………………………………………… 25
  习题 ………………………………………………………………………………………… 30

第2章 Servlet 基础 ……………………………………………………………………………… 31
  2.1 Servlet 简介 …………………………………………………………………………… 31
    2.1.1 什么是 Servlet ………………………………………………………………… 31
    2.1.2 Servlet 的特点 ………………………………………………………………… 31

2.2 编写第一个 Servlet ·········· 32
    2.2.1 编写 Servlet ·········· 32
    2.2.2 配置 Servlet ·········· 33
    2.2.3 运行 Servlet ·········· 34
    2.2.4 Servlet 的开发步骤与执行流程 ·········· 37
2.3 Servlet 编程 ·········· 40
    2.3.1 Servlet API ·········· 40
    2.3.2 Servlet 的生命周期 ·········· 45
    2.3.3 Servlet 获得初始化参数值 ·········· 50
    2.3.4 Servlet 处理表单 ·········· 52
    2.3.5 Servlet 中的跳转 ·········· 58
2.4 案例 ·········· 65
    2.4.1 案例设计 ·········· 65
    2.4.2 案例演示 ·········· 66
    2.4.3 代码实现 ·········· 66
习题 ·········· 70

第 3 章 状态管理与作用域对象 ·········· 73
3.1 Java Web 状态管理 ·········· 73
    3.1.1 Cookie ·········· 73
    3.1.2 Session ·········· 80
3.2 作用域对象 ·········· 89
    3.2.1 ServletContext ·········· 89
    3.2.2 HttpSession ·········· 91
    3.2.3 ServletRequest ·········· 91
    3.2.4 作用域对象的比较 ·········· 95
3.3 案例 ·········· 96
    3.3.1 案例设计 ·········· 96
    3.3.2 案例演示 ·········· 96
    3.3.3 代码实现 ·········· 96
习题 ·········· 99

第 4 章 JSP 语法基础 ·········· 103
4.1 JSP 基本概念 ·········· 103
4.2 JSP 元素 ·········· 103
    4.2.1 脚本元素 ·········· 103
    4.2.2 指令元素 ·········· 106
    4.2.3 动作元素 ·········· 111

4.3 JSP 内置对象 …………………………………………………………………… 114
　　4.3.1 内置对象概述 ………………………………………………………… 114
　　4.3.2 内置对象使用 ………………………………………………………… 115
4.4 JSP 注释 …………………………………………………………………………… 120
4.5 Java Web 中的路径问题 ………………………………………………………… 120
　　4.5.1 路径的基本概念 ……………………………………………………… 120
　　4.5.2 路径相关函数 ………………………………………………………… 121
　　4.5.3 Java Web 开发中经常涉及的路径问题 ……………………………… 122
4.6 案例 ……………………………………………………………………………… 122
　　4.6.1 案例设计 ……………………………………………………………… 123
　　4.6.2 案例演示 ……………………………………………………………… 123
　　4.6.3 代码实现 ……………………………………………………………… 126
习题 …………………………………………………………………………………… 143

第 5 章 过滤器和监听器 …………………………………………………………… 146

5.1 过滤器 …………………………………………………………………………… 146
　　5.1.1 过滤器概述 …………………………………………………………… 146
　　5.1.2 Filter 接口 …………………………………………………………… 147
　　5.1.3 过滤器开发步骤 ……………………………………………………… 147
　　5.1.4 过滤器应用举例 ……………………………………………………… 148
5.2 监听器 …………………………………………………………………………… 155
　　5.2.1 监听器简介 …………………………………………………………… 155
　　5.2.2 监听器接口 …………………………………………………………… 155
　　5.2.3 监听器开发 …………………………………………………………… 158
5.3 案例 ……………………………………………………………………………… 164
　　5.3.1 案例设计 ……………………………………………………………… 164
　　5.3.2 案例演示 ……………………………………………………………… 165
　　5.3.3 代码实现 ……………………………………………………………… 166
习题 …………………………………………………………………………………… 172

第 6 章 JavaBean 组件 ……………………………………………………………… 174

6.1 JavaBean 的概念 ………………………………………………………………… 174
6.2 JavaBean 的使用 ………………………………………………………………… 175
　　6.2.1 <jsp:useBean> ………………………………………………………… 175
　　6.2.2 <jsp:setProperty> …………………………………………………… 177
　　6.2.3 <jsp:getProperty> …………………………………………………… 178
6.3 JavaBean 的作用范围 …………………………………………………………… 180
6.4 案例 ……………………………………………………………………………… 185

	6.4.1 案例设计	185
	6.4.2 案例演示	186
	6.4.3 代码实现	187
习题		193

第 7 章 MVC 设计模式 ································ 196

7.1 JSP 开发模型 ································ 196
    7.1.1 JSP Model 1 ································ 196
    7.1.2 JSP Model 2 ································ 205
7.2 MVC 设计模式 ································ 206
7.3 案例 ································ 208
    7.3.1 案例设计 ································ 208
    7.3.2 案例演示 ································ 209
    7.3.3 代码实现 ································ 211
习题 ································ 225

第 8 章 EL 表达式 ································ 227

8.1 EL 简介 ································ 227
8.2 EL 访问数据 ································ 227
    8.2.1 访问作用域变量 ································ 228
    8.2.2 访问 JavaBean 属性 ································ 230
    8.2.3 访问集合元素 ································ 232
8.3 EL 内置对象 ································ 239
8.4 EL 运算符 ································ 247
    8.4.1 算术运算符 ································ 247
    8.4.2 关系运算符 ································ 248
    8.4.3 逻辑运算符 ································ 248
    8.4.4 条件运算符 ································ 248
    8.4.5 empty 运算符 ································ 249
8.5 在页面中禁止使用 EL 表达式 ································ 249
8.6 案例 ································ 251
    8.6.1 案例设计 ································ 251
    8.6.2 案例演示 ································ 252
    8.6.3 代码实现 ································ 254
习题 ································ 260

第 9 章 JSP 标签 ································ 263

9.1 自定义标签 ································ 263

9.1.1	自定义标签简介	263
9.1.2	标签接口和实现类	263
9.1.3	自定义标签开发	265
9.1.4	自定义标签应用举例	267

9.2 JSTL 标签库 276
9.3 JSTL 核心标签库 278

9.3.1	表达式标签	279
9.3.2	流程控制标签	283
9.3.3	循环标签	286
9.3.4	url 相关标签	290

9.4 案例 294

9.4.1	案例设计	294
9.4.2	案例演示	295
9.4.3	代码实现	297

习题 303

第 10 章 中文乱码处理 306

10.1 字符集和字符编码 306

10.1.1	字符编码目的	306
10.1.2	字符集与编码分类	306

10.2 Java Web 中的中文乱码处理 307

10.2.1	中文乱码产生原因	307
10.2.2	中文乱码问题解决方案	308

10.3 案例 315

10.3.1	案例设计	315
10.3.2	案例演示	316
10.3.3	代码实现	318

习题 329

第 11 章 Java Web 中的异常处理 330

11.1 Java Web 程序异常处理 330

11.1.1	Java Web 异常概述	330
11.1.2	Java Web 异常处理一般准则	331
11.1.3	Java Web 异常处理实例	331

11.2 Web 服务器中处理异常 338

11.2.1	HTTP 状态码拦截	339
11.2.2	Java 异常类型拦截	340

11.3 案例 342

11.3.1　案例设计……………………………………………………………342
　　11.3.2　案例演示……………………………………………………………343
　　11.3.3　代码实现……………………………………………………………343
习题……………………………………………………………………………………355

附录A　综合案例使用说明……………………………………………………………357
　A.1　数据库安装和导入…………………………………………………………357
　A.2　开发工具的安装和案例工程导入…………………………………………357
　A.3　综合案例运行………………………………………………………………358

参考文献………………………………………………………………………………359

Web 编程基础

随着 Internet 的迅速发展和普及,互联网得到了广泛的应用,这使得 Web 应用程序在社会的各个方面发挥着重要作用,Web 应用编程成为目前软件开发中使用最多的编程技术。本章主要介绍 Web 编程技术中相关基础内容,包括软件开发体系结构、HTTP、Web 应用程序的工作原理、Web 应用技术、Java Web 开发环境的搭建、Java Web 成功应用案例简介等相关内容。

1.1 软件开发体系结构

目前 Web 应用软件的开发体系结构有两种:C/S(Client/Server)结构和 B/S(Browser/Server)结构。

1.1.1 C/S 体系结构

C/S 软件开发体系结构的应用软件系统是一种两层结构的系统:第一层是在客户机系统上结合了表示逻辑与业务逻辑;第二层是通过网络结合了数据库服务器。它将多个复杂网络应用的用户交互界面 GUI 和业务应用处理与数据库访问及处理相分离,服务器与客户端之间通过消息传递机制进行对话,由客户端发出请求给服务器,服务器进行相应的处理后经传递机制送回客户端。腾讯 QQ 就是一种典型的 C/S 结构软件。

1.1.2 B/S 体系结构

B/S 软件开发体系结构的应用软件系统由浏览器(Browser)和服务器(Web Server、Other Server、MiddleWare)组成。数据和应用程序都存放在服务器上,浏览器可以通过下载服务器上的应用程序得到动态扩展。以 B/S 结构模式开发的系统维护工作集中在服务器上,客户端不用维护,操作风格比较一致,只要有浏览器的合法用户都可以十分容易地使用。B/S 结构是真正的开放系统架构,是随着 Internet 技术的兴起,对 C/S 结构的一种变化或者改进的结构。在这种结构下,用户界面完全通过浏览器实现,一部分事务逻辑在前端实现,但是主要事务逻辑在服务器端实现,客户端运行程序是通过浏览器软件(如 IE、NetScape 等)登录服务器进行的。该结构将应用逻辑全部置于服务器上,客户端利用 Web 浏览器下载应用,在浏览器上执行。B/S 结构主要是利用了不断成熟的浏览器技术,结合浏览器的多种 Script 语言(如 VBScript、JavaScript 等)和 ActiveX 技术,用通用浏览器就实现了原来需要复杂专用软件才能实现的强大功能,并节约了开发成本,是一种全新的软件系统开发结构。构建良好、稳定、容易扩展的基于 B/S 结构的 Web 应用软件已经成为目前软件开发中的研究热点。

1.1.3　C/S 和 B/S 体系结构的比较

B/S 软件开发体系结构作为目前使用最多、最流行的软件开发体系结构,与 C/S 结构相比具有很多优点。

(1) B/S 结构的优势在于首先它简化了客户端,它无须像 C/S 结构那样在不同的客户机上安装不同的客户应用程序,而只需安装通用的浏览器软件,这样不但可以节省客户机的硬盘空间与内存,而且使安装过程更加简便、网络结构更加灵活。

(2) B/S 结构简化了系统的开发和维护,系统的开发者无须再为不同级别的用户设计开发不同的客户应用程序,只需把所有的功能都实现在 Web 服务器上,并就不同的功能为各个组别的用户设置权限即可。各个用户通过 HTTP 请求在权限范围内调用 Web 服务器上的不同处理程序,从而完成对数据的操作。

(3) B/S 结构使用户的操作变得更简单,对于 C/S 结构,客户应用程序有自己特定的规格,使用者需要接受专门培训,而采用 B/S 结构时,客户端只是一个简单易用的浏览器软件,无论是决策层还是操作层的人员都无须培训,就可以直接使用。B/S 结构的这种特性,还使 MIS(Management Information System)维护的限制因素更少。

(4) B/S 结构的软件系统特别适用于网上信息发布,使得传统 MIS 的功能有所扩展。这是 C/S 结构所无法实现的,而这种新增的网上信息发布功能恰是现代企业所需的。这使得企业的大部分书面文件可以被电子文件取代,从而提高了企业的工作效率,使企业行政手续简化,节省了人力物力。此外,随着各种操作系统将浏览器技术植入操作系统内部,B/S 结构更成为当今应用软件的首选开发体系结构。

当然,C/S 结构的软件系统和 B/S 结构相比也有其优势,例如客户端除了和服务器端通信之外还可以处理一定的业务逻辑功能,这样可以减轻服务器的压力。

1.2　HTTP

HTTP(HyperText Transfer Protocol,超文本传输协议)是用于从 WWW 服务器传输超文本到本地浏览器的传送协议。HTTP 可以使浏览器更加高效,使网络传输减少。它不仅保证计算机正确快速地传输超文本文档,还确定传输文档中的哪一部分,以及哪一部分内容首先显示(如文本先于图形)。HTTP 是一个应用层协议,由请求和响应构成,是一个标准的客户/服务器模型。

HTTP 是由客户端发起请求,服务器端返回响应到客户端,这种机制限制了 HTTP 的使用,无法实现客户端在没有发送请求时,服务器将消息推送给客户端。HTTP 的请求响应模型如图 1-1 所示。

一次 HTTP 请求响应过程称为一个事务,其工作过程可分为 4 步,如图 1-1 所示。

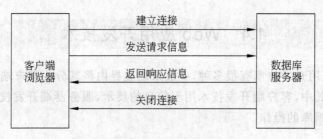

图 1-1　HTTP 请求响应模型

（1）首先客户端与服务器建立连接；

（2）客户机发送一个请求给服务器，请求方式的格式为：统一资源标识符（URL）、协议版本号，后边是 MIME 信息包括请求修饰符、客户机信息和其他可能的内容；

（3）服务器接到请求后，给予相应的响应信息，其格式为一个状态行，包括信息的协议版本号、一个成功或错误的代码，后边是 MIME 信息包括服务器信息、实体信息和其他可能的内容；

（4）客户端接收服务器所返回的信息并通过浏览器显示在用户的显示屏上，然后客户机与服务器断开连接。

HTTP 是一个无状态的协议，客户端的本次请求和上次请求没有对应关系。无状态是指协议对于事务处理没有记忆能力，缺少状态意味着如果后续处理需要前面的信息，则它必须重传，这样可能导致每次连接传送的数据量增大；另一方面，在服务器不需要先前信息时它的应答就较快。

1.3　Web 应用程序工作原理

目前大多数 Web 应用程序结构均采用最为流行的 B/S 软件开发体系结构。将 Web 应用程序部署在 Web 服务器上，只要 Web 服务器启动，用户便可以通过客户端浏览器发送 HTTP 请求到 Web 服务器，此时运行在 Web 服务器上对应的 Web 应用程序将处理客户端通过浏览器发来的请求，处理完成后对请求做出响应。Web 应用程序工作原理如图 1-2 所示。

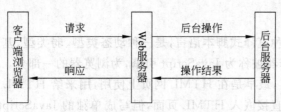

图 1-2　Web 应用程序工作原理

1.4　Web 应用开发技术

目前 Web 应用开发技术有很多种，一般来说业界内将其分为客户端开发技术和服务器端开发技术。其中，客户端开发技术用于信息的展示，服务器端开发技术用于实现业务逻辑的处理和数据库的操作。

1.4.1　客户端开发技术

Web 应用客户端开发技术为通常所说的 Web 静态页面技术，Web 应用程序的开发离不开客户端开发技术的支持。目前比较常用的客户端开发技术有 HTML、CSS、JavaScript 等，下面对这几种技术做简要介绍。

1. HTML

HTML(HyperText Mark-up Language，超文本标记语言)是 Web 应用程序客户端开发技术的基础，主要用于显示网页信息。HTML 不需要编译，由浏览器直接解释执行。HTML 作为一种标记语言，是由一些特定符号和语法组成的，所以理解和掌握都十分容易。HTML 中定义了很多标记，每个标记都是一条命令，它告诉浏览器如何显示文本。浏览器的功能是对这些标记进行解释，显示出文字、图像、动画、播放声音等。现在的静态网页制作，大都采用一些专门的网页制作工具，如 Dreamweaver、FrontPage 等。这些工具都是所见即所得，使用起来非常方便。但读者千万不要以为只懂这些工具就够了，在很多情况下，都需要开发者手动对 HTML 代码进行调整，才能达到更好、更专业的效果。

2. CSS

CSS(Cascading Style Sheet)称为级联样式表。在利用 HTML 制作网页时采用 CSS 样式，可以有效地对页面的布局、字体、颜色、背景和其他效果实现更加精确的控制，只要对相应的代码做一些简单的修改，就可以改变整个页面的风格。CSS 的使用极大地提高了开发者对信息展示格式的控制能力，特别是在目前比较流行的 CSS+DIV 布局的网站或者系统中，CSS 的作用更是具有举足轻重的地位。

3. JavaScript

JavaScript 是一种直译式脚本语言，是一种动态类型、弱类型、基于原型的语言，内置支持类型。它的解释器被称为 JavaScript 引擎，为浏览器的一部分。JavaScript 是广泛用于客户端的脚本语言，最早是在 HTML 网页上使用，用来给 HTML 网页增加动态功能。JavaScript 代码可以直接嵌入 HTML 页面，但写成单独的 JavaScript 文件更有利于结构和行为的分离。

JavaScript 语言不同于服务器端编程语言，例如 PHP、ASP 和 JSP，它主要作为客户端脚本语言在用户的浏览器上运行，不需要服务器的支持。所以在早期程序员比较喜欢 JavaScript 以减少对服务器的负担，而与此同时也带来另一个问题：作为直译语言，

JavaScript 语言的安全性比较差。而随着服务器的强壮,虽然现在的程序员更喜欢运行于服务端的脚本以保证安全,但 JavaScript 仍然以其跨平台、容易上手等优势大行其道。同时,有些特殊功能效果(如 AJAX)必须依赖 JavaScript 在客户端进行支持。随着引擎(如 V8)和框架(如 Node.js)的发展,及其事件驱动与异步 IO 等特性,JavaScript 逐渐被用来编写服务器端程序。

本书重点介绍 Java Web 开发技术,对于客户端常用开发技术 HTML、CSS 和 JavaScript 只做简要介绍,读者若不具备这方面的基础,请参阅其他相关书籍。

1.4.2 服务器端开发技术

开发 Web 应用系统或者 Web 动态网站,最主要的工作是服务器端代码的开发。目前常用的 Web 服务器端开发技术分为三大"流派":Java Web、.NET 技术和 PHP。三者作为主流的、成熟的 Web 服务器端开发技术,各自均有其优缺点。Java 语言具有众多优点,例如面向对象、跨平台、安全性高等,再加上很多成熟 Java 开源框架的出现,为 Java Web 开发极大地提高了开发效率,很多大型企业级应用系统均采用 Java Web 技术进行开发。下面对三者相关的开发技术做简要介绍。

1. PHP

PHP 原始为 Personal Home Page 的缩写,现在已经正式更名为 Hypertext Preprocessor,即超文本预处理器。PHP 是一种通用开源脚本语言,吸收了 C 语言、Java 和 Perl 的特点,易于学习,使用广泛,主要适用于 Web 开发领域。PHP 比 CGI 或者 Perl 更快速地执行动态网页,用 PHP 制作出的动态页面与其他的编程语言相比,PHP 是将程序嵌入到 HTML 文档中去执行,执行效率比完全生成 HTML 标记的 CGI 要高许多;PHP 还可以执行编译后代码,编译可以加密和优化代码运行,使代码运行更快。另外,PHP 可以跨平台运行,在 UNIX、Linux、Windows、Mac OS 下均可运行。

2. ASP.NET

ASP.NET 是.NET 框架的一部分,是一项微软公司的技术,是一种嵌入网页中的脚本,可以由因特网服务器执行的服务器端脚本技术。它可以在通过 HTTP 请求文档时再在 Web 服务器上动态创建。ASP 指 Active Server Pages(动态服务器页面),是运行于 IIS(Internet Information Server,即 Windows 开发的 Web 服务器)之中的程序。ASP.NET 是基于通用语言的编译运行的程序,其实现完全依赖于虚拟机,所以它拥有跨平台性。ASP.NET 构建的应用程序可以运行在几乎全部的平台上。ASP.NET 的网站或应用程序通常使用微软公司的 IDE(集成开发环境)产品 Visual Studio 进行开发。

3. Servlet

Servlet 是 Java 2.0 中新增的一个全新功能。Servlet 是一种独立于平台和协议的服务器端的 Java 应用程序,可以生成动态的 Web 页面。与传统的从命令行启动的 Java 应用程序不同,Servlet 由 Web 服务器进行加载,该 Web 服务器必须包含支持 Servlet 的

Java 虚拟机。Servlet 是 HTTP 服务器上的数据库或应用程序之间的中间层,用来接收 Web 浏览器或其他 HTTP 客户程序发来的请求。客户端访问 Servlet 时,Java 虚拟机用轻量级的 Java 线程处理每个请求,也就是同时有 N 个请求的情况下,Servlet 开启 N 个线程,但只有一个 Servlet 实例在内存中,所以 Servlet 运行效率高。Servlet 可以直接与 Web 服务程序对话,多个 Servlet 可以共享数据,Servlet 与数据库的连接比较简单。

4. JSP

JSP(Java Server Page,Java 服务器页面)是由 Sun 公司倡导、许多公司参与一起建立的一种动态网页技术标准。JSP 技术有点类似 ASP 技术,它是在传统的网页 HTML 文件中嵌入 Java 程序代码段和 JSP 标记,从而形成 JSP 文件,文件名的后缀为".jsp"。由于 Java 语言的跨平台特性,所以由 JSP 开发的 Web 应用也是跨平台的。JSP 的前身是 Servlet,由于 Servlet 的开发过程比较复杂,所以出现了 JSP 技术。

Servlet 和 JSP 作为 Java Web 开发技术的主要内容是本书的重点介绍内容,后面有十分详细的叙述,这里只做简要介绍。

1.5　Java Web 开发环境的搭建

1.5.1　Web 服务器

Web 服务器也称为 WWW(World Wide Web)服务器,主要功能是提供网上信息浏览服务。WWW 是 Internet 的多媒体信息查询工具,是 Internet 上近年才发展起来的服务,也是发展最快和目前用的最广泛的服务。下面介绍几种常用的 Web 服务器。

1. Resin

Resin 提供了最快的 JSP/Servlets 运行平台。如果选用 JSP 平台作为 Internet 商业站点的支持,那么速度、价格和稳定性都是要考虑到的,Resin 十分出色,表现更成熟,很具备商业软件的要求。Resin 是完全免费的。Resin3 之后已经不再是一个简单的 JSP 容器,开始支持 EJB,JTA 等企业功能。Resin4(2010 年后已经比较稳定)性能更是优秀。

2. JBoss

JBoss 是一个基于 JavaEE 的开放源代码的应用服务器。因为 JBoss 代码遵循 LGPL 许可,可以在任何商业应用中免费使用它,而不用支付费用。2006 年,JBoss 公司被 RedHat 公司收购。JBoss 是一个管理 EJB 的容器和服务器,支持 EJB1.1、EJB2.0 和 EJB3.0 的规范。但 JBoss 核心服务不包括支持 Servlet/JSP 的 Web 容器,一般与 Tomcat 或 Jetty 绑定使用。

3. WebSphere

WebSphere Application Server 是一种功能完善、开放的 Web 应用程序服务器,它基于 Java 的应用环境,用于建立、部署和管理 Internet 和 Intranet Web 应用程序。WebSphere 针对以 Web 为中心的开发人员,他们都是在基本 HTTP 服务器和 CGI 编程技术上成长起来的。IBM 提供 WebSphere 产品系列,通过提供综合资源、可重复使用的组件、功能强大并易于使用的工具,以及支持 HTTP 和 IIOP 通信的可伸缩运行时环境,来帮助这些用户从简单的 Web 应用程序转移到电子商务世界。WebSphere 适用于大型企业级应用,但是价格比较昂贵。

4. WebLogic

BEA WebLogic Server 是一种多功能、基于标准的 Web 应用服务器,为企业构建自己的应用提供了坚实的基础。各种应用开发、部署所有关键性的任务,无论是集成各种系统和数据库、还是提交服务、跨 Internet 协作,起始点都是 BEA WebLogic Server。由于它具有全面的功能、对开放标准的遵从性、多层架构、支持基于组件的开发,基于 Internet 的企业都选择它来开发、部署最佳的应用。BEA WebLogic Server 在使应用服务器成为企业应用架构的基础方面继续处于领先地位。BEA WebLogic Server 为构建集成化的企业级应用提供了稳固的基础,它们以 Internet 的容量和速度,在联网的企业之间共享信息、提交服务,实现协作自动化。WebLogic 适用于大型企业级应用,但是价格比较昂贵。

5. Tomcat

Tomcat 是一个开放源代码、运行 Servlet 和 JSP 的 Web 应用软件,它是基于 Java 的 Web 服务器。Tomcat Server 是根据 Servlet 和 JSP 规范进行工作的,因此可以说 Tomcat Server 实现了 Apache-Jakarta 规范且比绝大多数商业应用软件服务器要好。随着 Tomcat 版本的不断更新,再加上 Tomcat 是开源免费的 Web 服务器,Tomcat 成为众多开发者使用 Web 服务器的首选。本书介绍的 Java Web 应用程序均采用 Tomcat 作为 Web 服务器。

1.5.2　Tomcat 的安装与启动

Tomcat 以 JRE 为基础,因此需要首先安装 JDK。JDK 安装包需要从 Orcale 公司的网站 http://www.oracle.com/下载,目前的最新版本是 JDK7.0。读者可以根据需要下载对应的版本,并进行安装。

Windows 系统下安装完 JDK 后,需要配置相应的环境变量。

(1) 配置环境变量:右击"我的电脑"→"高级"→"环境变量"。

(2) 在系统变量里新建 JAVA_HOME 变量,变量值为:JDK 的安装路径。

(3) 新建 classpath 变量,值为:.;%JAVA_HOME%\lib;%JAVA_HOME%\lib\tools.jar。

(4) 修改 path 变量值,在值后添加:%JAVA_HOME%\bin;%JAVA_HOME%\jre\bin。

JDK 安装完成后,需要下载和安装 Tomcat。Tomcat 是免费开源的软件,提供了二进制版本和源代码版本,开发者可以直接去 Tomcat 官方网站 http://tomcat.apache.org

进行下载，Tomcat 目前最高版本为 Tomcat 8.0，这里使用 Tomcat 7.0。

访问 Tomcat 官方网站，在浏览器地址栏中输入"http://tomcat.apache.org"，进入如图 1-3 所示的界面。

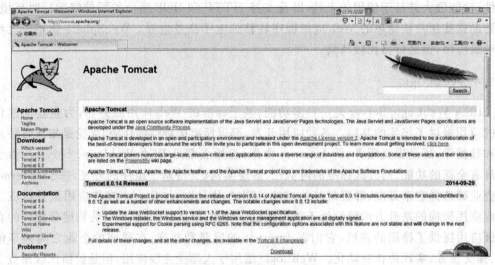

图 1-3　Tomcat 官方网站首页

在如图 1-3 所示的 Tomcat 官方网站上下载 apache-tomcat-7.0.55.exe 或 apache-tomcat-7.0.55.rar，其中 apache-tomcat-7.0.55.exe 为安装版本，apache-tomcat-7.0.55.rar 为解压版本。对于安装版本直接双击 apache-tomcat-7.0.55.exe 即可进行安装。安装过程就像安装 Windows 的其他软件一样简单，选择安装路径，单击"下一步"按钮直到完成即可安装成功。其中需要注意的是，在安装过程中会提示修改端口号，不修改将使用默认的端口号 8080，如果修改了请读者牢记，在访问 Web 应用程序时将使用该端口号。建议在端口号不冲突的情况下最好使用默认端口号 8080。对于解压版本直接将 apache-tomcat-7.0.55.rar 解压到本地硬盘上即可。

安装完成之后如果需要修改端口号，打开"Tomcat 安装目录\conf"文件夹下的 server.xml 文件，该文件是对 Web 服务器进行配置的文件，找到如下内容：

```
<Connector port="8080" protocol="HTTP/1.1"
          connectionTimeout="20000"
          redirectPort="8443" />
```

将"port＝"8080""中的 8080 改为你所希望的端口号即可。

Tomcat 安装配置完成后，双击"Tomcat 安装目录\bin"文件夹下的 startup.bat，如果弹出如图 1-4 所示的界面，说明 Tomcat 已启动。

如果要验证 Tomcat 是否启动成功，打开浏览器在地址栏中输入"http://localhost：8080/"，浏览器界面显示如图 1-5 所示，该界面说明 Tomcat 正常启动，即 Tomcat 安装成功。

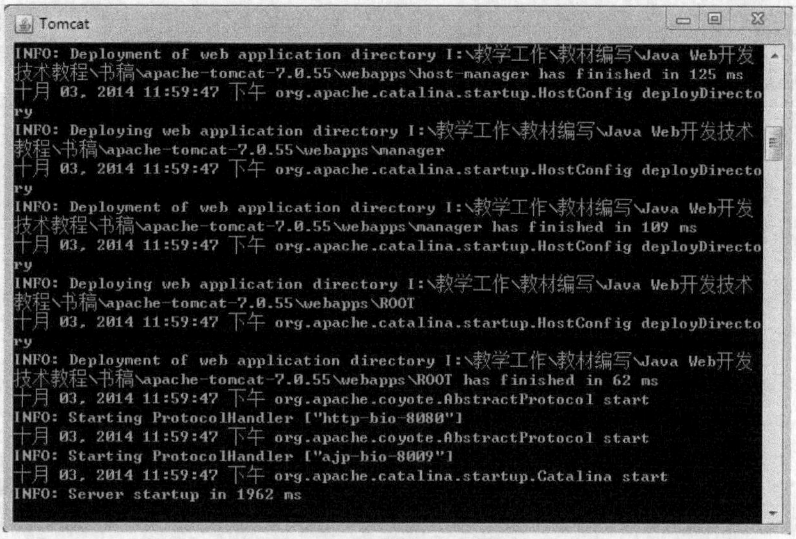

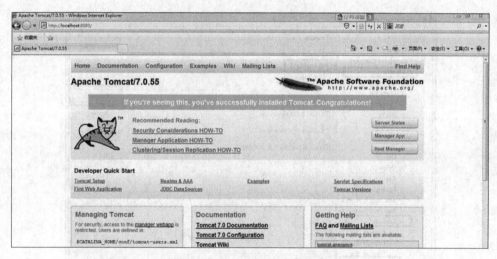

图 1-4　Tomcat 启动界面

图 1-5　Tomcat 正常启动界面

1.5.3　集成开发工具 MyEclipse 与 Tomcat 的集成

进行 Java Web 项目开发目前通常使用集成开发工具 MyEclipse，并将 Web 服务器 Tomcat 集成到 MyEclipse 中。下面介绍将 Tomcat 集成到 MyEclipse 的详细步骤。

下载并安装 Java 集成开发工具 MyEclipse 10.0。

（1）启动 MyEclipse，在 MyEclipse 中单击图标 右侧的倒三角符号，执行 Configure Server 命令，如图 1-6 所示。

（2）通过第（1）步进入如图 1-7 所示的界面，选择 Servers→Tomcat→Tomcat 7.x→将 Tomcat Server 选项由原来选中 Disable 改为 Enable，然后单击 Browse 按钮。

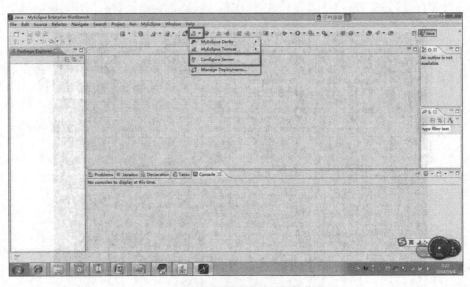

图 1-6　进入服务器配置对话框操作界面

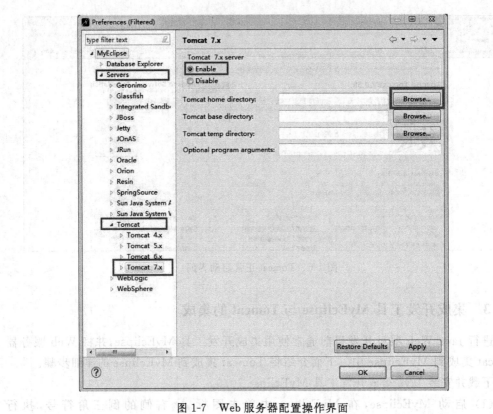

图 1-7　Web 服务器配置操作界面

（3）通过第（2）步进入如图 1-8 所示的界面，选择 Tomcat 目录，单击"确定"按钮即可。

第 1 章 Web 编程基础

图 1-8　Web 服务器目录配置操作界面

(4) 通过第(3)步进入如图 1-9 所示的界面,单击 OK 按钮即可。

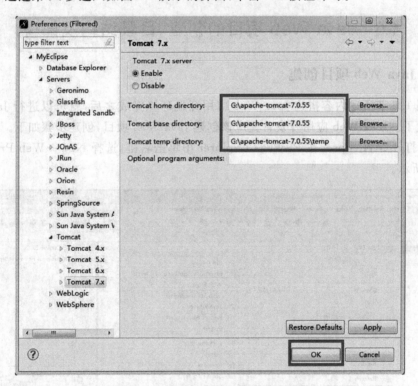

图 1-9　服务器目录配置确认操作界面

至此 Java Web 集成开发环境配置完成。配置完成后可以通过 MyEclipse 界面启动 Web 服务器 Tomcat,在如图 1-10 所示的界面中,单击图标　右侧的倒三角符号,选择 Tomcat 7.x→Start 命令,即启动了 Tomcat。此时,在 MyEclipse 控制台显示 Tomcat 启动信息。

Java Web 开发技术

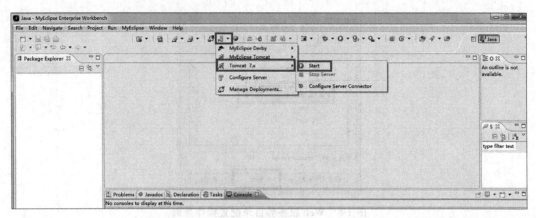

图 1-10　MyEclipse 中启动 Tomcat 操作界面

1.6　Java Web 项目的创建、目录结构及部署

1.6.1　Java Web 项目创建

按照 1.5 节所述内容搭建 Java Web 应用开发环境完成之后，就可以进行 Java Web 应用开发了。Java Web 应用开发首先应该创建 Java Web 项目，创建步骤如下。

（1）打开 MyEclipse，在 Package Explorer 中单击右键，选择 New→Web Project，如图 1-11 所示。

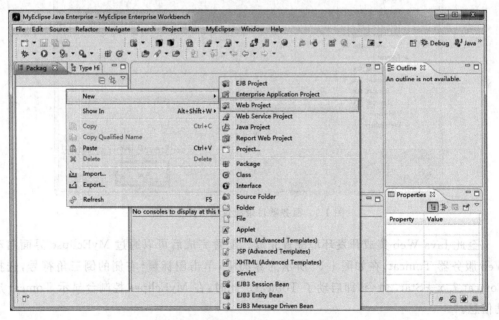

图 1-11　创建 Java Web 项目操作界面

（2）通过步骤（1）的操作进入如图1-12所示的对话框，用户在Project Name文本框中输入项目名称，此时Context root URL中自动填充为与Project Name相同的名称，该名称为Web应用名，默认与项目同名，用户可以改变它的值，但建议用户不要改动。其余各项选择默认值，然后单击Finish按钮，此时Java Web项目创建完成，如图1-13所示。

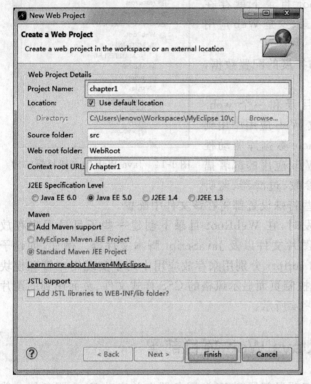

图1-12 New Web Project对话框

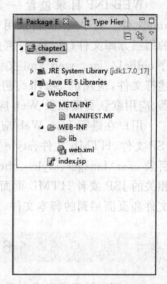

图1-13 Java Web项目初始概貌

1.6.2 Java Web项目目录结构

新建的Java Web项目初始目录结构如图1-13所示，从图中可以看出工程项目chapter1下有src和WebRoot两个子目录。一般情况下用户要根据具体业务需要，在遵照规范的目录结构基础上，为src和WebRoot目录合理地创建子目录，设计好Java Web应用程序的目录结构。

1. src目录

src目录下存放实现业务逻辑、数据操作、控制程序执行流程、描述实体对象、过滤器、监听器以及系统公用类的Java源代码文件，另外还包括资源文件等。

用户在进行Java Web应用开发时，根据Java源代码文件的作用创建不同的package，将不同的Java代码文件放在对应package中。例如，在src下创建包common、controller、dao、model、filter、resources分别用来存放实现系统公共功能、控制器、数据操作、业务逻辑操作、过滤功能的 Java 源代码文件和资源文件，如图 1-14 所示。当然，对于某些复杂的应用也可能要创建一些子包。

2. WebRoot 目录

新建项目初始目录中 WebRoot 目录下就存在 META-INF 和 WEB-INF 两个子目录，这是 Web 应用必需的两个目录。

META-INF 目录用来存放包和扩展的配置数据，如安全性、类加载器和版本信息。

WEB-INF 目录包含一个 lib 目录和一个 web.xml 文件。其中的 lib 目录用来存放 Web 应用中用到的第三方库文件(*.jar)，例如 Oracle 数据库驱动程序 ojdbc14.jar。web.xml 文件是 Web 应用的部署描述符文件，应用的 Servlet、初始化参数、过滤器、监听器、应用默认页面以及 Web 应用一些特殊设置都将在该文件中配置。

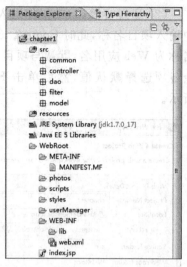

图 1-14 Java Web 的一般目录结构

用户在进行 Java Web 应用开发时，在 WebRoot 目录下创建一些子目录用来存放 JSP 文件、HTML 文件、style 文件、图片文件以及 JavaScript 脚本文件等。例如，创建子目录 userManager、styles、photos 和 scripts 分别用来存放应用中用于实现用户管理模块相关的 JSP 或者 HTML 页面文件、控制页面显示风格的 CSS 样式文件、页面用到的图片文件和页面用到的脚本文件，如图 1-14 所示。

1.6.3 Java Web 项目部署

Java Web 开发完成之后，必须把项目部署到 Web 服务器上并启动 Web 服务器才能被访问。项目部署操作如下。

(1) 在 MyEclipse 操作界面中单击"部署"按钮，如图 1-15 所示。

(2) 通过步骤(1)的操作，进入 Project Deployments 界面，如图 1-16 所示。在 Project 下拉框中选中 Java Web 工程项目 chapter1，然后单击 Add 按钮。

(3) 通过步骤(2)的操作，进入 New Deployment 界面，如图 1-17 所示。在 Server 下拉框中选中 Tomcat 7.x，然后单击 Finish 按钮。

(4) 通过步骤(3)的操作，进入 Project Deployments 界面，如图 1-18 所示。在 Deployments Status 栏中显示 Successfully deployed 说明项目部署成功，此时单击 OK 按钮即可，否则从步骤(1)重新开始或者检查项目是否有错误。

(5) 通过步骤(4)的操作将项目部署成功后，启动 Web 服务器 Tomcat，如图 1-19 所示。单击图标 右侧的倒三角，选择 Tomcat 7.x→Start，即启动了 Tomcat。此时，在 MyEclipse 控制台显示 Tomcat 启动信息。

Java Web 项目 chapter1 部署到服务器并启动服务器之后，用户可以通过浏览器访问该应用的资源。

第 1 章 Web 编程基础

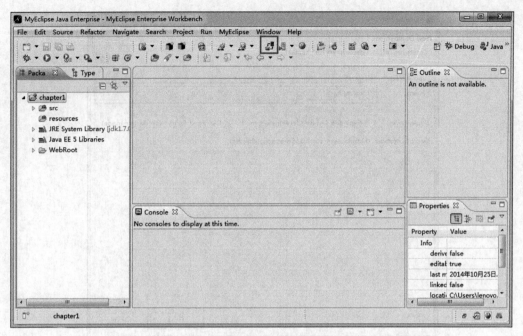

图 1-15　MyEclipse 操作界面

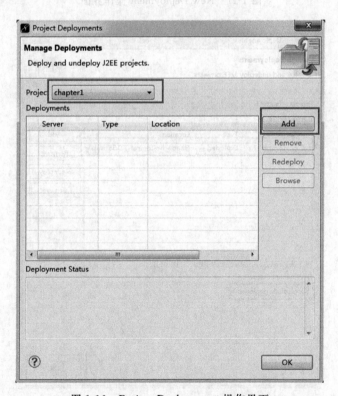

图 1-16　Project Deployments 操作界面

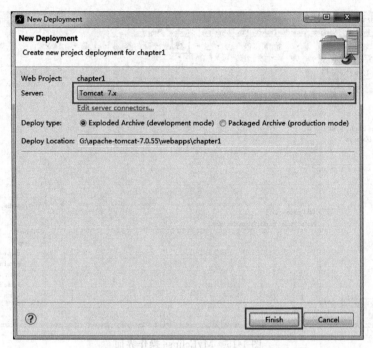

图 1-17　New Deployment 操作界面

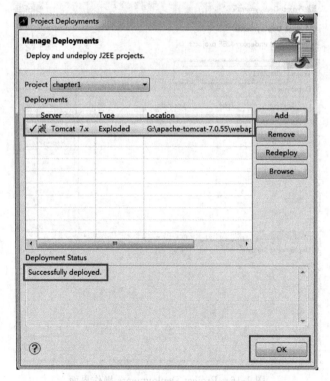

图 1-18　Project Deployments 操作界面

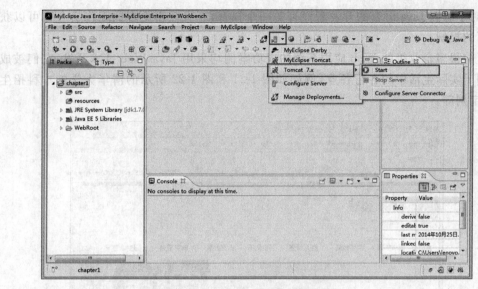

图 1-19　启动 Tomcat 操作界面

1.7　Java Web 应用成功案例简介

Java 语言本身技术先进,具有免费开源、跨平台、安全性高、面向对象等很多优点,在高校中 Java 已经成为许多学科研究、课程和计算的首选语言。在世界编程语言排行榜上一直居于前两位(多数年份居第一位)。如图 1-20 所示,2013 年世界编程语言排行榜,Java 语言居第一位,这充分说明了 Java 语言是目前最流行、使用人数最多的语言。不言

Position Mar 2013	Position Mar 2012	Delta in Position	Programming Language	Ratings Mar 2013	Delta Mar 2012	Status
1	1		Java	18.156%	+1.05%	A
2	2		C	17.141%	+0.05%	A
3	5	↑↑	Objective-C	10.230%	+2.49%	A
4	4		C++	9.115%	+1.07%	A
5	3	↓↓	C#	6.597%	-1.65%	A
6	6		PHP	4.809%	-0.75%	A
7	7		(Visual) Basic	4.607%	+0.24%	A
8	9	↑	Python	4.388%	+1.10%	A
9	13	↑↑↑↑	Ruby	2.150%	+0.74%	A
10	10		Perl	1.959%	-0.74%	A
11	8	↓↓↓	JavaScript	1.370%	-2.02%	A
12	48	↑↑↑↑↑↑↑↑↑↑↑↑	Bash	1.009%	+0.78%	A-
13	15	↑↑	Lisp	0.942%	+0.02%	A
14	12	↓↓	PL/SQL	0.921%	-0.50%	A--
15	11	↓↓↓↓	Delphi/Object Pascal	0.889%	-0.84%	A
16	16		Visual Basic .NET	0.888%	+0.10%	A
17	14	↓↓↓	Transact-SQL	0.836%	-0.09%	A-
18	17	↓	Pascal	0.697%	-0.07%	A--
19	21	↑↑	Lua	0.697%	+0.17%	B
20	26	↑↑↑↑↑	Assembly	0.633%	+0.21%	B

图 1-20　2013 年世界语言排行榜

而喻，对于掌握 Java 语言的开发者而言是一个福音，既可获得更多的工作机会也可以获得丰厚的报酬。

目前很多企业级的 Web 应用系统的成功案例均采用 Java Web 技术开发，它们被成功应用于实际生活中的各行各业，例如如图 1-21 和图 1-22 所示的清华大学的本科招生网、淘宝网等。

图 1-21　清华大学本科招生网站首页

图 1-22　淘宝网首页

还有涉及一些安全级别要求比较高或者跨平台运行的银行类项目应用，也是采用 Java Web 开发技术实现的，例如图 1-23 和图 1-24 所示的中国建设银行网站和中国农业银行网站。

第 1 章 Web 编程基础

图 1-23 中国建设银行网站首页

图 1-24 中国农业银行网站首页

1.8 案　　例

"学生成绩管理系统"作为贯穿全书的综合案例，与软件工程迭代开发相结合。根据工程开发实践，将每章内容与部分系统功能相结合进行组织，用于说明各章知识点和技术在实际工程实践项目中的应用,随着本书各章内容的讲解来逐步完成整个系统的实现。为了配合书中知识点的组织,案例在原系统的基础上进行了裁剪。读者可以根据自己学校的实际情况对所给案例进行修改,在课程结束时,设计出适合自己学校情况的学生成绩管理系统。

高校成绩管理工作是高校管理工作中的重要组成部分,成绩管理工作的好坏将直接影响学校的教学质量、教学秩序和学风的建设,同时,它也是学生在校各阶段的各种变化情况的风向指标,是学校了解、掌握学生基本情况的主要途径,更是对学生管理工作进行研究、决策的重要依据。

学生成绩管理系统能够为教师和学校管理者提供充足的信息和快捷的查询手段,但是长期以来人们使用传统人工的方式或单机版软件来管理学生成绩、填写各种表格,这种管理方式存在着许多缺点,如查询效率低、保密性差、保存不方便、数据不能共享等,更不利于快速查找、更新和维护。因此,结合本校实际开发一套基于 B/S 架构的成绩管理软件很有必要。

学生成绩管理系统的总体任务是:实现学生成绩管理信息的系统化、规范化和自动化,使得学生信息全校共享、成绩信息分权限管理。系统开发设计的思想是:尽量采用现有的软硬件环境及综合教务系统,引进先进的管理系统开发方案,提高系统开发水平和应用效果;符合学校成绩管理的规定,满足对学生成绩管理的需要,并达到操作过程直观、方便、实用、安全等要求;采用模块化程序设计方法,便于系统功能的各种组合和修改;根据学校管理者需求进行数据的添加、删除、修改和查询等操作。

1.8.1 案例设计

1. 系统功能分析与设计

学生成绩管理系统共有教务管理员、教师、学生三类用户,不同用户具有操作系统功能的不同权限。其中,教务管理员可以对管理员信息、学生信息、教师信息、课程信息和班级信息进行管理;负责对课程进行安排;另外,可以对学生成绩录入、修改、查询以及个人信息修改等功能进行操作。教师可以对本人所带课程的成绩录入、修改、查询以及个人信息修改等功能进行操作。学生可以对本人所修的课程成绩进行查看以及对个人信息进行修改等。

系统主要包括用户管理、基础数据维护、课程安排、成绩管理、个人信息管理 5 大模块。功能模块图如图 1-25 所示。

学生成绩管理系统各模块功能具体描述如下。

用户管理包括用户信息初始化、用户登录等功能。用户信息初始化:系统管理员通过添加管理员、学生、教师相关信息来进行用户的初始化,学生、教师可以修改本人相关信息,例如登录密码等,管理员可以修改系统所有信息。用户登录:管理员、学生和教师利用用户名和密码来登录系统,登录成功后根据各自权限访问系统。

基础数据维护包括管理员信息维护、学生信息维护、教师信息维护、课程信息维护、班级信息维护。学生信息维护:管理员可以对学生进行管理,查看所有学生信息,添加学生信息(新生入校),更改学生信息(学籍异动)以及删除学生(学生退学、留级等)。教师信息

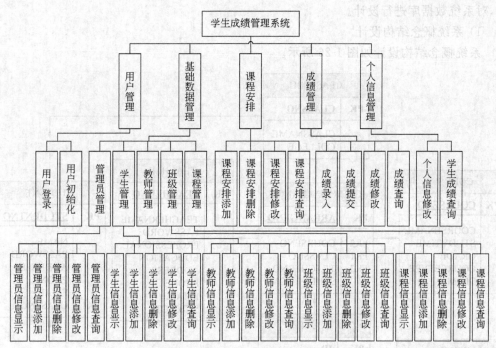

图 1-25 系统功能模块图

维护:管理员可以对教师进行管理,查看所有教师信息,添加教师信息(新教师调入),更改教师信息以及删除教师(教师调离、退休等)。课程信息维护:管理员可以对课程进行管理,查看所有课程信息,添加新课程,更改课程信息以及删除课程。班级信息维护:管理员可以对班级进行管理,查看所有班级信息,添加班级信息,更改班级信息以及删除班级(学生毕业)。

课程安排:管理员根据教学计划给各班级安排课程、教师以及授课地点,并且可以对相关信息进行维护,系统提供多条件查询课程安排。

成绩管理:管理员和任课教师可以对学生成绩进行录入、修改、查询。其中,管理员可以录入、修改、查询所有课程成绩,教师只能录入、修改、查询自己所带课程的成绩。

个人信息管理:已经登录的管理员、教师和学生可以查看及修改本人的基本信息,例如登录密码、电话、邮箱等。

学生成绩查询:登录成功的学生用户可以查看已修完所有课程的成绩及相关信息。

另外,所有用户必须登录后才能使用该系统的功能;系统管理员、教师、学生具有不同的权限,不能进行越权操作;系统用户界面必须方便快捷并提供页面验证,具有友好出错提示;系统应具有较好的可移植性;系统应具有良好的可维护性,以便以后对系统进行扩充和修改。

2. 系统数据库分析与设计

通过上述对系统功能的分析和设计,可以从中抽象出系统所要管理的数据实体有管理员、教师、学生、课程、班级、课程安排和成绩等 7 个实体。根据系统需求分析和功能设计,对系统数据库进行设计。

1) 系统概念结构设计

系统概念结构设计如图1-26所示。

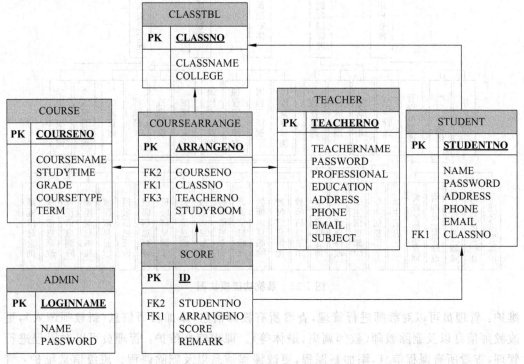

图1-26 系统E-R图

2) 系统数据库表的逻辑设计

根据数据库的概念结构设计,进行数据表的逻辑结构设计,如表1-1~表1-7所示。

表1-1 管理员表(admin)逻辑结构设计

序号	中文名称	字段名称	数据类型	外键	主键	说 明
1	登录名	LoginName	VARCHAR2(30)		PK	唯一标识
2	姓名	Name	VARCHAR2(30)			管理员真实姓名
3	密码	Password	VARCHAR2(30)			管理员用户登录时的用户密码

表1-2 班级表(classTbl)逻辑结构设计

序号	中文名称	字段名称	数据类型	外键	主键	说 明
1	班级编号	ClassNo	VARCHAR2(30)		PK	唯一标识
2	班级名称	ClassName	VARCHAR2(30)			班级名称
3	所在院系	College	VARCHAR2(50)			所在院系

表 1-3 教师表（teacher）逻辑结构设计

序号	中文名称	字段名	数据类型	外键	主键	说明
1	教师编号	TeacherNo	VARCHAR2(30)		PK	唯一标识；教师登录名
2	教师姓名	TeacherName	VARCHAR2(30)			教师的真实姓名
3	密码	Password	VARCHAR2(30)			教师用户登录时的用户密码
4	职称	Professional	Number			标识教师的职称： 0 教授或正高工 1 副教授或副高工 2 讲师 3 助教
5	学历学位	Education	VARCHAR2(30)			教师最后学历学位
6	家庭住址	Address	VARCHAR2(100)			教师家庭住址
7	联系电话	Phone	VARCHAR2(50)			联系电话
8	邮箱	Email	VARCHAR2(30)			邮箱
9	专业研究方向	Subject	VARCHAR2(50)			所学专业、学术专长

表 1-4 课程表（course）逻辑结构设计

序号	中文名称	字段名	数据类型	外键	主键	说明
1	课程编号	CourseNo	VARCHAR2(30)		PK	唯一标识
2	课程名称	CourseName	VARCHAR2(50)			课程名称
3	学时	StudyTime	Number			课程学时
4	学分	Grade	Number			课程学分
5	课程类别	CourseType	Number			课程类别： 0 基础课 1 专业基础课 2 专业课 3 选修课
6	学期	Term	Number			课程的开课学期

表 1-5 学生表（student）逻辑结构设计

序号	中文名称	字段名	数据类型	外键	主键	说明
1	学号	StudentNo	VARCHAR2(30)		PK	唯一标识；学号编写规则：年度＋学院标识＋专业标识＋流水号，共12位；学生登录名
2	姓名	Name	VARCHAR2(30)			学生的真实姓名

续表

序号	中文名称	字段名称	数据类型	外键	主键	说明
3	密码	Password	VARCHAR2(30)			学生用户登录时的用户密码
4	家庭住址	Address	VARCHAR2(100)			学生家庭住址
5	联系电话	Phone	VARCHAR2(50)			联系电话
6	邮箱	Email	VARCHAR2(30)			邮箱
7	所在班级	ClassNo	VARCHAR2(30)	FK (classTbl)		所在班级,关联班级表

表1-6 课程安排表(courseArrange)逻辑结构

序号	中文名称	字段名称	数据类型	外键	主键	说明
1	安排编号	ArrangeNo	VARCHAR2(30)		PK	唯一标识
2	课程编号	CourseNo	VARCHAR2(30)	FK (course)		课程编号,关联课程表
3	班级编号	ClassNo	VARCHAR2(30)	FK1 (classTbl)		班级编号,关联班级表
4	教师编号	TeacherNo	VARCHAR2(30)	FK (teacher)		教师编号,关联教师表
5	上课教室	StudyRoom	VARCHAR2(30)			上课地点

表1-7 成绩表(score)逻辑结构设计

序号	中文名称	字段名称	数据类型	外键	主键	说明
1	ID	ID	Number		PK	唯一标识
2	学号	StudentNo	VARCHAR2(30)	FK (student)		学生编号,关联学生表
3	安排编号	ArrangeNo	VARCHAR2(30)	FK (courseArrange)		教师编号,关联教师表
4	成绩	Score	Number			课程得分
5	备注	Remark	VARCHAR2(5)			有关成绩的备注信息,例如A代表正常,B代表缺考、作弊,C代表其他等

  本系统的数据库管理系统采用Oracle 11g,根据上述数据库设计创建数据库student以及相关的数据库用户和数据库表。具体创建数据库和数据库表的SQL可参见随书电子资源。

  本章的案例程序主要是利用客户端开发技术(HTML、CSS和JavaScript等)设计并实现系统的静态页面。

1.8.2 案例演示

在浏览器地址栏中输入"http://localhost:8080/ch01/login.html",运行效果如图 1-27 所示,输入"http://localhost:8080/ch01/index.html",运行效果如图 1-28 所示。事实上,直接使用浏览器打开这两个文件也是完全可以运行的。

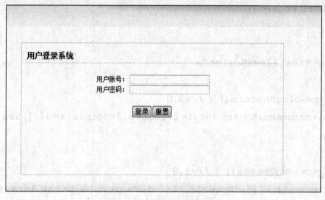

图 1-27 用户登录页面

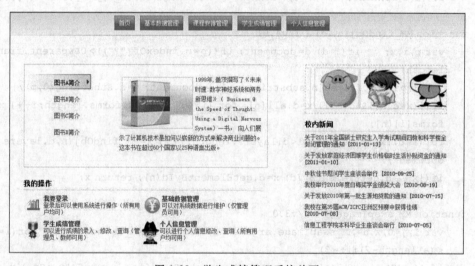

图 1-28 学生成绩管理系统首页

1.8.3 代码实现

创建工程 ch01,在根目录下编写登录页面 login.html 和首页 index.html 的代码。login.html 的源代码如下:

```
<!DOCTYPE html PUBLIC "-//W3C//DTD XHTML 1.0 Transitional//EN" "http://www.w3.org/TR/xhtml1/DTD/xhtml1-transitional.dtd">
<html xmlns="http://www.w3.org/1999/xhtml">
<head>
```

```html
<title>学生成绩管理-用户登录</title>
<link rel="stylesheet" type="text/css" id="css" href="style/main.css" />
<link rel="stylesheet" type="text/css" id="css" href="style/style1.css" />
<script src="js/main.js" type="text/javascript"></script>
<style type="text/css">
<!--
.STYLE1 {color: #CC6600}
-->
</style>
<script type="text/JavaScript">
<!--
function MM_swapImgRestore() { //v3.0
    var i,x,a=document.MM_sr; for(i=0;a&&i<a.length&&(x=a[i])&&x.oSrc;i++) x.src
    =x.oSrc;
}
function MM_preloadImages() { //v3.0
    var d=document; if(d.images){ if(!d.MM_p) d.MM_p=new Array();
    var i,j=d.MM_p.length,a=MM_preloadImages.arguments; for(i=0; i<a.length; i++)
    if (a[i].indexOf("#")!=0){ d.MM_p[j]=new Image; d.MM_p[j++].src=a[i];}}
}
function MM_findObj(n, d) { //v4.01
    var p,i,x;    if(!d) d=document; if((p=n.indexOf("?"))>0&&parent.frames.
    length) {
        d=parent.frames[n.substring(p+1)].document; n=n.substring(0,p);}
    if(!(x=d[n])&&d.all) x=d.all[n]; for (i=0;!x&&i<d.forms.length;i++) x=d.
    forms[i][n];
    for(i=0;!x&&d.layers&&i<d.layers.length;i++) x=MM_findObj(n,d.layers[i].
    document);
    if(!x && d.getElementById) x=d.getElementById(n); return x;
}
function MM_swapImage() { //v3.0
    var i,j=0,x,a=MM_swapImage.arguments; document.MM_sr=new Array; for(i=0;i
    <(a.length-2);i+=3)
    if ((x=MM_findObj(a[i]))!=null){document.MM_sr[j++]=x; if(!x.oSrc) x.oSrc
    =x.src; x.src=a[i+2];}
}
//-->
</script>
<script type="text/JavaScript">
    /*判断是否为数字*/
    function isNumber(str) {
        var Letters="1234567890";
        for (var i=0; i<str.length; i=i+1) {
            var CheckChar=str.charAt(i);
```

```
            if (Letters.indexOf(CheckChar)==-1) {
                return false;
            }
        }
        return true;
    }
    /*判断是否为Email*/
    function isEmail(str) {
        var myReg=/^[-_A-Za-z0-9]+@([_A-Za-z0-9]+\.)+[A-Za-z0-9]{2,3}$/;
        if (myReg.test(str)) {
            return true;
        }
        return false;
    }
    /*判断是否为空*/
    function isEmpty(value) {
        return /^\s*$/.test(value);
    }
    function check(){
        if(isEmpty(document.myForm.loginName.value)){
            alert("登录名不能为空!");
            document.myForm.loginName.focus();
            return false;
        }
        if(isEmpty(document.myForm.password.value)){
            alert("密码不能为空!");
            document.myForm.password.focus();
            return false;
        }
        return true;
    }
</script>
</head>
<body onload="MM_preloadImages('images/login-21.gif')">
<div id="btm">
<div id="main">
    <div id="header">
        <div id="top"></div>
        <div id="logo"><h1>用户登录</h1></div>
        <div id="mainnav"><span></span></div>
    </div>
    <div id="content">
        <div id="left">
            <div id="ltd" align="center">
```

```html
<h2>用户登录系统</h2>
<form name="myForm" action="" method="post" >
<table align="center">
    <tr>
        <td width="30%" align="right">
            用户账号:
        </td>
        <td width="70%" align="left" >
            < input type="text" id="loginName" name="loginName"
            style="width:150px"/>
        </td>
    </tr>
    <tr>
        <td width="30%" align="right">
            用户密码:
        </td>
        <td width="70%" align="left">
            < input type="password" id="password" name="password"
            style="width:150px"/>
        </td>
    </tr>
    <tr>
        <td width="30%" align="right">
              
        </td>
        <td width="70%" align="left">
              
        </td>
    </tr>
    <tr>
        <td colspan="2" align="center"><input type="submit" value=
        "登录" onclick="return check();"/><input type="reset" value
        ="重置"/></td>
    </tr>
    <tr>
        <td width="30%" align="right">
              
        </td>
    </tr>
    <tr>
        <td width="30%" align="right">
              
        </td>
    </tr>
```

```
                <tr>
                    <td width="30%" align="right">
                          
                    </td>
                </tr>
                <tr>
                    <td width="30%" align="right">
                          
                    </td>
                </tr>
                <tr>
                    <td width="30%" align="right">
                          
                    </td>
                </tr>
            </table>
        </form>
    </div>
</div>
<div class="clear"></div>
</div>
<div id="footer">
    <div id="copyright">
        <div id="copy">
            <p>CopyRight&copy;2014</p>
            <p>内蒙古工业大学信息工程学院软件工程系</p>
        </div>
    </div>
    <div id="bgbottom"></div>
</div>
</div>
</div>
</body>
</html>
```

【代码分析】程序 login.html 中使用了 HTML 显示内容,使用 CSS 设置内容的显示样式,使用 JavaScript 设置验证等动态行为。本案例使用的所有界面均使用以上三种静态页面技术编写,读者可以参考相关资料进行学习。

程序 login.html 中使用了一个＜form＞表单标记,其中添加了用户账号和用户密码两个＜input＞文本框标记,一个登录＜submit＞提交标记和一个重置＜reset＞重置标记。JavaScript 脚本定义了判断文本框内容是否为空的函数 isEmpty(value),定义了 check() 函数利用 isEmpty(value) 函数判断用户名和密码文本框是否为空来对页面进行验证。限于篇幅,系统涉及的其他静态页面源代码请参见随书电子资源。

习 题

1. 选择题

(1) 超文本标记语言的英文缩写为(　　)。
　　A. HTML　　　　B. HTTP　　　　C. URL　　　　D. CSS
(2) 以下选项中不属于 B/S 结构的优点的是(　　)。
　　A. 简化了客户端　　　　　　　　B. 减轻了服务器端压力
　　C. 方便了操作　　　　　　　　　D. 便于发布网络信息
(3) 下列开发技术中不属于常用客户端开发技术的是(　　)。
　　A. HTML　　　　B. JavaScript　　C. CSS　　　　D. Servlet
(4) HTTP 的全称是(　　)。
　　A. Hypertext Transport Protocol　　B. Hyper text Transport Protocol
　　C. Hypertext Transfer Protocol　　 D. Hyper text Transfer Protocol
(5) HTTP 哪个请求方式，请求参数会出现在网址列上？(　　)
　　A. get　　　　B. post　　　　C. delete　　　D. 以上都不是
(6) Servlet 必须基于哪一类型的容器才能提供服务？(　　)
　　A. Applet 容器　　　　　　　　　B. 应用程序客户端容器
　　C. Web 容器　　　　　　　　　　D. EJB 容器
(7) 以下描述中错误的内容是(　　)。
　　A. Web 服务器主要功能是提供网上信息浏览服务
　　B. Tomcat 是一个非免费的基于 Java 的 Web 应用软件容器
　　C. Tomcat 以 JRE 为基础，使用前提是必须安装 JDK
　　D. Tomcat 是一个支持 Java Web 最小的 Web 容器，由 Apache 提供

2. 填空题

(1) 目前常用的 Web 服务器端开发技术分为三大"流派"：＿＿＿＿、＿＿＿＿和＿＿＿＿。
(2) HTTP 永远都是＿＿＿＿发起请求，＿＿＿＿发回响应。
(3) Tomcat 默认使用的端口号为＿＿＿＿。

3. 程序设计题

(1) 搭建 Java Web 开发环境，并写一个 Servlet 程序使浏览器输出"开始学习 Java Web 程序设计！"。
(2) 编写 Servlet 程序实现页面表单中文本框数字内容的相加，并把结果输出。

4. 简答题

(1) 简述 B/S 和 C/S 体系结构有什么区别与联系。
(2) 请详细描述 Web 应用处理的过程。

Servlet 基础

Servlet 是 Java 语言应用于 Java Web 服务器端开发的扩展技术,它的产生为 Java Web 开发奠定了基础,它是 Java Web 应用中最核心的组件。随着 Web 开发技术的不断发展,Servlet 也在不断发展与完善,并凭借其安全性高、跨平台等诸多优点,深受广大 Java 编程人员的青睐。本章以理论与实践相结合的方式介绍 Servlet 的定义与特点、Servlet 的编写与配置、Servlet 的工作流程、Servlet 的基本结构与生命周期、Servlet 的编程接口以及 Servlet 的应用编程等内容。

2.1 Servlet 简介

Servlet 是基于 Java 技术的 Web 组件,它是 JSP 组件的前身,是 Java Web 开发技术的基础和核心组件。

2.1.1 什么是 Servlet

Servlet 是在服务器端运行的小程序,Servlet 一词由 Java Server Applet 而来。Servlet 是一个独立于平台的 Java 类,实际就是按照 Servlet 规范编写的一个 Java 类。Servlet 被编译成为平台独立的字节码文件,可以被 Web 服务器加载和运行,可以生成动态的 Web 页面。Servlet 可以通过 Web 服务器(Web 容器)接收客户端发来的请求,执行某一特定的功能,然后返回结果到客户端浏览器。Servlet 是 Java Web 服务器组件,它的运行离不开 Web 服务器。

2.1.2 Servlet 的特点

1. 简单、实用的 API 方法

Servlet 对象对 Web 应用进行了封装,针对 HTTP 请求提供了丰富的 API 方法,它可以处理表单提交数据、会话跟踪、读取和设置 HTTP 头信息等,对 HTTP 请求数据的处理非常方便,只需要调用相应的 API 方法即可。

2. 高效率

对于 N 个并发的请求,Servlet 将启动 N 个线程来处理,但是只需要创建一个

Servlet 实例；而对于传统的 CGI 程序，将创建 N 个实例来处理并发的请求。所以，与传统的 CGI 以及其他许多类似 CGI 的技术相比，Servlet 具有更高的效率。

3. 功能强大

Servlet 能够直接和 Web 服务器交互，Servlet 还能够在各个程序之间共享数据，使得数据库连接池之类的功能也很容易实现，这些作为传统的 CGI 程序是无法做到的。

4. 可移植性

Servlet 用 Java 语言编写，Servlet API 具有完善的标准。因此，为 IPlanet Enterprise Server 写的 Servlet 无须任何实质上的改动即可移植到 Apache、Microsoft IIS 或者 WebStar，几乎所有的主流服务器都直接或通过插件支持 Servlet。

2.2 编写第一个 Servlet

2.2.1 编写 Servlet

Servlet 本质上就是一个 Java 类。创建一个 Servlet 很简单，就是定义一个 Java 类，不过这个类要遵循 Servlet 规范，即要继承 javax.servlet.http.HttpServlet 类，覆盖其中的 doGet 或者 doPost 方法，在 doGet 或者 doPost 方法中编写处理请求的代码。

例程 2-1：演示 Servlet 的编写。程序为：FirstServlet.java。

```java
package com.ch02;
import java.io.IOException;
import java.io.PrintWriter;
import javax.servlet.ServletException;
import javax.servlet.http.HttpServlet;
import javax.servlet.http.HttpServletRequest;
import javax.servlet.http.HttpServletResponse;
public class FirstServlet extends HttpServlet{
                                            //FirstServlet 类继承 HttpServlet
    protected void doGet(HttpServletRequest req, HttpServletResponse resp)
            throws ServletException, IOException {
        resp.setContentType("text/html;charset=UTF-8");
                                            //设置响应的文本类型和编码方式
        PrintWriter out =resp.getWriter();   //通过输出流向客户端浏览器做出响应
        out.println("<HTML><HEAD><TITLE>hello world</TITLE></HEAD>");
        out.println("<font size=5 color=red>hello, world!!!");
        out.println("</font></BODY></HTML>");
        out.close();
    }
    protected void doPost(HttpServletRequest req, HttpServletResponse resp)
            throws ServletException, IOException {
```

```
        doGet(req, resp);                    //调用同类的doGet方法
    }
}
```

程序FirstServlet.java中定义了一个名字为FirstServlet的Java类,该类继承了HttpServlet,即遵循Servlet规范,所以FirstServlet是一个Servlet。在FirstServlet类中覆盖了从HttpServlet继承的doGet或者doPost方法,分别用来处理以Get方式或Post方式从客户端发来的Web请求。Web请求方式除了Get方式和Post方式实际上还有很多种,例如Put方式、Delete方式等,只不过Get方式和Post方式最为常用,其他方式基本不用,所以通常只覆盖doGet或者doPost方法。程序语句"resp.setContentType("text/html;charset=UTF-8");"是用来设置响应的文本类型和编码方式,通过响应对象获得的输出流对象out用来向客户端浏览器输出响应内容。通过代码可以看出响应向客户端输出的内容为HTML标记,这实际上是一个动态的Web页面。

2.2.2 配置Servlet

Servlet编写完成后,必须在工程WebRoot→WEB-INF中的web.xml中进行配置后才能使用。web.xml是Web应用的主配置文件,包含Web应用配置的主要信息,提供根据访问路径和配置信息定位到具体某一个Servlet、Filter或Listener的功能。

在web.xml中根元素<web-app>下配置Servlet(配置后Web Server能够进行定位),每个servlet对应一组<servlet></sevlet>和<servlet-mapping></servlet-mapping>:

```
<servlet>
        <servlet-name>servlet名</servlet-name>
        <servlet-class>servlet的class的全名</servlet-class>
</sevlet>
<servlet-mapping>
        <servlet-name>servlet名</servlet-name>
        <url-pattern>Servlet的访问路径</url-pattern>
</servlet-mapping>
```

在配置文件中<servlet>元素的子元素<servlet-name>中配置Servlet名称,该名称可以起任意名字,只要和<servlet-mapping>中的<servlet-name>值匹配即可,但必须保证不和其他<servlet-name>元素配置值重复;<servlet>元素的子元素<servlet-class>配置Servlet类的全名,即包名+类名;<servlet-mapping>元素的子元素<servlet-name>中的Servlet名必须和<servlet>元素的子元素<servlet-name>中配置的Servlet名相同;在<servlet-mapping>元素的子元素<url-pattern>中配置的Servlet访问路径指的是Servlet在浏览器访问时使用的路径,该路径必须以符号/开头(/为应用的根路径),它是访问一个Servlet的唯一路径。<url-pattern>中配置的Servlet访问路径是一个逻辑值,其值是虚拟的。

例程2-1中的FirstServlet主要配置信息如下:

```xml
<servlet>
    <servlet-name>first</servlet-name>
    <servlet-class>com.ch02.FirstServlet</servlet-class>
</servlet>
<servlet-mapping>
    <servlet-name>first</servlet-name>
    <url-pattern>/firstServlet</url-pattern>
</servlet-mapping>
```

2.2.3 运行 Servlet

运行 Servlet 即通过客户端浏览器访问 Servlet。在一个 Java Web 工程中编写并配置好 Servlet 之后,还不能运行 Servlet,此时还要把 Servlet 所在工程项目部署到 Web 服务器上并启动 Web 服务器后,才能运行该 Servlet。Java Web 项目部署与启动 Web 服务器的详细操作步骤参见 1.6.3 节。

Servlet 配置完成之后,把 Servlet 所在工程项目部署到 Web 服务器上并启动 Web 服务器,此时 Servlet 将可以被访问,其访问地址格式为"协议://服务器地址:服务器端口号/WEB 应用名/Servlet 的访问路径"。在访问 Servlet 的过程中,Web Server 负责定位具体访问哪个 Servlet,其过程如下。

(1) 查找 web.xml 中 Servlet 配置信息中的＜url-pattern＞值与客户端请求的路径相匹配的项;

(2) 通过查找结果(即＜servlet-mapping＞)所对应的＜servlet-name＞值查找匹配的＜servlet＞标记(元素＜servlet＞中的子元素＜servlet-name＞的值与＜servlet-mapping＞中的子元素＜servlet-name＞的值相同);访问查找结果中的＜servlet-class＞值指定的 Servlet 类。Web 服务器将在第一次访问 Servlet 时实例化该 Servlet 的一个对象,为客户端提供响应服务。

访问 Servlet 的方式有以下三种。

(1) 直接在浏览器地址栏中输入访问路径来访问 Servlet。

直接在浏览器地址栏中输入访问路径来访问 Servlet,这种请求提交方式为 GET,将调用 Servlet 类的 doGet 方法。例如,例程 2-1 中的 FirstServlet.java,在浏览器地址栏中输入"http://localhost:8080/chapter2/firstServlet",其中 http 为协议,localhost 为服务器地址(本机主机名称,也可使用本机回路地址 127.0.0.1),8080 为服务器端口号,chapter2 为应用名(Java Web 工程项目的应用名默认和项目相同),firstServlet 为 FirstServlet 在 web.xml 配置文件中 url-pattern 的值。运行效果如图 2-1 所示。

直接在浏览器地址栏中输入访问路径来访问 FirstServlet,执行其中的 doGet 方法,服务器向客户端浏览器输出"hello,world!!!"。

(2) 通过超链接访问 Servlet。

Servlet 可以接收客户端发来的请求,在 HTML 或者 JSP 中常常使用超链接向 Servlet 发出访问请求。通过超链接访问 Servlet 可以在超链接的 href 属性中指定值为

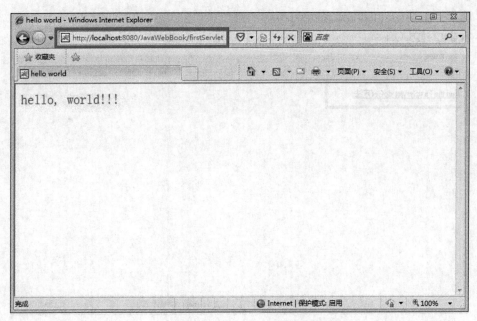

图 2-1　FirstServlet.java 运行效果

Servlet 的 url-pattern 值，这种请求提交方式为 GET，响应将执行 Servlet 中的 doGet 方法。

下面介绍通过超链接访问 Servlet 的方法，首先在 chapter2 工程中的 WebRoot 目录下创建 index.html 文件，该文件包含访问例程 2-1 中 FirstServlet 的超链接，源代码如下：

```
<!DOCTYPE html>
<html>
    <head>
        <title>index.html</title>
        <meta http-equiv="keywords" content="keyword1,keyword2,keyword3">
        <meta http-equiv="description" content="this is my page">
        <meta http-equiv="content-type" content="text/html; charset=UTF-8">
    </head>
    <body>
        <a href="firstServlet">Run the FirstServlet doGet 方法</a>
    </body>
</html>
```

在 index.html 文件中加入一个超链接"Run the FirstServlet doGet 方法"，href 属性值为"firstServlet"，单击超链接将执行 FirstServlet 的 doGet 方法。在浏览器的地址栏中输入"http://localhost:8080/chapter2/index.html"将运行 index.html，效果如图 2-2 所示。

单击图 2-2 中的"Run the FirstServlet doGet 方法"超链接将执行 FirstServlet 中

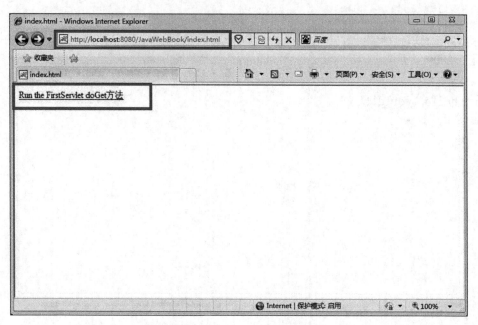

图 2-2 index.html 运行效果

doGet 方法,运行效果与直接在浏览器地址栏中输入路径进行 Servlet 访问一样,如图 2-1 所示。

(3) 通过表单提交访问 Servlet。

在 HTML 或者 JSP 中除了使用超链接向 Servlet 发出访问请求外,还可以通过表单提交的方式向 Servlet 提交访问请求。表单标记<form>有一个属性 method,当 method 的值为 get 时,则表单提交方式为 GET,调用 Servlet 的 doGet 方法;当 method 的值为 post 时,则表单提交方式为 POST,调用 Servlet 的 doPost 方法。

使用表单提交方式访问例程 2-1 中的 FirstServlet,在 index.html 文件中增加如下代码:

```
<form action="firstServlet" method="post">
    input your name:<input type="text"  name="loginname"><br>
    input your password:<input type="password" name="pwd"><br>
    <input type="submit" value="login">
</form>
```

上述代码在 index.html 中创建了一个<form>表单,表单 action 的属性值 "firstServlet"表示表单提交请求的目标资源为 FirstServlet,method 属性值为 post 表明表单提交方式为 POST,将执行 Servlet 的 doPost 方法。表单中添加了两个文本标记和一个提交按钮。在浏览器的地址栏中输入"http://localhost:8080/chapter2/index.html"将运行 index.html,效果如图 2-3 所示。

单击图 2-3 中的 login 按钮将执行 FirstServlet 中的 doPost 方法,因为在 FirstServlet 的 doPost 方法中代码直接调用 doGet 方法,所以运行效果与前两种访问方式进行

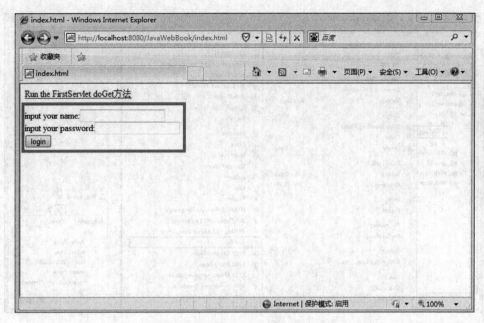

图 2-3　index.html 运行效果图

Servlet 访问一样,如图 2-1 所示。

2.2.4　Servlet 的开发步骤与执行流程

通过上述 Servlet 的编写、配置与运行,可以总结出开发一个 Servlet 的一般步骤如下。

(1) 编写 Servlet:编写一个 Java 类使其继承 HttpServlet 类并覆盖 doGet 和 doPost 方法。

(2) 在配置文件 web.xml 中配置 Servlet。

(3) 将 Servlet 所在 Java Web 项目部署到 Web 服务器(Tomcat)上。

(4) 启动 Web 服务器。

(5) 发送请求访问 Servlet。

在上述 Servlet 的开发步骤中,第(1)、(2)步可以通过 MyEclipse 直接定义 Servlet 并自动生成 Servlet 配置信息。具体操作步骤如下(以创建和配置例程 2-1 中 FirstServlet.java 为例)。

(1) 选中 com.ch02 包(FirstServlet.java 所在包)单击右键,选择 New→Servlet,如图 2-4 所示。单击 Servlet 进入创建 Servlet 对话框,如图 2-5 所示。

(2) 在 Name 文本框中输入 Servlet 的名字;在 Superclass 文本框中默认父类为 javax.servlet.http.HttpServlet,建议读者不要改动;在 Which method stubs would you like to create? 复选框中选择 Servlet 要创建的方法,一般只需选择 doGet 和 doPost 方法。输入完成后,单击 Next 按钮进入如图 2-6 所示的 Servlet 配置对话框。

(3) 当进入如图 2-6 所示的 Servlet 配置对话框时,MyEclipse 自动为 Servlet 生成了

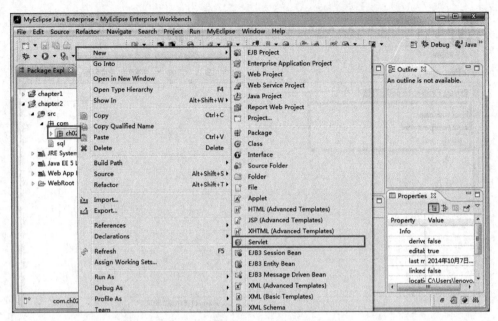

图 2-4 直接创建 Servlet 操作界面

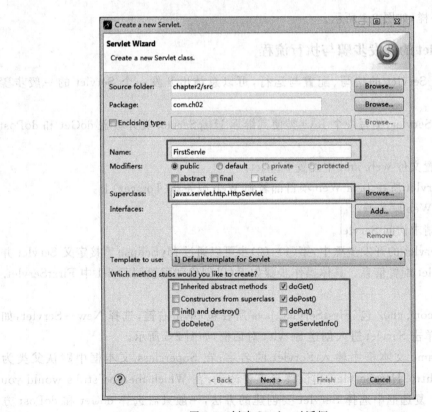

图 2-5 创建 Servlet 对话框

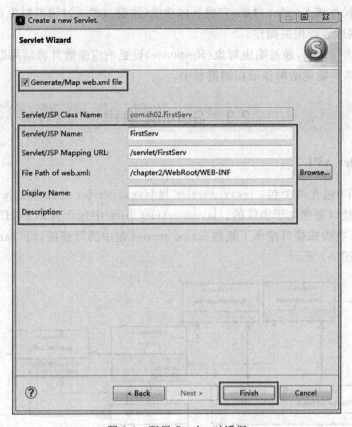

图 2-6 配置 Servlet 对话框

配置信息,除了 Servlet 的 class 名字外,Servlet 的其他配置信息用户均可修改。

Generate/Map web.xml file 复选框用户可以不选择,然后手动在 web.xml 中配置 Servlet。

Servlet/JSP Name 文本框中输入配置文件 web.xml 中 Servlet 的名字。

Servlet/JSP Mapping URL 文本框中输入配置文件的 url-pattern。

File Path of web.xml 文本框中的值为 web.xml 文件所在目录,用户不要改变该值,否则会影响 Web 应用的运行。

Display Name 和 Description 文本框中是配置文件 web.xml 中所配置 Servlet 的显示名称和描述信息,运行时不起任何作用,可以不填。

(4)在如图 2-6 所示的 Servlet 配置对话框中单击 Finish 按钮,即完成了 FirstServlet 的创建和配置。用户在创建的 FirstServlet.java 中编写代码覆盖 doGet 和 doPost 即可。

客户端发送请求访问 Servlet,Servlet 的执行流程如下。

(1)客户端浏览器向 Web 服务器发送请求访问某一 Servlet,例如注册。

(2)Web 服务器根据配置信息定位到具体的 Servlet,如完成注册功能的 Servlet。

(3)如果该 Servlet 是第一次被访问,此时 Servlet 对象在内存中不存在,则创建该 Servlet 的对象;如果该 Servlet 已经被访问过,则 Servlet 的对象已经存在于内存。然后创建一个线程操作该 Servlet 对象,完成具体功能(注册),即通过请求对象(Request)获得客户端数据,然后进行相关操作。

(4) 获得运行结果,通过响应对象(Response)设置响应参数并将结果返回到客户端。

(5) 客户端将响应结果显示在浏览器中。

2.3 Servlet 编程

2.3.1 Servlet API

Servlet API 包含两个包:javax.servlet 与 javax.servlet.http 包。javax.servlet 包中定义的类和接口是独立于协议的。javax.servlet.http 中包含具体于 HTTP 的类和接口,该包中的某些类或接口继承了某些 javax.servlet 包中的类或接口。Servlet API 的接口与类结构如图 2-7 所示。

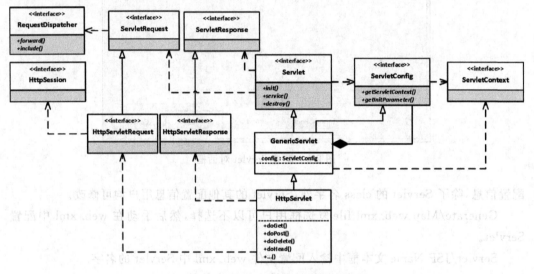

图 2-7　Servlet API 接口与类结构

Servlet 的运行需要 Web 服务器的支持,Web 服务器通过调用 Servlet 对象提供的标准 API,对客户端请求做出处理。下面对 Servlet 编程中常用的接口和类分别做简要介绍。

1. Servlet 接口

javax.servlet.Servlet 接口是任何一个 Servlet 都要直接或者间接实现的,上一小节在定义 FirstServlet 时直接继承了 HttpServlet 类,通过图 2-7 可以看出 HttpServlet 类间接实现了 Servlet 接口,所以 FirstServlet 也间接实现了 Servlet 接口。Servlet 接口中定义的方法如表 2-1 所示。

表 2-1 Servlet 接口方法和说明

方法	说明
public void init(ServletConfig config)	Servlet 实例化后,Servlet 容器调用此方法完成初始化工作
public ServletConfig getServletConfig()	获取 Servlet 对象的配置信息,返回 ServletConfig 对象
public void service(ServletRequest request,ServletResponse response)	处理 request 对象封装的客户端请求,并使用 response 对象返回请求结果
public String getServletInfo()	返回描述 Servlet 的一个字符串
public void destory()	当 Servlet 将要卸载时由 Servlet 引擎调用,以释放资源

2. ServletConfig 接口

Servlet 接口的 init(ServletConfig config)方法有一个 ServletConfig 类型的参数。当 Servlet 容器初始化一个 Servlet 对象时,会为这个 Servlet 实例对象创建一个 ServletConfig 对象。在 ServletConfig 对象中包含 Servlet 的初始化参数信息,此外,ServletConfig 对象还与当前 Web 应用的 ServletContext 对象关联。Servlet 容器在调用 Servlet 对象的 init(ServletConfig config)方法时,会把 ServletConfig 对象作为参数传给 Servlet 对象,init(ServletConfig config)方法会使当前 Servlet 对象与 ServletConfig 对象之间建立关联关系。ServletConfig 接口中的方法介绍如表 2-2 所示。

表 2-2 ServletConfig 接口方法和说明

方法	说明
public String getServletName()	该方法返回一个 Servlet 实例的名称,该名称由服务器管理员提供
public ServletContext getServletContext()	返回一个 ServletContext 对象的引用
public String getInitParameter(String name)	返回一个由参数 name 决定的初始化变量的值,如果该变量不存在,返回 null
public Enumeration getInitParameterNames()	返回一个存储所有初始化变量的枚举函数。如果 Servlet 没有初始化变量,返回一个空枚举函数

3. GenericServlet 类

GenericServlet 类是一种与协议无关的 Servlet,它直接实现了 Servlet 接口,是一个抽象类。该类实现了 Servlet 接口中除 service()方法之外的所有抽象方法,但是是默认实现,这意味着在定义一个 Servlet 时可以直接继承 GenericServlet 类来简化程序中的代码。GenericServlet 类除了实现了 Servlet 接口外,还实现了 ServletConfig 接口。GenericServlet 类在实际开发中使用较少,这里不再详细介绍。

4. HttpServlet 类

HttpServlet 类继承了 GenericServlet 类,也是一个抽象类。HttpServlet 类为

Servlet 接口提供了 HTTP 相关的通用实现。在 Java Web 开发中,自定义 Servlet 时通常都直接继承 HttpServlet 类。HTTP 把客户端请求方式分为 GET、POST、PUT、DELETE 等多种方式,在 HttpServlet 类中针对每一种请求方式都提供了相应的服务方法,如 doGet、doPost、doPut 和 doDelete 方法等。

HttpServlet 类实现了 Servlet 接口中的 service(ServletRequest request, ServletResponse response)方法,该方法用来处理客户端发来的请求。该方法实际上去调用它的重载方法 service(HttpServletRequest request, HttpServletResponse response),该重载方法首先调用 request 参数的 getMethod()方法来获得客户端的请求方式,然后根据请求方式去匹配对应的服务方法。例如,如果是 GET 请求方式,则调用 doGet 方法,以此类推。

HttpServlet 类为所有针对特定请求方式的 doXxxx()方法提供了默认实现,在 HttpServlet 类的默认实现中,doGet、doPost、doPut 和 doDelete 方法等都会返回一个对应 HTTP 中状态代码为 405 或者 400 的错误。所以,在开发中定义一个 Servlet 继承 HttpServlet 类时一定要根据响应不同方式的请求来覆盖相应的 doXxxx()方法。

5. ServletRequest 和 ServletResponse 接口

Servlet 由 Servlet 容器(Web 服务器)来管理,当客户请求到来时,容器创建一个 ServletRequest 对象,封装请求数据,同时创建一个 ServletResponse 对象,封装响应数据。这两个对象将被容器作为 service()方法的参数传递给 Servlet,Serlvet 利用 ServletRequest 对象获取客户端发来的请求数据,利用 ServletResponse 对象发送响应数据。

ServletRequest 接口中定义了很多有用的方法,在实际开发中经常用到,如表 2-3 所示。

表 2-3 ServletRequest 接口方法和说明

方 法	说 明
public Object getAttribute(String name)	返回请求范围对象内以 name 为名字的属性的值,如果该属性不存在,这个方法将返回 null
public void setAttribute(String name, Object o)	将对象 o 保存在请求范围对象内名字为 name 的属性,如果第二个参数 o 为 null,那么相当于调用 removeAttribute(name)
public void removeAttribute(String name)	移除请求范围中名字为 name 的属性
public String getCharacterEncoding()	返回请求正文使用的字符编码的名字。如果请求没有指定字符编码,这个方法将返回 null
public String getContentType()	返回请求正文的 MIME 类型。如果类型不可知,这个方法将返回 null
public String getParameter(String name)	返回请求范围对象中 name 属性的值。如果 name 参数有多个值,那么这个方法将返回值列表中的第一个值。如果在请求中没有找到这个参数,这个方法将返回 null

方　　法	说　　明
public String[] getParameterValues(String name)	返回请求范围对象中 name 参数所有的值。如果这个参数在请求中并不存在，这个方法将返回 null。该方法常用来获取客户端复选框的一组值
public RequestDispatcher getRequestDispatcher (String path)	返回指定资源名的 RequestDispatcher 对象，作为 path 所定位的资源的封装

ServletResponse 接口中定义了很多有用的方法，在实际开发中经常用到，如表 2-4 所示。

表 2-4　ServletResponse 接口方法和说明

方　　法	说　　明
public String getCharacterEncoding()	返回响应使用字符解码的名字。除非显式设置，否则为 ISO-8859-1
public OutputStream getOutputStream()	返回用于将返回的二进制输出写入客户端的流，此方法和 getWrite()方法二者只能调用其一
public Writer getWriter()	返回用于将返回的文本输出写入客户端的一个字符写入器，此方法和 getOutputStream()二者只能调用其一
public void setContentType(String type)	设置字符编码方式和内容类型

6. HttpServletRequest 和 HttpServletResponse 接口

HttpServletRequest 是专用于 HTTP 的 ServletRequest 子接口。客户端浏览器发出的请求被封装成为一个 HttpServletRequest 对象。所有的信息包括请求的地址、请求的参数、提交的数据、上传的文件、客户端的 IP 甚至客户端操作系统都包含在其内。HttpServletResponse 继承了 ServletResponse 接口，并提供了与 HTTP 有关的方法，这些方法的主要功能是设置 HTTP 状态码和管理 Cookie。

HttpServletRequest 接口中定义了很多有用的方法，在实际开发中经常用到，如表 2-5 所示。

表 2-5　HttpServletRequest 接口方法和说明

方　　法	说　　明
public Cookie[] getCookies()	返回一个数组，该数组包含这个请求中当前的所有 cookie。如果这个请求中没有 cookie，返回一个空数组
public String getMethod()	返回这个请求使用的 HTTP 请求方式(例如：GET、POST、PUT)
public String getRequestedSessionId()	返回这个请求相应的 session id。如果由于某种原因客户端提供的 session id 是无效的，这个 session id 将与在当前 session 中的 session id 不同，与此同时，将建立一个新的 session

续表

方法	说明
public HttpSession getSession()	返回与这个请求关联的当前有效的 session。在没有 session 与这个请求关联的情况下,将会新建一个 session
public HttpSession getSession(boolean create)	返回与这个请求关联的当前有效的 session。在没有 session 与这个请求关联的情况下,如果参数 create 为真时将新建一个 session;如果参数 create 为假,这个方法会返回空值。为了确保 session 能够被完全维持,Servlet 开发者必须在响应被提交之前调用该方法

HttpServletResponse 接口中定义了很多有用的方法,在实际开发中经常用到,如表 2-6 所示。

表 2-6　HttpServletResponse 接口方法和说明

方法	说明
public void addCookie(Cookie cookie)	将一个 Cookie 加入到响应中
public void sendRedirect(String location)	重定向到某一 Web 组件
public void sendError(int status)	设置响应状态码为指定值(可选的状态信息)

7. ServletContext 接口

一个 ServletContext 接口对象表示一个 Web 应用程序的上下文,运行在 Java 虚拟机中的每个 Web 应用程序都有一个与之相关的 Servlet 上下文,Servlet API 提供了一个 ServletContext 接口用来表示上下文。它提供了一些方法,Servlet 可以利用这些方法与 Servlet 容器进行通信。ServletContext 接口中的方法如表 2-7 所示。

表 2-7　ServletContext 接口方法和说明

方法	说明
public Object getAttribute(String name)	返回 Servlet 上下文中名字为 name 的对象。名字为 name 的对象是全局对象,因为它们可以被同一 Servlet 在另外某一任意时刻访问,或上下文中任意其他的 Servlet 访问
public void setAttribute(String name,Object obj)	将对象 obj 保存在 Servlet 上下文中,名字为 name
public String getInitParameter(String name)	返回指定上下文范围的初始化参数值

8. HttpSession 接口

HttpSession 接口被 Servlet 容器用来实现在 HTTP 客户端和 HTTP 会话两者的关联。这种关联可能在多个连接和请求中持续一段给定的时间。Session 用来在无状态的 HTTP 下越过多个请求页面来维持状态和识别用户。HttpSession 接口中的方法如表 2-8 所示。

表 2-8　HttpSession 接口方法和说明

方　　法	说　　明
public long getCreationTime()	返回建立 Session 的时间,这个时间表示为自 1970 年 1 月 1 日 0 时 0 分 0 秒 0 毫秒(GMT)至 Session 创建所经历的毫秒数
public String getId()	返回分配给这个 Session 的标识符。一个 HTTP Session 的标识符是一个由服务器来建立和维持的唯一字符串
public long getLastAccessedTime()	返回客户端最后一次发出与这个 Session 有关的请求的时间,如果这个 Session 是新建立的,返回−1。这个时间表示为自 1970 年 1 月 1 日 0 时 0 分 0 秒 0 毫秒(GMT)至最后一次访问所经历的毫秒数
public int getMaxInactiveInterval()	从最后一次与客户端发生交互开始计算,返回会话存活的最大秒数
public void setMaxInactiveInterval(int seconds)	从最后一次与客户端发生交互开始计算,设置会话存活的最大秒数
public Object getAttribute(String name)	返回会话中名字为 name 的对象,如果不存在返回 null
public void setAttribute(String name, Object value)	将对象 value 保存在会话中,指定名字为 name
public void removeAttribute(String name)	删除会话中此前保存的名称为 name 的对象

2.3.2　Servlet 的生命周期

Servlet 的生命周期是由 Servlet 容器(即 Web 服务器)来控制的,通过简单的概括可以分为 4 步:Servlet 类加载→实例化 Servlet→Servlet 提供服务→销毁 Servlet。

(1) Servlet 类加载,该阶段仅执行一次。

Servlet 容器在 Servlet 被第一次请求时执行。主要操作内容包括将 Servlet 对应的 class 文件载入内存,该过程开发人员无法参与。

(2) 实例化 Servlet,该阶段仅执行一次。

Servlet 容器创建 ServletConfig 对象,ServletConfig 对象包含 Servlet 的初始化配置信息,此外 Servlet 容器还会使得 ServletConfig 对象与当前 Web 应用的 ServletContext 对象关联。Servlet 容器创建 Servlet 对象,Servlet 容器依次调用 Servlet 对象的 init (ServletConfig config)和 init()方法。通过实例化步骤,创建了 Servlet 对象和 ServletConfig 对象,并且 Servlet 对象与 ServletConfig 对象关联,而 ServletConfig 对象又与当前对象的 ServletContext 对象关联。开发人员可以将需要执行的初始化代码放入 init()方法中来完成特定的初始化工作,init(ServletConfig config)方法不要重写,留给 Servlet 容器使用。

一般情况下,Servlet 类加载和实例化 Servlet 这两个阶段在 Servlet 被第一次请求时由 Web 容器执行。如果在 web.xml 文件中的 Servlet 配置信息中设置了＜load-on-startup＞元素,那么情况就并非如此了。例如,配置代码如下:

```xml
<servlet>
    <servlet-name>first</servlet-name>
    <servlet-class>com.ch02.FirstServlet</servlet-class>
    <load-on-startup>0</load-on-startup>
</servlet>
```

上述代码在配置文件中的＜servlet＞元素中添加了子元素＜load-on-startup＞并指定其值为0，那么 Web 应用启动时就加载并实例化这个 Servlet。

在 Servlet 的配置当中，＜load-on-startup＞0＜/load-on-startup＞的含义是表示容器是否在启动的时候就加载和实例化这个 Servlet。当＜load-on-startup＞值为 0 或者大于 0 时，表示容器在应用启动时就加载和实例化这个 Servlet；当是一个负数时或者没有指定时，则指示容器在该 Servlet 被访问时才加载和实例化。正数的值越小，启动该 Servlet 的优先级越高。当 Web 应用被重新启动时，Web 应用中的所有 Servlet 会在特定的时间被重新加载和实例化。

（3）Servlet 提供服务，该阶段客户端请求一次执行一次，具体执行几次取决于客户端的请求次数。

在这个阶段 Servlet 可以随时响应客户端的请求。当 Servlet 容器接到访问特定的 Servlet 请求时，Servlet 容器会创建针对这个请求的 ServletRequest 和 ServletResponse 对象，ServletRequest 对象用来封装请求信息，ServletResponse 对象用来生成响应信息，然后调用 service(ServletRequest request, ServletResponse response)方法，并把这两个对象当作参数传递给 service(ServletRequest request, ServletResponse response)方法。service(ServletRequest request, ServletResponse response)方法将 ServletRequest 类型的对象 request 和 ServletResponse 类型的 response 对象转化为对应的 HttpServletRequest 类型和 HttpServletResponse 类型对象，然后调用 service(HttpServletRequest request, HttpServletResponse response)方法，根据请求的提交方式调用 doGet(HttpServletRequest request, HttpServletResponse response)或者 doPost(HttpServletRequest request, HttpServletResponse response)。实际上，不管是 post 方式还是 get 方式提交请求，都会在 service 方法中处理，然后由 service 方法来交由相应的 doPost 或 doGet 方法处理，如果重写 service 方法，就不会再处理 doPost 或 doGet 了，所以开发人员在实际开发过程中不要重写 service 方法。当 Servlet 容器把 Servlet 生成的响应结果发送到客户端后，Servlet 容器会销毁 ServletRequest 和 ServletResponse 对象。

（4）销毁 Servlet，该阶段执行一次。

当 Web 应用被终止时，Servlet 容器会先调用 Web 应用中所有的 Servlet 对象的 destroy()方法，然后再销毁 Servlet 对象。此外，容器还会销毁与 Servlet 对象关联的 ServletConfig 对象。在 destroy()方法的实现中，可以编写代码释放 Servlet 所占用的资源，例如，关闭文件输入输出流、关闭与数据库的连接等。

在 Sevlet 的生命周期中，Servlet 的加载、实例化和销毁只会发生一次，因此 init()和 destroy()方法只能被 Servlet 容器调用一次，而 service()方法的执行次数取决于 Servlet 被客户端访问的次数。

为了使读者对 Servlet 的生命周期有一个深入的理解,下面来看一个 Servlet 的生命周期的实例。

例程 2-2:演示 Servlet 的生命周期。程序为 LiftServlet.java。

```
package com.ch02;
import java.io.IOException;
import javax.servlet.ServletConfig;
import javax.servlet.ServletException;
import javax.servlet.ServletRequest;
import javax.servlet.ServletResponse;
import javax.servlet.http.HttpServlet;
import javax.servlet.http.HttpServletRequest;
import javax.servlet.http.HttpServletResponse;
public class LiftServlet extends HttpServlet {
    //用来响应客户端发来的 GET 方式请求
    public void doGet(HttpServletRequest request, HttpServletResponse response)
            throws ServletException, IOException {
        System.out.println("doGet()...");
    }
    //用来响应客户端发来的 POST 方式请求
    public void doPost(HttpServletRequest request, HttpServletResponse response)
            throws ServletException, IOException {
        System.out.println("doPost()...");
    }
    //不能被覆盖
    protected void service(HttpServletRequest arg0, HttpServletResponse arg1)
            throws ServletException, IOException {
        System.out.println("service(HS,HS)...");
        super.service(arg0, arg1);
    }
    //不能被覆盖
    public void service(ServletRequest arg0, ServletResponse arg1)
            throws ServletException, IOException {
        System.out.println("service(S,S)...");
        super.service(arg0, arg1);
    }
    //编写代码释放所占用的资源
    public void destroy() {
        System.out.println("destroy()...");
        //super.destroy();
    }
    //写特定的初始化代码
    public void init() throws ServletException {
        System.out.println("init()...");
```

```
    }
    //不能被覆盖,留做 Web 服务器做 Servlet 初始化
    public void init(ServletConfig config) throws ServletException {
        System.out.println("init(ServletConfig)...");
        super.init(config);
    }
}
```

在 web.xml 文件配置 LiftServlet 代码如下：

```
<servlet>
    <servlet-name>lift</servlet-name>
    <servlet-class>com.ch02.LiftServlet</servlet-class>
</servlet>
<servlet-mapping>
    <servlet-name>lift</servlet-name>
    <url-pattern>/liftServlet</url-pattern>
</servlet-mapping>
```

在 MyEclipse 中重新部署工程，在 MyEclipse 中单击"部署"按钮进入如图 2-8 所示的 Project Deployments 界面。在 Project Deployments 对话框中，Project 下拉框中选中 LiftServlet 所在工程 chapter2，Deployments 列表中选中对应服务器，然后单击 Redeploy 按钮，此时在 Deployment Status 框中如果显示 Successfully deployed 说明重部署成功，单击 OK 按钮即可。

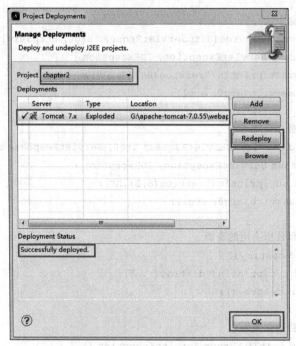

图 2-8　MyEclipse Project Deployments 界面

Java 项目工程 chapter2 重新部署完成后，启动 Web 服务器。在 MyEclipse 控制台打印信息中全部是有关 Web 服务器 Tomcat 的启动信息，说明此时 LiftServlet 中的任何方法均没有执行。

在浏览器地址栏中输入"http://localhost:8080/chapter2/liftServlet"访问 LiftServlet，此时在 MyEclipse 控制台打印信息如图 2-9 所示。

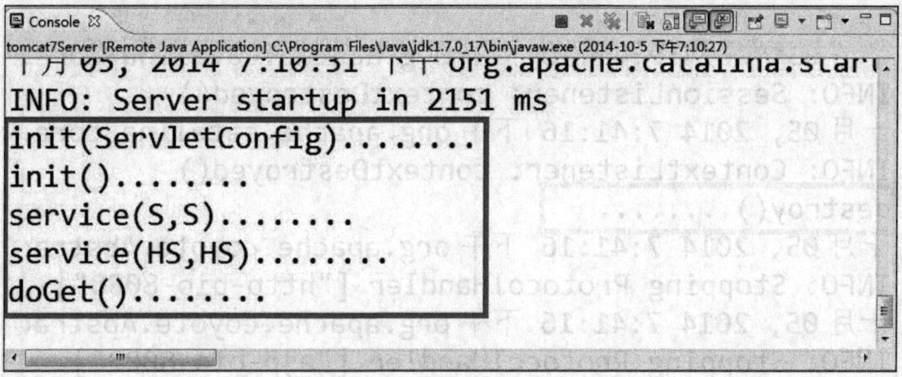

图 2-9　第一次访问 LiftServlet 结果界面

如图 2-9 所示的程序打印结果说明，LiftServlet 第一次被访问时，依次执行了该 Servlet 的 init(ServletConfig config)、init()、service(ServletRequest arg0，ServletResponse arg1)、service(HttpServletRequest arg0，HttpServletResponse arg1)、doGet(HttpServletRequest request，HttpServletResponse response)方法。

通过刷新浏览器再次访问 LiftServlet，此时在 MyEclipse 控制台打印信息如图 2-10 所示。

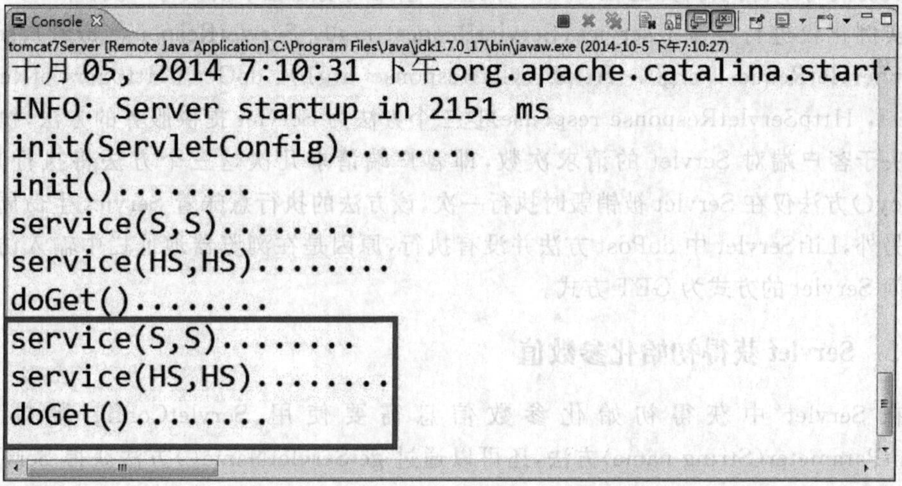

图 2-10　第二次访问 LiftServlet 结果界面

如图 2-10 所示的程序打印结果说明，LiftServlet 第二次被访问时，依次执行了该 Servlet 的 service(ServletRequest arg0，ServletResponse arg1)、service(HttpServletRequest arg0，

HttpServletResponse arg1）、doGet（HttpServletRequest request，HttpServletResponse response)这三个方法。读者可以验证在第三次、第四次等以后若干次访问该Servlet时都是执行这三个方法。init(ServletConfig config)、init()这两个方法并没有执行，这说明这两个方法只有在第一次访问Servlet时执行，以后不再执行。

将Web服务器正常终止，此时在MyEclipse控制台打印信息如图2-11所示。

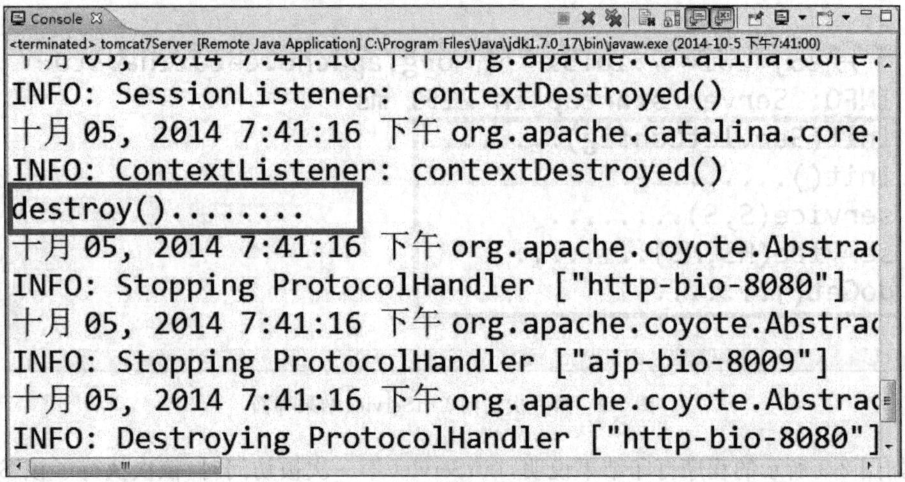

图2-11 Web服务器停止结果界面

如图2-10所示的程序打印结果说明，Web服务器终止将销毁Servlet，此时执行Servlet的destroy()方法。

综上所述，init(ServletConfig config)、init()方法在Servlet第一次被访问时执行，以后每次访问将不再执行，并且在Servlet的整个生命周期过程中仅执行一次，即Servlet加载和实例化仅进行一次。service(ServletRequest arg0，ServletResponse arg1）、service（HttpServletRequest arg0，HttpServletResponse arg1）、doGet（HttpServletRequest request，HttpServletResponse response)这三个方法为Servlet提供服务的方法，执行次数取决于客户端对Servlet的请求次数，即客户端请求几次这三个方法将执行几次。destroy()方法仅在Servlet被销毁时执行一次，该方法的执行意味着Servlet生命周期结束。另外，LiftServlet中doPost方法并没有执行，原因是在浏览器地址栏中输入访问路径访问Servlet的方式为GET方式。

2.3.3 Servlet获得初始化参数值

在Servlet中获得初始化参数信息需要使用ServletConfig接口中的getInitParameter(String name)方法，还可以通过getServletName()方法获得Servlet名字。Servlet名字即web.xml文件中相应＜servlet＞元素的＜servlet-name＞子元素的值，如果没有为Servlet配置＜servlet-name＞子元素，则返回Servlet类的名字。初始化参数包括参数名和参数值，在web.xml文件中配置一个Servlet时，可以通过＜init-param＞元素来设置初始化参数，＜init-param＞元素的＜param-name＞子元素设定参数名,<param-value>子元素设定参数值。

自定义的 Servlet 中可以直接使用 ServletConfig 接口中定义的方法。因为 HttpServlet 类继承 GenericServlet 类,而 GenericServlet 类实现了 ServletConfig 接口,所以在 HttpServlet 类或 GenericServlet 类及子类中都可以直接调用 ServletConfig 接口中的方法。

例程 2-3:演示 Servlet 获得初始化参数信息。程序为 GetInitParamServlet.java。

程序 GetInitParamServlet.java 代码如下:

```java
package com.ch02;
import java.io.IOException;
import java.io.PrintWriter;
import javax.servlet.ServletException;
import javax.servlet.http.HttpServlet;
import javax.servlet.http.HttpServletRequest;
import javax.servlet.http.HttpServletResponse;
public class GetInitParamServlet extends HttpServlet{
    protected void doGet(HttpServletRequest req, HttpServletResponse resp)
            throws ServletException, IOException {
        //通过 getInitParameter 方法获得配置文件中设置的初始化参数值
        String name=this.getInitParameter("name");
        String password=this.getInitParameter("password");
        //通过 getServletName 方法获得配置文件中设置 Servlet 名字
        String servletName=this.getServletName();
        resp.setContentType("text/html;charset=UTF-8");
                                                //设置响应的文本类型和编码方式
        PrintWriter out=resp.getWriter();//通过输出流向客户端做出响应
        out.println("<HTML><HEAD><TITLE>hello world</TITLE></HEAD>");
        out.println("<font size=5 color=red>ServletName="+servletName+
            "</font><br>");
        out.println("<font size=5 color=blue>name="+name+",password="+
            password+"</font>");
        out.println("</BODY></HTML>");
        out.close();
    }
    protected void doPost(HttpServletRequest req, HttpServletResponse resp)
            throws ServletException, IOException {
        doGet(req, resp);
    }
}
```

GetInitParamServlet 配置信息如下:

```
<servlet>
    <servlet-name>getInitParam</servlet-name>
```

```xml
        <servlet-class>com.ch02.GetInitParamServlet</servlet-class>
        <init-param>
            <param-name>name</param-name>
            <param-value>zhangsan</param-value>
        </init-param>
        <init-param>
            <param-name>password</param-name>
            <param-value>zs123</param-value>
        </init-param>
    </servlet>
    <servlet-mapping>
        <servlet-name>getInitParam</servlet-name>
        <url-pattern>/getInitParamServlet</url-pattern>
    </servlet-mapping>
```

重新部署 Servlet 所在工程,启动 Web 服务器,通过浏览器访问 GetInitParamServlet,运行结果如图 2-12 所示。

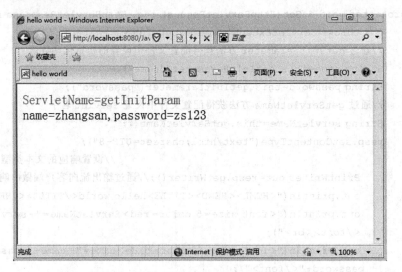

图 2-12　访问 GetInitParamServlet 运行结果

程序 GetInitParamServlet.java 通过 ServletConfig 提供的 getInitParameter(String name) 和 getServletName() 方法获得了配置文件中设置的初始化参数值和 GetInitParamServlet 的名字,并通过响应对象输出到客户端浏览器。

2.3.4　Servlet 处理表单

Servlet 处理 HTML 表单在实际开发中经常用到。在 HTML 的表单标记＜form＞的 action 属性中设置表单提交的目标,提交的目标可以是某一个 Servlet,那么 action 属性值是这个 Servlet 的 url-pattern,这里的 url-pattern 是配置文件 web.xml 中指定＜url-pattern＞值中去除/的部分。＜form＞的 method 属性中设置提交方式,如 GET 或

POST,提交后将去执行 Servlet 中对应的 doGet 或者 doPost 方法。表单提交的信息包括 HTML 中<form>和</form>标记之间的所有输入控件和选择控件,如文本框、单选按钮、复选框、文本域等。表单提交的信息将被封装在 ServletRequest 对象中,在 Servlet 中 ServletRequest 对象相关方法可以通过 HTML 控件的名字获得表单信息。

例程 2-4:演示 Servlet 处理表单数据。程序为 register.html、RegisterServlet.java。
HTML 文件 register.html 代码如下:

```
<!DOCTYPE HTML PUBLIC "-//W3C//DTD HTML 4.01 Transitional//EN">
<html>
    <head>
        <title></title>
        <meta http-equiv="keywords" content="keyword1,keyword2,keyword3">
        <meta http-equiv="description" content="this is my page">
        <meta http-equiv="content-type" content="text/html; charset=UTF-8">
    </head>
    <body>
        <form action="registerServlet" method="post">
        <center><h3>注册页面</h3></center>
        <table align="center" border="1">
        <tr><td>姓名:</td><td><input type="text" name="name" /></td></tr>
        <tr><td>密码:</td><td><input type="password" name="password"/></td></tr>
        <tr><td>确认密码:</td><td><input type="password" name="qpassword"/>
        </td></tr>
        <tr><td rowspan="2">性别</td><td><input type="radio" name="sex"
        value="男"/>男</td></tr>
        <tr><td align="left"><input type="radio" name="sex" value="女"/>女
        </td></tr>
        <tr><td rowspan="4">爱好</td><td><input type="checkbox" name="
        feature" value="运动" />运动</td></tr>
        <tr><td align="left"><input type="checkbox" name="feature" value="音
        乐"/>音乐</td></tr>
        <tr><td align="left"><input type="checkbox" name="feature" value="书
        法" checked/>书法</td></tr>
        <tr><td align="left"><input type="checkbox" name="feature" value="绘
        画"/>绘画</td></tr>
        <tr><td>省份:</td><td>
                <select name="privence">
                    <option value="北京">北京</option>
                    <option value="上海">上海</option>
                    <option value="天津" selected>天津</option>
                    <option value="内蒙古自治区">内蒙古自治区</option>
                    <option value="山东省">山东省</option>
                </select>
        </td></tr>
```

```html
                <tr><td>自我简介:</td><td><textarea rows="5" cols="10" name="author">
                </textarea></td></tr>
                <tr>
                    <td><input type="submit" value="提交"/></td>
                    <td><input type="reset" value="重置"/></td>
                </tr>
            </table>
        </form>
    </body>
</html>
```

程序 RegisterServlet.java 代码如下:

```java
package com.ch02;
import java.io.IOException;
import java.io.PrintWriter;
import javax.servlet.ServletException;
import javax.servlet.http.HttpServlet;
import javax.servlet.http.HttpServletRequest;
import javax.servlet.http.HttpServletResponse;
public class RegisterServlet extends HttpServlet {
    public void doPost ( HttpServletRequest request, HttpServletResponse response)
            throws ServletException, IOException {
        //设置请求编码方式
        request.setCharacterEncoding("UTF-8");
        //设置响应编码方式
        response.setCharacterEncoding("UTF-8");
        response.setContentType("text/html;charset=UTF-8");
        //通过 HttpServletRequest 中的 getParameter 方法获得客户端提交参数值
        String name=request.getParameter("name");
        String password=request.getParameter("password");
        String qpassword=request.getParameter("qpassword");
        String sex=request.getParameter("sex");
        //通过 HttpServletRequest 中的 getParameterValues 方法获得客户端提交的名
        //字为"feature"的一组参数值,通常用来获得复选框的所有选项值
        String[] feature=request.getParameterValues("feature");
        String privence=request.getParameter("privence");
        String author=request.getParameter("author");
        PrintWriter out=response.getWriter();
        //判断两次输入的密码是否一致,如果不一致输出出错信息,一致将输入的信息在客户
        //端显示
        if(! password.equals(qpassword)){
            out.println("<html>");
            out.println("<head><title>RegisterServlet</title></head>");
```

```
            out.println("<body><h1>两次输入的密码不一致,请重新输入!!!");
            out.println("</h1></body>");
            out.println("</html>");
        }else{
            out.println("<html>");
            out.println("<head><title>RegisterServlet</title></head>");
            out.println("<body><h1>个人注册详细信息如下:<br>");
            out.println("姓名:"+name+"<br>");
            out.println("密码:"+password+"<br>");
            out.println("确认密码:"+qpassword+"<br>");
            out.println("性别:"+sex+"<br>");
            out.println("爱好:");
            for(int i=0;i<feature.length;i++){
                out.println(feature[i]+" ");
            }
            out.println("<br>");
            out.println("省份:"+privence+"<br>");
            out.println("个人简介:"+author+"<br>");
            out.println("</h1></body>");
            out.println("</html>");
        }
        out.close();
    }
    protected void doGet(HttpServletRequest req, HttpServletResponse resp)
            throws ServletException, IOException {
        doPost(req, resp);
    }
}
```

RegisterServlet 配置代码如下：

```
<servlet>
    <servlet-name>register</servlet-name>
    <servlet-class>com.ch02.RegisterServlet</servlet-class>
</servlet>
<servlet-mapping>
    <servlet-name>register</servlet-name>
    <url-pattern>/registerServlet</url-pattern>
</servlet-mapping>
```

Servlet 所在工程重新部署,启动 Web 服务器,先通过浏览器访问 register.html,运行结果如图 2-13 所示。

在如图 2-13 所示的 HTML 页面中输入对应的值,然后单击"提交"按钮,将执行 RegisterServlet 的 doPost 方法,运行结果如图 2-14 所示。

在 register.html 中代码"<form action="registerServlet" method="post">"设置了表单数据提交的目标是配置为"/registerServlet"的 Servlet 即 RegisterServlet,提交方

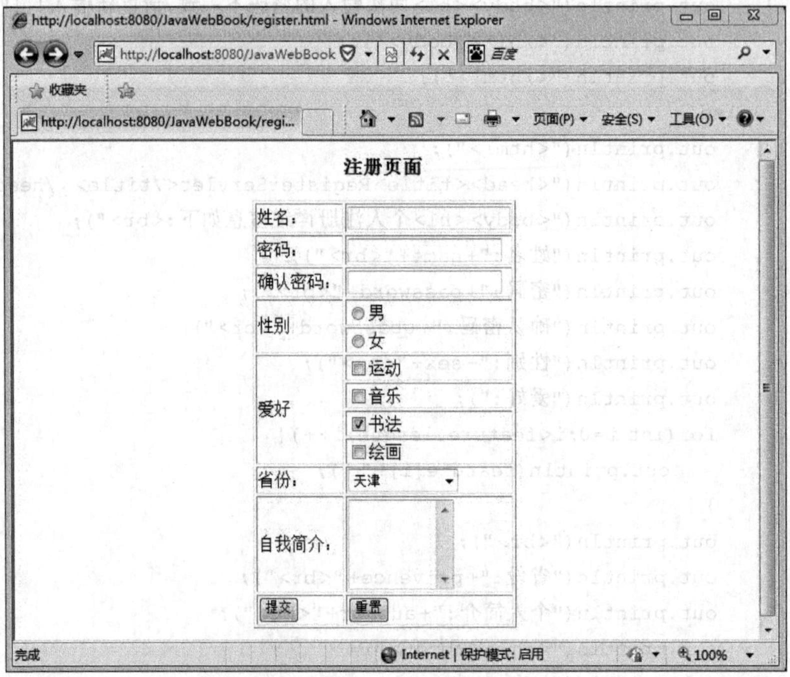

图 2-13 register.html 运行结果

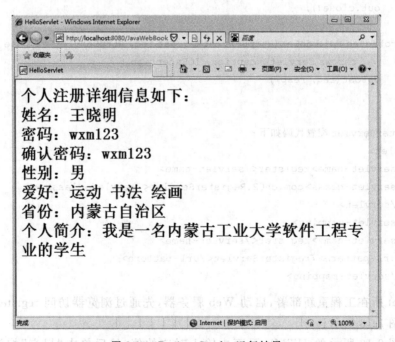

图 2-14 RegisterServlet 运行结果

式为 POST，所以 register.html 运行后输入数据提交后将执行 RegisterServlet 中的 doPost 方法。在 RegisterServlet 的 doPost 方法中，request 请求对象调用方法

getParameter(String name) 和 getParameterValues(String name)来获得表单提交的输入控件和选择控件的值,getParameter(String name)可以获得名字为指定参数的控件的一个值,如文本框、文本域、单选按钮或下拉框,通常用来获得单个数值。getParameterValues(String name)可以获得名字为指定参数的控件的一组值,如复选框。程序将从前台表单获取的信息通过 response 对象的输出流返回到客户端浏览器。

在 register.html 中"<form action="registerServlet" method="post">"中的代码"method="post""设置了表单属性 method 值为 post,则客户端页面给 Web 服务器端 Servlet 传递参数的提交方式为 POST 方式,将执行 Servlet 中的 doPost 方法。如果代码改为"method="get"",则表单传递参数的提交方式为 GET 方式,此时将执行 Servlet 中的 doGet 方法。另外,通过 URL 携带传递参数的提交方式也为 GET 方式,例如"http://localhost:8080/chapter10/registerServlet? name=zhangsan"是在访问 RegisterServlet 的同时并以 GET 方式向 RegisterServlet 传递参数 name,其值为 zhangsan。"http://localhost:8080/chapter10/registerServlet? name=zhangsan&password=zs123"是在访问 RegisterServlet 的同时并以 GET 方式向 RegisterServlet 传递两个参数 name 和 password,其值分别为 zhangsan 和 zs123。

例程 2-5:演示通过 URL 传递参数。程序为 URLDeliverParamServlet.java。

URLDeliverParamServlet.java 程序代码如下:

```
package com.ch02;
import java.io.IOException;
import java.io.PrintWriter;
import javax.servlet.ServletException;
import javax.servlet.http.HttpServlet;
import javax.servlet.http.HttpServletRequest;
import javax.servlet.http.HttpServletResponse;
public class URLDeliverParamServlet extends HttpServlet {
    public void doGet(HttpServletRequest request, HttpServletResponse response)
            throws ServletException, IOException {
        //设置响应的文本类型和编码方式
        response.setContentType("text/html;charset=UTF-8");
        String name=request.getParameter("name");
        String password=request.getParameter("password");
        PrintWriter out=response.getWriter();
        out.println("name:"+name);
        out.println("password:"+password);
        out.flush();
        out.close();
    }
}
```

URLDeliverParamServlet 配置代码如下:

<servlet>

```
        <servlet-name>URLDeliverParamServlet</servlet-name>
        <servlet-class>com.ch02.URLDeliverParamServlet</servlet-class>
    </servlet>
    <servlet-mapping>
        <servlet-name>URLDeliverParamServlet</servlet-name>
        <url-pattern>/URLDeliverParamServlet</url-pattern>
    </servlet-mapping>
```

URLDeliverParamServlet 所在工程重新部署，启动 Web 服务器，在浏览器地址栏中输入 http://localhost:8080/chapter10/registerServlet? name = zhangsan & password = zs123 访问 URLDeliverParamServlet，运行结果如图 2-15 所示。

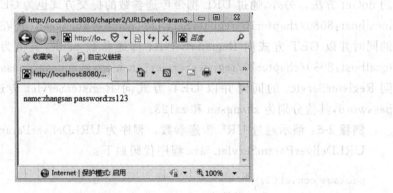

图 2-15　URLDeliverParamServlet 运行结果

例程 2-5 通过 URL 为 URLDeliverParamServlet 传递参数 name 和 password，这种参数传递方式为 GET 方式，所以执行 URLDeliverParamServlet 的 doGet 方法。该 Servlet 的 doGet 方法通过 request 对象的 getParameter 方法获得 name 和 password 参数的值，然后输出到客户端浏览器。

2.3.5　Servlet 中的跳转

1. 转发与包含

Web 应用在响应客户端的一个请求时，有可能响应过程比较复杂，需要多个 Web 组件(Servlet、JSP 或者 HTML)共同协作，才能生成响应结果。尽管一个 Servlet 对象无法去直接调用另一个 Servlet 对象的 service()方法，但是 Servlet 规范为 Web 组件之间的协作提供了两种途径：请求转发和请求包含。请求转发是 Servlet(源组件)先对客户端请求做一些预处理操作，然后把请求转发给其他 Web 组件(目标组件)来完成生成响应结果在内的后续操作。请求包含是 Servlet(源组件)把其他 Web 组件(目标组件)生成的响应结果包含在自身生成的响应结果中。请求转发和请求包含的共同点是：源组件和目标组件处理的都是同一个客户端请求，共享同一个 ServletRequest 对象和 ServletResponse 对象。

Servlet API 中定义了 javax.servlet.RequestDispatcher 接口，该接口中定义的

forward()和 include()方法分别实现了请求转发和请求包含。在 Servlet 中,当调用 forward()和 include()方法时,要把当前的 ServletRequest 对象和 ServletResponse 对象作为参数传给 forward()和 include()方法,这使得源组件和目标组件共享同一个 ServletRequest 对象和 ServletResponse 对象。在 Servlet 中可以通过两种方式获得 RequestDispatcher 对象:调用 ServletContext 的 getRequestDispatcher(String path)或者调用 ServletRequest 的 getRequestDispatcher(String path)方法,path 参数为指定目标组件的路径,区别在于前者 path 参数必须是绝对路径,而后者 path 参数既可以是绝对路径,也可以是相对路径。

2. 重定向

重定向是服务器对客户端的请求做出响应,响应的结果是让客户端浏览器去重新发出对另外一个 URL 的访问请求。

HttpServletResponse 对象的 sendRedirect()方法实现了 Web 组件的重定向。在 Servlet 中调用 sendRedirect()方法,源组件和目标组件使用各自不同的 ServletRequest 对象和 ServletResponse 对象,源组件和目标组件属于两个独立的访问请求和响应过程。

例程 2-6:演示请求转发和重定向的使用。程序为 ConnectionFactory.java、ResourceClose.java、login.html、LoginServlet.java、SuccessServlet.java 和 ErrorServlet.java。例程登录页面 login.html 提交登录信息至 LoginServlet,LoginServlet 查询数据库 XE 中的表 UserTbl 中是否存在该用户,如果存在请求转发至 SuccessServlet,否则客户端重定向至 ErrorServlet。例程 2-6 涉及数据库操作,首先应创建数据库以及数据库用户,然后创建 UserTbl 表。另外,涉及数据库操作例程所在项目工程一定要导入 Oracle 驱动程序 ojdbc14.jar 文件(参见随书电子资源)。创建表 UserTbl 的语句如下:

```
drop table userTbl;
create table usertbl(
    name varchar2(30),
    password varchar2(20),
    sex varchar2(10),
    privence varchar2(20),
    author varchar2(200)
);
insert into usertbl values('zhangsan','zs123','男','内蒙古自治区','我是一名大学生');
commit;
```

程序 ConnectionFactory.java、ResourceClose.java 是分别用来获取数据库连接、关闭数据库操作对象的类文件,本书凡是涉及数据库操作的例程均使用这两个类文件。

程序 ConnectionFactory.java 代码如下:

```
package com.ch02;
import java.sql.Connection;
import java.sql.DriverManager;
```

```java
public class ConnectionFactory {
    private static String driver="oracle.jdbc.driver.OracleDriver";
    private static String url="jdbc:oracle:thin:@127.0.0.1:1521:XE";
    private static String userName="webbook";
    private static String password="webbook";
    public static Connection getConnection(){
        try{
            Class.forName(driver);
            return DriverManager.getConnection(url,userName,password);
        }catch(Exception e){
            e.printStackTrace();
            return null;
        }
    }
}
```

程序 ResourceClose.java 代码如下：

```java
package com.ch02;
import java.sql.Connection;
import java.sql.ResultSet;
import java.sql.Statement;
public class ResourceClose {
    public static void close(ResultSet rs,Statement stmt,Connection conn){
        try{
            if(rs!=null)rs.close();
            if(stmt!=null)stmt.close();
            if(conn!=null)conn.close();
        }catch(Exception e){
            e.printStackTrace();
        }
    }
    public static void close(ResultSet rs,Statement stmt){
        close(rs,stmt,null);
    }
    public static void close(Statement stmt,Connection conn){
        close(null,stmt,conn);
    }
    public static void close(ResultSet rs,Connection conn){
        close(rs,null,conn);
    }
    public static void close(Connection conn){
        close(null,null,conn);
    }
    public static void close(ResultSet rs){
```

```
            close(rs,null,null);
        }
        public static void close(Statement stmt){
            close(null,stmt,null);
        }
    }
```

程序 login.html 代码如下：

```html
<!DOCTYPE HTML PUBLIC "-//W3C//DTD HTML 4.01 Transitional//EN">
<html>
    <head>
        <title>login.html</title>
        <meta http-equiv="keywords" content="keyword1,keyword2,keyword3">
        <meta http-equiv="description" content="this is my page">
        <meta http-equiv="content-type" content="text/html; charset=UTF-8">
    </head>
    <body>
        <form action="loginServlet" method="post">
            用户名:<input type="text" name="name"/><br>
            密码:<input type="password" name="password"/><br>
            <input type="submit" value="登录"/>
            <input type="reset" value="清空"/>
        </form>
    </body>
</html>
```

程序 LoginServlet.java 代码如下：

```java
package com.ch02;
import java.io.IOException;
import java.sql.Connection;
import java.sql.PreparedStatement;
import java.sql.ResultSet;
import javax.servlet.ServletException;
import javax.servlet.http.HttpServlet;
import javax.servlet.http.HttpServletRequest;
import javax.servlet.http.HttpServletResponse;
public class LoginServlet extends HttpServlet{
    protected void doGet(HttpServletRequest req, HttpServletResponse resp)
            throws ServletException, IOException {
        doPost(req, resp);
    }
    public void doPost (HttpServletRequest request,HttpServletResponse response)
        throws ServletException,IOException{
        request.setCharacterEncoding("UTF-8");
```

```java
        response.setCharacterEncoding("UTF-8");
        response.setContentType("text/html;charset=UTF-8");
        //通过请求对象从客户端获取 name 和 password 值
        String name=request.getParameter("name");
        String password=request.getParameter("password");
        PreparedStatement pstmt=null;
        ResultSet rs=null;
        Connection conn=null;
        try{
            String sql="select * from usertbl where name=? and password=?";
            conn=ConnectionFactory.getConnection();
            pstmt=conn.prepareStatement(sql);
            pstmt.setString(1, name);
            pstmt.setString(2, password);
            rs=pstmt.executeQuery();
            if(rs.next()){
                //将 name 信息保存在 request 对象中
                request.setAttribute("name", name);
                //请求转发
        request. getRequestDispatcher (" successServlet "). forward (request,
        response);
            }else{
                //将 name 信息保存在 request 对象中
                request.setAttribute("name", name);
                //客户端重定向
                response. sendRedirect("http://localhost:8080/chapter2/errorServlet");
            }
        }catch(Exception e){
            e.printStackTrace();
        }finally{
            ResourceClose.close(rs, pstmt, conn);
        }
    }
}
```

程序 SuccessServlet.java 代码如下:

```java
package com.ch02;
import java.io.IOException;
import java.io.PrintWriter;
import java.sql.Connection;
import java.sql.PreparedStatement;
import java.sql.ResultSet;
import javax.servlet.ServletException;
import javax.servlet.http.HttpServlet;
```

```java
import javax.servlet.http.HttpServletRequest;
import javax.servlet.http.HttpServletResponse;
public class SuccessServlet extends HttpServlet{
    protected void doGet(HttpServletRequest req, HttpServletResponse resp)
            throws ServletException, IOException {
        doPost(req, resp);
    }
    public void doPost (HttpServletRequest request,HttpServletResponse response)
        throws ServletException,IOException{
        request.setCharacterEncoding("UTF-8");
        response.setCharacterEncoding("UTF-8");
        response.setContentType("text/html;charset=UTF-8");
        PrintWriter out=response.getWriter();
        out.println("<HTML><HEAD><TITLE>登录成功</TITLE></HEAD>");
            out.println("<font size=5 color=red>恭喜您"+request.getAttribute("name")+",登录成功!!! </font>");
            out.println("</BODY></HTML>");
        out.close();
    }
}
```

程序 ErrorServlet.java 代码如下:

```java
package com.ch02;
import java.io.IOException;
import java.io.PrintWriter;
import java.sql.Connection;
import java.sql.PreparedStatement;
import java.sql.ResultSet;
import javax.servlet.ServletException;
import javax.servlet.http.HttpServlet;
import javax.servlet.http.HttpServletRequest;
import javax.servlet.http.HttpServletResponse;
public class ErrorServlet extends HttpServlet{
    protected void doGet(HttpServletRequest req, HttpServletResponse resp)
            throws ServletException, IOException {
        doPost(req, resp);
    }
    public void doPost (HttpServletRequest request,HttpServletResponse response)
        throws ServletException,IOException{
        request.setCharacterEncoding("UTF-8");
        response.setCharacterEncoding("UTF-8");
        response.setContentType("text/html;charset=UTF-8");
        PrintWriter out=response.getWriter();
        out.println("<HTML><HEAD><TITLE>登录失败</TITLE></HEAD>");
```

```
            out.println("<font size=5 color=red>对不起"+request.getAttribute
            ("name")+",登录失败!!!</font>");
            out.println("</BODY></HTML>");
            out.close();
        }
}
```

LoginServlet、SuccessServlet 和 ErrorServlet 在 web.xml 中的配置信息如下：

```
<servlet>
    <servlet-name>login</servlet-name>
    <servlet-class>com.ch02.LoginServlet</servlet-class>
</servlet>
<servlet-mapping>
    <servlet-name>login</servlet-name>
    <url-pattern>/loginServlet</url-pattern>
</servlet-mapping>
<servlet>
    <servlet-name>success</servlet-name>
    <servlet-class>com.ch02.SuccessServlet</servlet-class>
</servlet>
<servlet-mapping>
    <servlet-name>success</servlet-name>
    <url-pattern>/successServlet</url-pattern>
</servlet-mapping>
<servlet>
    <servlet-name>error</servlet-name>
    <servlet-class>com.ch02.ErrorServlet</servlet-class>
</servlet>
<servlet-mapping>
    <servlet-name>error</servlet-name>
    <url-pattern>/errorServlet</url-pattern>
</servlet-mapping>
```

部署 Java Web 项目 chapter2，启动 Web 服务器，访问登录页面 login.html，运行结果如图 2-16 所示。在如图 2-16 所示的登录页面中输入用户名 zhangsan 和密码 zs123，登录成功进入如图 2-17 所示的页面，否则跳转到如图 2-18 所示的页面。

程序 login.html 实现了登录页面，在登录页面上输入用户名 zhangsan、密码 zs123（创建表时添加了一条用户名为 zhangsan、密码为 zs123 的记录）提交到 LoginServlet。

图 2-16　登录界面

图 2-17 登录成功界面

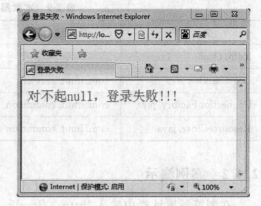
图 2-18 登录失败界面

在 LoginServlet 中,通过 request 对象获得登录页面提交的用户名和密码值,根据用户名和密码做数据库查询操作,存在该用户则将用户名 name 值通过代码"request.setAttribute("name", name);"保存在 request 对象中,并通过代码"request.getRequestDispatcher("successServlet").forward(request, response);"将请求转发至 SuccessServlet,由于请求转发时,LoginServlet 和 SuccessServlet 共享 request 对象,在 SuccessServlet 中可以通过"request.getAttribute("name")"获取 request 对象中保存的 name 值,所以在图 2-17 中可以显示 name 的值 zhangsan。

在 LoginServlet 中做数据库查询操作,不存在该用户则将用户名 name 值通过代码"request.setAttribute("name", name);"保存在 request 对象中,并通过代码"response.sendRedirect("http://localhost:8080/chapter2/errorServlet");"将客户端重定向至 ErrorServlet,由于客户端重定向时,LoginServlet 和 ErrorServlet 属于两个不同请求响应过程,不共享 request 对象,在 ErrorServlet 中通过"request.getAttribute("name")"并不能获取 request 对象中保存的 name 值,所以在图 2-18 中显示为 null。

2.4 案 例

本章案例实现"学生成绩管理系统"中登录的部分功能,由于课程进度原因无法实现系统中完整的登录功能,案例暂时先实现接收用户登录的信息,如果正确则显示名称信息,否则通过超链接返回到登录页面。

2.4.1 案例设计

在本模块中编写 login.html 供用户输入登录信息,编写 LoginServlet 接收用户登录的信息并在当前 Servlet 中显示该信息。将连接数据库和关闭数据库的操作分别使用 ConnectionFactory.java 和 ResourceClose.java 类来实现。本章案例使用的主要源文件如表 2-9 所示。

表 2-9　本章案例使用的文件

文　件	所在包/路径	功　能
login.html	/	登录界面
LoginServlet.java	com.imut.servlet	接收登录信息,访问数据库,根据结果显示信息
ConnectionFactory.java	com.imut.commmon	连接数据库
ResourceClose.java	com.imut.commmon	关闭数据库

2.4.2　案例演示

在浏览器地址栏中输入"http://localhost:8080/ch02/login.html",显示登录页面,效果如图 1-18 所示。在该页面中输入用户名和密码,单击"登录"按钮,如果用户名和密码正确则会显示欢迎信息,运行效果如图 2-19 所示,否则提示登录失败,并通过超级链接的形式提供返回登录页面的功能,运行效果如图 2-20 所示。

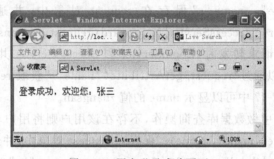

图 2-19　用户登录成功页面

图 2-20　用户登录失败页面

2.4.3　代码实现

创建工程 ch02,根据案例设计描述分别给出各部分具体实现。本章主要给出数据库连接类 ConnectionFactory.java、数据库关闭类 ResourceClose.java 以及 LoginServlet.java 的具体代码。

程序 ConnectionFactory.java 代码如下:

```java
package com.imut.commmon;
import java.sql.Connection;
import java.sql.DriverManager;
public class ConnectionFactory {
    /*
     * 数据库连接需要的四大参数:driver 代表操作 Oracle 数据库的驱动名称;url 代表数据
       库的 URL;
     * userName 代表连接数据库的用户名;password 代表连接数据库的用户密码。
     * 读者一定要注意参数值的拼写。
     */
    private static String driver="oracle.jdbc.driver.OracleDriver";
    private static String url="jdbc:oracle:thin:@localhost:1521:XE";
    //url 格式:协议名+@+数据库所在主机 IP 地址+数据库端口号+数据库名字;中间使用":"
      隔开。
    private static String userName="webbook";
    private static String password="webbook";
    /*
     * 该方法用来获得数据库连接,其返回类型为 Connection 引用类型;
     * 定义为 static 方法可以不用实例化类(ConnectionFactory)对象直接通过类名使用该
       方法。
     */
    synchronized public static Connection getConnection(){
        try{
            Class.forName(driver);//注册数据库驱动
            return DriverManager.getConnection(url,userName,password);
            //使用 DriverManager 的 getConnection 方法来获得连接,作为方法的返回值。
        }catch(Exception e){
            //如果程序出现异常,打印异常的栈信息,方法的返回值为 null。
            e.printStackTrace();
            return null;
        }
    }
}
```

程序 ResourceClose.java 代码如下:

```java
package com.imut.commmon;
import java.sql.Connection;
import java.sql.ResultSet;
import java.sql.Statement;
public class ResourceClose {
    public static void close(ResultSet rs,Statement stmt,Connection conn){
        try{
            if(rs!=null)rs.close();
            if(stmt!=null)stmt.close();
```

```java
            if(conn!=null)conn.close();
        }catch(Exception e){
            e.printStackTrace();
        }
    }
    public static void close(ResultSet rs,Statement stmt){
        close(rs,stmt,null);
    }
    public static void close(Statement stmt,Connection conn){
        close(null,stmt,conn);
    }
    public static void close(Connection conn){
        close(null,null,conn);
    }
    public static void close(ResultSet rs){
        close(rs,null,null);
    }
    public static void close(Statement stmt){
        close(null,stmt,null);
    }
}
```

程序 LoginServlet.java 代码如下:

```java
package com.imut.servlet;
import java.io.IOException;
import java.io.PrintWriter;
import java.sql.Connection;
import java.sql.PreparedStatement;
import java.sql.ResultSet;
import java.sql.SQLException;
import javax.servlet.ServletException;
import javax.servlet.http.HttpServlet;
import javax.servlet.http.HttpServletRequest;
import javax.servlet.http.HttpServletResponse;
import com.imut.commmon.ConnectionFactory;
import com.imut.commmon.ResourceClose;
public class LoginServlet extends HttpServlet {
    public void doGet(HttpServletRequest request, HttpServletResponse response)
            throws ServletException, IOException {
        this.doPost(request, response);
    }
    public void doPost(HttpServletRequest request, HttpServletResponse response)
            throws ServletException, IOException {
        //设置响应的 MIME 类型以及编码格式
```

```java
response.setContentType("text/html;charset=UTF-8");
//设置请求的编码格式
request.setCharacterEncoding("UTF-8");
//接收用户登录传递过来的用户名和密码
String loginName=request.getParameter("loginName");
String password=request.getParameter("password");
Connection conn=null;
PreparedStatement pstmt=null;
ResultSet rs=null;
String name=null;
try{
conn=ConnectionFactory.getConnection();
String sql="select * from admin where loginName=? and password=?";
pstmt=conn.prepareStatement(sql);
pstmt.setString(1, loginName);
pstmt.setString(2, password);
rs=pstmt.executeQuery();
while(rs.next()){
    name=rs.getString(2);
}
}catch(SQLException e){
    e.printStackTrace();
}finally{
    ResourceClose.close(rs, pstmt, conn);
}
PrintWriter out=response.getWriter();
out.println ( " <!DOCTYPE  HTML  PUBLIC  \" -//W3C//DTD  HTML  4. 01
Transitional//EN\">");
out.println("<HTML>");
out.println("<HEAD><TITLE>A Servlet</TITLE></HEAD>");
out.println("<BODY>");
if(name!=null){
    out.println("登录成功,欢迎您:"+name+"<br>");
}else{
    out.println("用户名或密码错误,请<a href='login.html'>重新登录</a><
    br>");
}
out.println("    </BODY>");
out.println("</HTML>");
out.flush();
out.close();
}
}
```

程序 LoginServlet.java 在 web.xml 中的配置信息如下:

```
<servlet>
    <servlet-name>loginServlet</servlet-name>
    <servlet-class>com.imut.servlet.LoginServlet</servlet-class>
</servlet>
<servlet-mapping>
    <servlet-name>loginServlet</servlet-name>
    <url-pattern>/loginServlet</url-pattern>
</servlet-mapping>
```

【代码分析】程序 LoginServlet.java 使用 HttpServletRequest 对象的 getParameter() 方法接收用户登录的信息,并使用接收到的用户名和密码作为 SQL 查询的条件。如果查询到的结果集 ResultSet 对象不为空,证明用户名和密码正确,则把得到的结果集中的第二列即真实姓名取出赋给 name 变量,提示登录成功,并显示登录者的真实姓名。如果结果集为空,表明用户名和密码错误,打印错误提示信息,同时提供返回到登录页面的超级链接。

习　题

1. 选择题

(1) 下列哪一项不是 Servlet 使用的方法?(　　)

　　A. doGet()　　　　B. doPost()　　　　C. service()　　　　D. close()

(2) 下列哪个方法在 Servlet 载入时执行一次,而且只执行一次,负责对 Servlet 进行初始化?(　　)

　　A. service()　　　　B. doGet()　　　　C. init()　　　　D. destroy()

(3) 下面是 Servlet 部署文件片段

```
<servlet>
    <servlet-name>Hello</servlet-name>
    <servlet-class>com.chap.HelloServlet</servlet-class>
</servlet>
<servlet-mapping>
    <servlet-name>Hello</servlet-name>
    <url-pattern>/helpHello</url-pattern>
</servlet-mapping>
```

Servlet 的类名是(　　)。

　　A. Hello　　　　B. HelloServlet　　　　C. helpHello　　　　D. /helpHello

(4) 下面哪一项对 Servlet 描述错误?(　　)

　　A. Servlet 是一个特殊的 Java 类,它必须直接或间接实现 Servlet 接口

　　B. Servlet 接口定义了 Servlet 的生命周期方法

　　C. 当多个客户请求一个 Servlet 时,服务器为每一个客户开启一个进程

D. Servlet 客户线程调用 service 方法响应客户请求

(5) 在 Web 容器中,以下哪个类的实例代表 HTTP 的请求?（　　）
    A. HttpRequest　　　　　　　　B. HttpServletRequest
    C. HttpServletResponse　　　　　D. HttpPrintWriter

(6) Servlet 必须基于哪一个类型的容器才能提供服务?（　　）
    A. Applet 容器　　　　　　　　B. 应用程序客户端容器
    C. Web 容器　　　　　　　　　D. EJB 容器

(7) HttpServlet 是定义在以下哪个套件中?（　　）
    A. javax.servlet　　　　　　　　B. javax.servlet.http
    C. java.http　　　　　　　　　　D. javax.http

(8) 在 Web 应用程序的目录结构中,web.xml 是直接放置在以下哪个目录中?（　　）
    A. WEB-INF 目录　　　　　　　B. conf 目录
    C. lib 目录　　　　　　　　　　D. classes 目录

(9) HttpServlet 的子类要从 HTTP 请求中获得请求参数,应该调用哪个方法?（　　）
    A. 调用 HttpServletRequest 对象的 getAttribute() 方法
    B. 调用 ServletContext 对象的 getAttribute() 方法
    C. 调用 HttpServletRequest 对象的 getParameter() 方法
    D. 调用 HttpServletRequest 对象的 getHeader() 方法

(10) 对于自己编写的 Servlet1,以下对 Servlet1 的定义正确的是（　　）。
    A. class Servlet1 implements javax.servlet.servlet
    B. class Servlet1 extends javax.servlet.GenericServlet
    C. class Servlet1 extends javax.servlet.http.HttpServlet
    D. class Servlet1 extends javax.servlet.ServletRequest

2. 填空题

(1) 一般编写一个 Servlet 就是编写一个_____的子类。

(2) Web 应用关闭时,_____就被销毁。

(3) 使用 Servlet 处理表单提交时,两个最重要的方法是_____和_____。

(4) HTTP 把客户端的请求方式分为_____、_____、_____、_____等多种方式,在 HttpServlet 类中对每一种请求方式都提供相应的服务方法,如_____、_____、_____和_____方法等。

(5) Servlet API 包含两个包:_____与_____包。

(6) 在 web.xml 文件中配置一个 Servlet 时,可以通过_____元素来设置初始化参数,该元素有两个子元素分别实现参数名和参数值的设置,这两个子元素分别为_____和_____。

3. 简答题

(1) 试述 Servlet 的生命周期。

(2) 如何编写、配置及运行一个 Servlet 程序。

(3) 请描述 Servlet 如何处理表单请求。

4. 程序设计题

(1) 编写一个 Servlet 程序,令其在被访问时在页面上显示一个九九乘法表。

(2) 编写 Servlet 程序,实现如下功能:先通过浏览器访问如下书籍信息 HTML 页面,运行"单击"提交按钮,运行完成的 Servlet 程序,获取书籍表单中相关控件的信息,然后通过 Response 对象的输出流返回到客户端浏览器,最终显示效果如图 2-21 所示。

图 2-21　浏览书籍信息

状态管理与作用域对象

状态管理和作用域对象在 Java Web 应用程序开发中占有十分重要的地位,任何一个 Java Web 工程应用都要用到状态管理和作用域对象相关内容。Web 应用中客户端与服务器端的交互状态往往需要保存下来,由于 HTTP 是无状态的,所以某些状态信息的保存必须依靠协议外的机制实现。目前,在 Java Web 开发中常采用 Cookie 和 Session 来实现状态信息的保存。作用域对象的主要作用是充当容器将某些有用的属性信息保存下来,供整个 Web 应用中不同的 Servlet 或者 JSP 等组件共享。本章将以理论与实践相结合的方式介绍状态存储技术 Cookie 和 Session、作用域对象 ServletContext、HttpSession、ServletRequest 以及三者之间的比较等内容。

3.1 Java Web 状态管理

HTTP 是个无状态的协议。在 HTTP 中,客户端打开一个连接,并发送请求到服务器,服务器响应请求到客户端,最后关闭连接。在关闭连接之后 Web 服务器不会保存这次请求响应过程的任何状态信息。当下一次请求发起时,Web 服务器会把这次请求看成一个新的连接,和前面的请求无关。HTTP 的无状态是指协议对于事务处理没有记忆能力。不保存状态信息意味着如果后续处理需要前面的状态信息,则它必须重传,这样可能导致每次连接传送的数据量增大;另一方面,在服务器不需要先前信息时它的应答就较快。

在 Web 程序开发中,保持状态是非常重要的,一个具有状态的协议或者存储状态的协议外机制可以用来帮助在多个请求和响应之间实现复杂的应用逻辑,例如常见的自动登录、网络购物车功能的实现。于是,两种用于保持 HTTP 连接状态的技术 Cookie 和 Session 就应运而生了。

3.1.1 Cookie

1. Cookie 介绍

Cookie 是通过客户端保持状态的解决方案。从定义上讲,Cookie 是 Web 服务器通过浏览器保存在客户端硬盘上的一个文本文件,其中保存的是以名/值(name/value)形式存储的文本信息。服务器在接收到客户端请求之后,开发人员可以通过请求对象访问这个文件,用来解决浏览器用户和服务器之间的无状态通信。很多 Web 应用系统登录时可以选择"保存一周"、"两周内自动登录"或者选择"记住我"等功能都是通过 Cookie 技术实现的。

2. Cookie 编程

Cookie 编程要用到 Servlet API 中提供的类 javax.servlet.http.Cookie、接口 javax.servlet.http.HttpServletRequest 和 javax.servlet.http.HttpServletResponse 中的部分方法。一般步骤如下。

1) 创建 Cookie 对象

通过调用 Cookie 类的构造方法即可创建一个 Cookie 对象,代码如下:

```
Cookie cookie=new Cookie(String name,String value);
```

2) 设置 Cookie 的属性

通过 Cookie 类的方法 setMaxAge(int expiry)来设置 Cookie 的生命周期或者存活时间,expiry 单位为秒。代码如下:

```
cookie.setMaxAge(3600);              //设置 Cookie 的存活时间为 1 小时
```

3) 发送 Cookie 对象

在服务器端创建了 Cookie 对象,通过响应对象 response 的 add(Cookie cookie)方法将 Cookie 对象发送到客户端浏览器,保存在客户端计算机的硬盘上。代码如下:

```
response.addCookie(cookie);
```

4) 读取 Cookie

通过请求对象 request 的 getCookies()方法来获得一个 Cookie 数组,然后遍历 Cookie 数组,通过 name 查找所要的 cookie。代码如下:

```
Cookie[] cookies=req.getCookies();
for(int i=0;i<cookies.length;i++){
    if(cookies[i].getName().equals("username")){
        username=cookies[i].getValue();
        out.println("welcome,"+username+"!");
        break;
    }
}
```

5) 修改 Cookie

通过 Cookie 类的方法 setValue(String newValue)来修改 Cookie 的值。代码如下:

```
cookie.setValue("zhangsan");
```

6) 删除 Cookie

Servlet API 并没有提供删除 Cookie 对象的方法,但是可以通过 Cookie 类的方法

setMaxAge(int expiry)设置其存活时间为 0 来使 Cookie 失效,达到和删除一样的效果。代码如下:

```
cookie.setMaxAge(0);
```

在使用 Cookie 时必须注意,要保证浏览器接收所有 Cookie。具体操作为:打开浏览器,选择"工具"→"Internet 选项"命令,进入"Internet 选项"对话框,如图 3-1 所示。在"Internet 选项"对话框中选择"隐私"选项卡,将设置区域中滚动滑块置于最底端,此时浏览器将接受所有 Cookie。如果将设置区域中滚动滑块置于最顶端,此时浏览器将阻止所有 Cookie 即禁用 Cookie。

图 3-1 设置浏览器接收所有 Cookie

例程 3-1:演示 Cookie 技术的使用。程序为 input.html、SetCookieServlet.java、GetCookieServlet.java。例程中 input.html 提交表单数据由 SetCookieServlet 获取并利用 Cookie 保存到客户端,GetCookieServlet 来读取 Cookie 数据并显示。

程序 input.html 代码如下:

```
<!DOCTYPE HTML PUBLIC "-//W3C//DTD HTML 4.01 Transitional//EN">
<html>
    <head>
        <title>input.html</title>
        <meta http-equiv="keywords" content="keyword1,keyword2,keyword3">
        <meta http-equiv="description" content="this is my page">
        <meta http-equiv="content-type" content="text/html; charset=UTF-8">
    </head>
    <body>
        <form action="setCookieServlet" method="post">
        <table>
            <tr>
                <td align="right">用户名:</td>
                <td ><input type="text" name="username"></td>
            </tr>
            <tr>
                <td align="center" colspan="2"><input type="submit" value="提交"><input type="reset" value="重置"></td>
            </tr>
        </table>
        </form>
    </body>
```

```
</html>
```

程序 SetCookieServlet.java 代码如下：

```java
package com.ch03;
import java.io.IOException;
import java.io.PrintWriter;
import java.util.Date;
import javax.servlet.ServletException;
import javax.servlet.http.Cookie;
import javax.servlet.http.HttpServlet;
import javax.servlet.http.HttpServletRequest;
import javax.servlet.http.HttpServletResponse;
public class SetCookieServlet extends HttpServlet {
    public void doPost(HttpServletRequest request, HttpServletResponse response)
        throws ServletException, IOException {
    response.setContentType("text/html;charset=UTF-8");
    String username=request.getParameter("username");
    PrintWriter out=response.getWriter();
    String responseContent=null;
    if(username!=null&&!username.equals("")){
        //创建 Cookie 对象
        Cookie c1=new Cookie("username",username);
        Date date=new Date();
        Cookie c2=new Cookie("lastCall",date.toString());
        //设置 Cookie 的存活时间
        c1.setMaxAge(60*60*24*30);
        c2.setMaxAge(60*60*24*30);
        //将 Cookie 发送到客户端保存
        response.addCookie(c1);
        response.addCookie(c2);
        responseContent="本次登录的用户名和时间已成功写入 Cookie.<br><a href
            ='/chapter3/getCookieServlet'>读取 Cookie 信息</a>";
    }else{
        responseContent="用户名为空,请重新输入.<br><a href='/chapter3/
            input.html'>重新输入</a>";
    }
    out.println("<!DOCTYPE HTML PUBLIC \"-//W3C//DTD HTML 4.01 Transitional//
        EN\">");
    out.println("<HTML>");
    out.println("    <HEAD><TITLE>A Servlet</TITLE></HEAD>");
    out.println("    <BODY>");
    out.println("<h2><font color=red>"+responseContent+"</font></h2>");
    out.println("    </BODY>");
    out.println("</HTML>");
```

```java
        out.flush();
        out.close();
    }
}
```

程序 GetCookieServlet 代码如下:

```java
package com.ch03;
import java.io.IOException;
import java.io.PrintWriter;
import javax.servlet.ServletException;
import javax.servlet.http.Cookie;
import javax.servlet.http.HttpServlet;
import javax.servlet.http.HttpServletRequest;
import javax.servlet.http.HttpServletResponse;
public class GetCookieServlet extends HttpServlet {
    protected void doGet(HttpServletRequest req, HttpServletResponse resp)
            throws ServletException, IOException {
        doPost(req, resp);
    }
    public void doPost(HttpServletRequest request, HttpServletResponse response)
            throws ServletException, IOException {
        response.setContentType("text/html;charset=UTF-8");
        PrintWriter out=response.getWriter();
        out.println("<!DOCTYPE HTML PUBLIC \"-//W3C//DTD HTML 4.01 Transitional//EN\">");
        out.println("<HTML>");
        out.println("<HEAD><TITLE>A Servlet</TITLE></HEAD>");
        out.println("<BODY>");
        out.println("<h2>从 Cookie 中读取的用户名和上次登录时间</h2>");
        //获得客户端保存的 Cookie 数组
        Cookie[] cookies=request.getCookies();
        Cookie cookie=null;
        //读取 Cookie
        for(int i=0;i<cookies.length;i++){
            cookie=cookies[i];
            if(cookie.getName().equals("username")){
                out.println("用户名:"+cookie.getValue()+"<br>");
            }
            if(cookie.getName().equals("lastCall")){
                out.println("上次登录时间:"+cookie.getValue()+"<br>");
            }
        }
        out.println("</BODY>");
        out.println("</HTML>");
```

```
            out.flush();
            out.close();
        }
    }
```

SetCookieServlet 和 GetCookieServlet 配置代码如下：

```
<servlet>
    <servlet-name>setCookie</servlet-name>
    <servlet-class>com.ch03.SetCookieServlet</servlet-class>
</servlet>
<servlet-mapping>
    <servlet-name>setCookie</servlet-name>
    <url-pattern>/setCookieServlet</url-pattern>
</servlet-mapping>
<servlet>
    <servlet-name>getCookie</servlet-name>
    <servlet-class>com.ch03.GetCookieServlet</servlet-class>
</servlet>
<servlet-mapping>
    <servlet-name>getCookie</servlet-name>
    <url-pattern>/getCookieServlet</url-pattern>
</servlet-mapping>
```

部署 Java Web 项目，启动 Web 服务器，先通过浏览器访问 input.html，运行结果如图 3-2 所示。

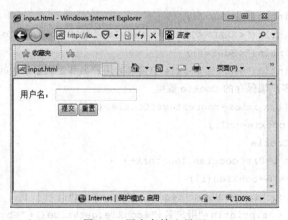

图 3-2　用户名输入界面

在如图 3-2 所示界面中，"用户名"文本框没有输入值时单击"提交"按钮，程序进入如图 3-3 所示界面，此时可以单击"重新输入"链接重新进入如图 3-2 所示界面进行输入。在如图 3-2 所示界面中"用户名"文本框中输入"zhangsan"，然后单击"提交"按钮，程序进入如图 3-4 所示界面。在如图 3-4 所示界面中单击"读取 Cookie 信息"链接，程序运行结果界面如图 3-5 所示。

第 3 章 状态管理与作用域对象

图 3-3　重新输入界面

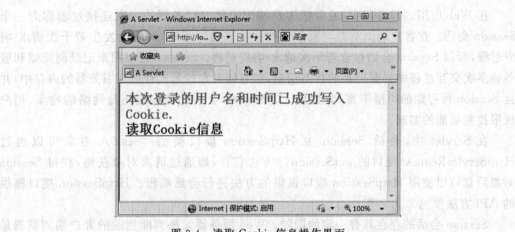

图 3-4　读取 Cookie 信息操作界面

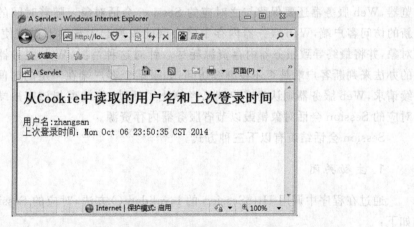

图 3-5　显示 Cookie 信息界面

程序 SetCookieServlet.java 类从表单读取客户端提交的用户名信息,如果用户名为空向客户端浏览器返回一个带有重新输入超链接的信息页面,让用户重新输入用户名。

用户名不为空则创建 Cookie 对象将用户名信息和当前时间保存，将 Cookie 对象发送到客户端保存，并向客户端浏览器返回一个带有"读取 Cookie 信息"（调用 GetCookieServlet）超链接的信息页面。单击"读取 Cookie 信息"超链接调用 GetCookieServlet，GetCookieServlet.java 类实现了从客户端读取 Cookie 信息的功能，并将读取的 Cookie 信息显示在浏览器中。

3. Cookie 的缺点

Cookie 的缺点首先是安全性差，Cookie 在个人计算机硬盘上所保存的文本信息是明文格式存储的，没有任何的加密措施，可以被任意篡改和删除；其次，用户可以让浏览器禁止使用 Cookie，所以通过 Cookie 保存的信息不应是服务器程序运行必需的信息；最后，Cookie 存储数据大小及数量是有限制的。

3.1.2 Session

在 Web 应用运行过程中，通常把客户端浏览器与服务器的一次连接过程称为一个 Session 会话。在客户端浏览器与服务器的一次连接过程中，可能会发生若干次请求-响应过程，所以 Session 会话包含若干次请求-响应过程。Session 可以用来记录浏览器和服务器多次交互过程中需要记录的状态信息，信息以对象的形式存放在服务器的内存中，并且 Session 可存储的类型丰富、数据量大。Session 会话的典型应用是网络购物车、用户权限控制功能的实现。

在 Servlet 中，会话 Session 为 HttpSession 接口类型。Session 对象可以通过 HttpServletRequest 接口的 getSession()方法获得，即通过请求对象获得，获得 Session 对象后就可以使用 HttpSession 接口提供的方法进行会话编程。HttpSession 接口提供的 API 方法参见 2.3.1 节。

Session 会话的存在具有一定的期限。Web 服务器无法判断当前的客户端浏览器是否还会继续访问，也无法检测客户端浏览器是否关闭，所以即使客户已经离开或关闭了浏览器，Web 服务器还要保留与之对应的 Session 会话对象。随着时间的推移而不断增加新的访问客户端，Web 服务器内存中将会因此积累起大量的不再被使用的 HttpSession 对象，并将最终导致服务器内存资源耗尽。针对这种情况，Web 服务器采用"超时限制"的办法来判断客户端是否还在继续访问，如果某个客户端在一定的时间之内没有发出后续请求，Web 服务器则认为客户端已经停止了活动，将结束与该客户端的会话并将与之对应的 Session 会话对象销毁以节省服务器内存资源。

Session 会话结束有以下三种方式。

1. 主动关闭

通过在程序中调用 HttpSession 的 invalidate()方法，对应的 Session 将销毁。代码如下：

```
session.invalidate();
```

2. 会话超时

Session 会话默认的超时时间为 30 分钟。超时时间从 Session 不活动开始计时,如果在计时过程中有新的活动则重新开始计时。

修改超时时间有以下两种方式。

方法 1:在 web.xml 中配置,该配置对 Web 应用中的所有 Session 均有效。配置代码如下:

```
<session-config>
    <session-timeout>10</session-timeout>        //这里设置的时间单位为分钟
</session-config>
```

方法 2:通过在程序中调用 HttpSession 的 setMaxInactiveInterval(int interval) 方法设置超时时间,interval 单位为秒。该操作只对调用方法的当前 Session 对象有效。代码如下:

```
session.setMaxInactiveInterval(3600);           //设置超时时间为 1 小时
```

3. 浏览器关闭

关闭 IE 浏览器可以销毁当前 Session 会话,对于其他浏览器不一定适用。

例程 3-2:演示 Session 会话的使用。程序为 login.html、LoginServlet.java、IndexServlet.java、View1Servlet.java、View2Servlet.java。程序中用到的 ConnectionFactory.java 和 ResourceClose.java 与例程 2-6 相同,详细参见随书电子资源。

程序 login.html 代码如下:

```html
<!DOCTYPE HTML PUBLIC "-//W3C//DTD HTML 4.01 Transitional//EN">
<html>
    <head>
        <title>login.html</title>
        <meta http-equiv="keywords" content="keyword1,keyword2,keyword3">
        <meta http-equiv="description" content="this is my page">
        <meta http-equiv="content-type" content="text/html; charset=UTF-8">
    </head>
    <body>
        <form action="login" method="post">
            用户名:<input type="text" name="name"/><br>
            密码:<input type="password" name="password"/><br>
            <input type="submit" value="登录"/>
            <input type="reset" value="清空"/>
        </form>
    </body>
</html>
```

程序 LoginServlet.java 代码如下：

```java
package com.ch03;
import java.io.IOException;
import java.sql.Connection;
import java.sql.PreparedStatement;
import java.sql.ResultSet;
import javax.servlet.ServletException;
import javax.servlet.http.HttpServlet;
import javax.servlet.http.HttpServletRequest;
import javax.servlet.http.HttpServletResponse;
import javax.servlet.http.HttpSession;
public class LoginServlet extends HttpServlet{
    /*
     * 验证登录状态，如果登录成功，将用户名信息保存到
     * HttpSession级的attribute中
     **/
    public void doPost (HttpServletRequest request,HttpServletResponse response)
        throws ServletException,IOException{
        request.setCharacterEncoding("UTF-8");
        response.setCharacterEncoding("UTF-8");
        response.setContentType("text/html;charset=UTF-8");
        String name=request.getParameter("name");
        String password=request.getParameter("password");
        /*
         * 获得相应的 Session:request.getSession()
         * 如果当前用户已经有 Session 则返回
         * 如果当前用户没有 Session 则创建新的 Session 对象返回；
         **/
        HttpSession session=request.getSession();
        PreparedStatement pstmt=null;
        ResultSet rs=null;
        Connection conn=null;
        try{
            String sql="select * from usertbl1 where name=? and password=?";
            conn=ConnectionFactory.getConnection();
            pstmt=conn.prepareStatement(sql);
            pstmt.setString(1, name);
            pstmt.setString(2, password);
            rs=pstmt.executeQuery();
            if(rs.next()){
                //将name信息保存在HttpSession对象中
                session.setAttribute("name", name);
                request. getRequestDispatcher ( " index "). forward ( request,
```

```
                response);
            }else{
                request.getRequestDispatcher("/login.html").forward(request,
                response);
            }
        }catch(Exception e){
            e.printStackTrace();
        }finally{
            ResourceClose.close(rs, pstmt, conn);
        }
    }
}
```

程序 IndexServlet.java 代码如下：

```
package com.ch03;
import java.io.IOException;
import java.io.PrintWriter;
import javax.servlet.ServletException;
import javax.servlet.http.HttpServlet;
import javax.servlet.http.HttpServletRequest;
import javax.servlet.http.HttpServletResponse;
import javax.servlet.http.HttpSession;
public class IndexServlet extends HttpServlet{
    public void doGet(HttpServletRequest request,HttpServletResponse response)
        throws ServletException,IOException{
        doPost(request,response);
    }
    public void doPost (HttpServletRequest request,HttpServletResponse response)
        throws ServletException,IOException{
        //通过 request 获得 Session 对象
        HttpSession session=request.getSession();
        response.setContentType("text/html;charset=UTF-8");
        PrintWriter out=response.getWriter();
        out.println("<html><body><h1>");
        //获得 Session 的 ID
        out.println("session ID:"+session.getId()+"<br>");
        //从 Session 对象中获得 name 值
        out.println("Session Attribute:"+session.getAttribute("name")+"<br>");
        //session.getMaxInactiveInterval():获得 Session 默认的过期时间
        out.println(" Session timeout:"+ session. getMaxInactiveInterval () +
        "s"+"<br>");
        /*
        *设置当前 Session 的过期时间为 300 秒,注:该设置只对当前 Session 有效对 Web 应
         用中其他 Session 无效
```

```
        **/
        session.setMaxInactiveInterval(300);
        out.println("Session timeout:"+
            session.getMaxInactiveInterval()+"s"+"<br>");
        /*
        *强制销毁当前Session
        **/
        session.invalidate();
        out.println("<a href=\"view1\">View1</a><br>");
        out.println("<a href=\"view2\">View2</a><br>");
        out.println("</h1></body></html>");
        out.close();
    }
}
```

程序 View1Servlet.java 代码如下:

```
package com.ch03;
import java.io.IOException;
import java.io.PrintWriter;
import javax.servlet.ServletException;
import javax.servlet.http.HttpServlet;
import javax.servlet.http.HttpServletRequest;
import javax.servlet.http.HttpServletResponse;
import javax.servlet.http.HttpSession;
public class View1Servlet extends HttpServlet{
    public void doGet(HttpServletRequest request,
            HttpServletResponse response)
        throws ServletException,IOException{
        doPost(request,response);
    }
    public void doPost(HttpServletRequest request,HttpServletResponse response)
        throws ServletException,IOException{
        HttpSession session=request.getSession();
        response.setContentType("text/html;charset=UTF-8");
        PrintWriter out=response.getWriter();
        out.println("<html><body><h1>");
        out.println("This is view1.....<br>");
        out.println("session ID:"+session.getId()+"<br>");
        out.println("Session Attribute:"+session.getAttribute("name")+"<br>");
        out.println("</h1></body></html>");
        out.close();
    }
}
```

程序 View2Servlet.java 代码如下：

```java
package com.ch03;
import java.io.IOException;
import java.io.PrintWriter;
import javax.servlet.ServletException;
import javax.servlet.http.HttpServlet;
import javax.servlet.http.HttpServletRequest;
import javax.servlet.http.HttpServletResponse;
import javax.servlet.http.HttpSession;
public class View2Servlet extends HttpServlet{
    public void doGet(HttpServletRequest request,
            HttpServletResponse response)
    throws ServletException,IOException{
        doPost(request,response);
    }
    public void doPost (HttpServletRequest request, HttpServletResponse response)
    throws ServletException,IOException{
        HttpSession session=request.getSession();
        response.setContentType("text/html;charset=UTF-8");
        PrintWriter out=response.getWriter();
        out.println("<html><body><h1>");
        out.println("This is view2.....<br>");
        out.println("session ID:"+session.getId()+"<br>");
        out.println("Session Attribute:"+session.getAttribute("name")+"<br>");
        out.println("</h1></body></html>");
        out.close();
    }
}
```

LoginServlet.java、IndexServlet.java、View1Servlet.java、View2Servlet.java 配置代码如下：

```xml
<servlet-mapping>
    <servlet-name>login</servlet-name>
    <url-pattern>/login</url-pattern>
</servlet-mapping>
<servlet>
    <servlet-name>index</servlet-name>
    <servlet-class>com.ch03.IndexServlet</servlet-class>
</servlet>
<servlet-mapping>
    <servlet-name>index</servlet-name>
    <url-pattern>/index</url-pattern>
```

```xml
        </servlet-mapping>
        <servlet>
            <servlet-name>view1</servlet-name>
            <servlet-class>com.ch03.View1Servlet</servlet-class>
        </servlet>
        <servlet-mapping>
            <servlet-name>view1</servlet-name>
            <url-pattern>/view1</url-pattern>
        </servlet-mapping>
        <servlet>
            <servlet-name>view2</servlet-name>
            <servlet-class>com.ch03.View2Servlet</servlet-class>
        </servlet>
        <servlet-mapping>
            <servlet-name>view2</servlet-name>
            <url-pattern>/view2</url-pattern>
        </servlet-mapping>
```

部署 Java Web 项目，启动 Web 服务器，先通过 IE 浏览器访问登录页面 login.html，运行结果如图 3-6 所示。在如图 3-6 所示登录页面中输入用户名"zhangsan"和密码"zs123"，登录成功进入如图 3-7 所示页面，否则跳转到登录页面进行重新输入。

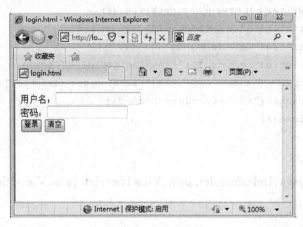

图 3-6 登录界面

在如图 3-7 所示页面中分别单击 View1 和 View2 链接，程序分别运行如图 3-8 和图 3-9 所示界面。

如果从其他客户端计算机的浏览器或者同一客户端重新打开其他浏览器页面直接输入 http://localhost:8080/chapter3/view1 和 http://localhost:8080/chapter3/view2 来访问 View1Servlet 和 View2Servlet，程序分别运行如图 3-10 和图 3-11 所示界面。

用户通过登录页面 login.html 提交登录名和密码到 LoginServlet，在 LoginServlet 中访问数据库判断是否存在该登录用户，如果存在将登录用户名保存在 Session 中并跳转到 IndexServlet，否则页面直接跳转到 login.html 供用户重新输入登录信息。在 IndexServlet

第 3 章 状态管理与作用域对象

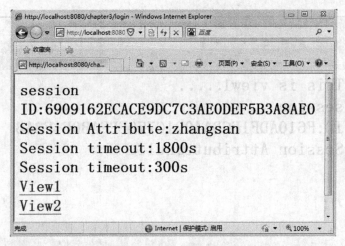

图 3-7　IndexServlet 显示界面

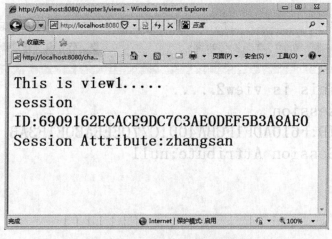

图 3-8　View1Servlet 显示界面

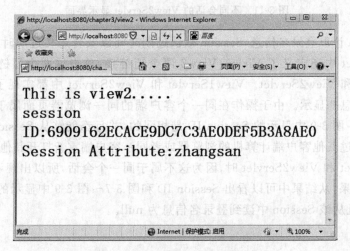

图 3-9　View2Servlet 显示界面

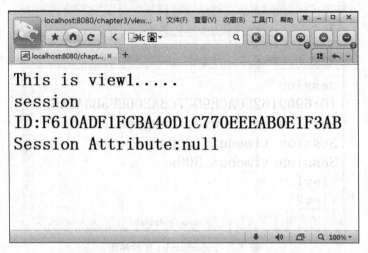

图 3-10 不同会话的 View1Servlet 显示界面

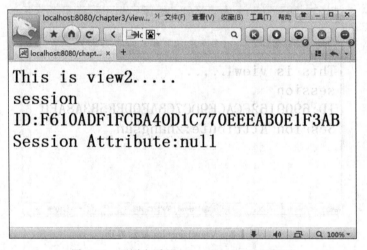

图 3-11 不同会话的 View2Servlet 显示界面

中通过 Session 读取登录名信息、Session 默认过期时间以及自定义过期时间并显示。另外，IndexServlet 中还定义了 View1、View2 两个超链接，单击这两个超链接，分别执行 View1Servlet 和 View2Servlet。View1Servlet 和 View2Servlet 中都实现了通过 Session 读取登录名信息并显示。由于操作在同一个客户端的同一浏览器页面属于同一个会话，所以在图 3-7~图 3-9 中显示的 Session ID 是相同的，并且都能够从 Session 中获得登录名信息。当通过其他客户端计算机的浏览器或者同一客户端重新打开其他浏览器页面访问 View1Servlet 和 View2Servlet 时，因为这不属于同一个会话，所以出现了图 3-10 和图 3-11 所示的结果，从结果中可以看出 Session ID 和图 3-7~图 3-9 中显示的 SessionID 是不相同的，并且从该 Session 中读到登录名信息为 null。

3.2 作用域对象

在 Java Web 编程中常见的作用域对象有 ServletContext、HttpSession 和 ServletRequest。使用作用域对象的目的是可以在不同的 Servlet、JSP 等 Web 组件之间进行数据的传送,作用域对象的主要作用是充当容器对象存放任何对象信息。作用域对象的主要方法是存取作用域对象中的数据,即保存信息方法 setAttribute(String name, Obejct value)和获取信息方法 getAttribute(String name)。

3.2.1 ServletContext

ServletContext 对象在 Web 应用启动时创建,Web 应用停止时销毁,其中保存的信息存在范围为整个 Web 应用。ServletContext 对象对于一个 Web 应用来说是唯一的,ServletContext 对象中一般存放一些 Web 应用中所有用户都要使用,并且在整个 Web 应用运行过程中一直要使用的数据。

例程 3-3:演示利用 ServletContext 实现统计页面被访问次数。程序为 CountServlet.java。

程序 CountServlet.java 代码如下:

```
package com.ch03;
import java.io.IOException;
import java.io.PrintWriter;
import javax.servlet.ServletException;
import javax.servlet.http.HttpServlet;
import javax.servlet.http.HttpServletRequest;
import javax.servlet.http.HttpServletResponse;
public class CountServlet extends HttpServlet {
    public void doGet(HttpServletRequest request, HttpServletResponse response)
            throws ServletException, IOException {
        doPost(request,response);
    }
    public void doPost(HttpServletRequest request, HttpServletResponse response)
            throws ServletException, IOException {
        response.setContentType("text/html;charset=UTF-8");
        Integer count = (Integer) this.getServletContext().getAttribute("count");
        if(count==null){
            count=1;
            this.getServletContext().setAttribute("count", count);
        }else{
            count=count+1;
            this.getServletContext().setAttribute("count", count);
        }
```

```
            PrintWriter out=response.getWriter();
            out.println ( " <! DOCTYPE HTML PUBLIC \" -//W3C//DTD HTML 4. 01
        Transitional//EN\">");
            out.println("<HTML>");
            out.println("    <HEAD><TITLE>A Servlet</TITLE></HEAD>");
            out.println("    <BODY>");
            out.println("<h2><font size=5 color=red>页面被第"+count+"次被访问!!</
        font></h2>");
            out.println("    </BODY>");
            out.println("</HTML>");
            out.flush();
            out.close();
        }
    }
```

CountServlet 配置代码如下：

```
<servlet>
    <servlet-name>count</servlet-name>
    <servlet-class>com.ch03.CountServlet</servlet-class>
</servlet>
<servlet-mapping>
    <servlet-name>count</servlet-name>
    <url-pattern>/count</url-pattern>
</servlet-mapping>
```

部署 Java Web 项目，启动 Web 服务器，通过浏览器访问 CountServlet，运行结果如图 3-12 所示。

图 3-12　CountServlet 运行结果

CountServlet.java 首先从 ServletContext 对象中获取访问次数变量 count 的值，如果 CountServlet 是第一次被访问 count 值为 null，此时将 count 的值置为 1，并将 count 保

存到ServletContext对象中。如果CountServlet不是第一次被访问，则从ServletContext对象中获取的count为某一数值而不为null，此时将count的值置为count＋1，并更新ServletContext对象count的值为新值。从任意客户端浏览器向CountServlet发送访问请求都属于同一个应用，所以共享ServletContext对象，count即为CountServlet的访问次数。

3.2.2 HttpSession

HttpSession对象在客户端发起一个新的会话时创建，在会话结束时销毁，其中保存的信息存在范围为整个会话中。HttpSession对象中不宜存放过多数据，在整个会话中经常需要使用的数据以及在进行客户端重定向时需要传递数据时，一般将数据放在HttpSession对象中。

3.2.3 ServletRequest

ServletRequest对象在客户端浏览器发起请求时创建，服务器端向客户端做出响应时结束，其中保存的信息存在范围为这次请求响应过程中。在进行服务器内部跳转时，一般都会将需要传递的数据放在ServletRequest对象中。

例程3-4：演示ServletContext、HttpSession、ServletRequest的使用比较。本程序为对例程3-2做部分代码改动。

登录页面login.html代码不做改动。
程序LoginServlet.java代码改动如下：

```
...
if(rs.next()){
    //将name信息保存在ServletRequest对象中
    request.setAttribute("name", name);            //新增代码
    //将name信息保存在HttpSession对象中
    session.setAttribute("name", name);
    //将name信息保存在ServletContext对象中
    ServletContext sc=this.getServletContext();    //新增代码
    sc.setAttribute("name", name);                 //新增代码
    request.getRequestDispatcher("index").forward(request,response);
}else{
...
```

程序IndexServlet.java代码改动如下：

```
...
out.println("<html><body><h1>");
//获得Session的ID
out.println("session ID:"+session.getId()+"<br>");
//从request对象中获得name值
out.println("request Attribute:"+request.getAttribute("name")+"<br>");
```

```
                                                                  //新增代码
//从 Session 对象中获得 name 值
out.println("Session Attribute:"+session.getAttribute("name")+"<br>");
//从 ServletContext 对象中获得 name 值
out.println("ServletContext Attribute:" + this.getServletContext().
getAttribute("name")+"<br>");                                     //新增代码
//session.getMaxInactiveInterval():获得 Session 默认的过期时间
out.println("Session timeout:" + session.getMaxInactiveInterval() +
"s"+"<br>");
...
```

程序 View1Servlet.java 代码改动如下：

```
...
out.println("<html><body><h1>");
out.println("This is view1.....<br>");
out.println("session ID:"+session.getId()+"<br>");
out.println("request Attribute:"+request.getAttribute("name")+"<br>");
                                                                  //新增代码
out.println("Session Attribute:"+session.getAttribute("name")+"<br>");
out.println("ServletContext Attribute:" + getServletContext().getAttribute
("name")+"<br>");                                                 //新增代码
out.println("</h1></body></html>");
...
```

程序 View2Servlet.java 代码改动如下：

```
...
out.println("<html><body><h1>");
out.println("This is view2.....<br>");
out.println("session ID:"+session.getId()+"<br>");
out.println("request Attribute:"+request.getAttribute("name")+"<br>");
                                                                  //新增代码
out.println("Session Attribute:"+session.getAttribute("name")+"<br>");
out.println("ServletContext Attribute:" + getServletContext().getAttribute
("name")+"<br>");                                                 //新增代码
out.println("</h1></body></html>");
...
```

Servlet 的配置代码不用做任何改动。

部署 Java Web 项目，启动 Web 服务器，先通过 IE 浏览器访问登录页面 login.html，运行结果如图 3-13 所示。在如图 3-13 所示登录页面中输入用户名"zhangsan"和密码"zs123"，登录成功进入如图 3-14 所示页面，否则跳转到登录页面重新进行输入。

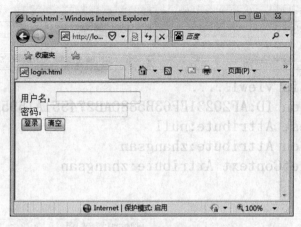

图 3-13 登录界面

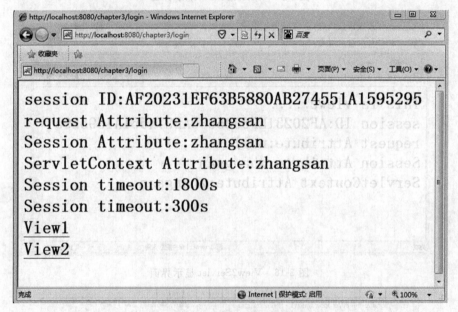

图 3-14 IndexServlet 显示界面

在如图 3-14 所示页面中分别单击 View1 和 View2 链接,程序分别运行如图 3-15 和图 3-16 所示。

如果从其他客户端计算机的浏览器或者同一客户端重新打开猎豹浏览器页面直接输入 http://localhost:8080/chapter3/view1 和 http://localhost:8080/chapter3/view2 来访问 View1Servlet 和 View2Servlet,程序分别运行如图 3-17 和图 3-18 所示。

用户通过登录页面 login.html 提交登录名和密码到 LoginServlet,在 LoginServlet 中访问数据库是否存在该登录用户,如果存在将登录用户名信息分别保存在 request、Session 和 ServletContext 中并通过服务器内部跳转方式跳转到 IndexServlet,否则页面直接跳转到 login.html 让用户重新输入登录信息。

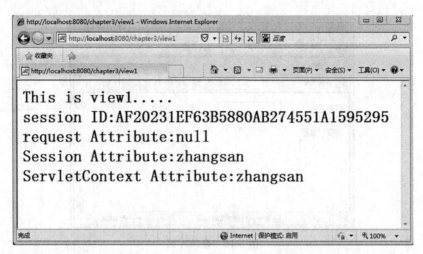

图 3-15　View1Servlet 显示界面

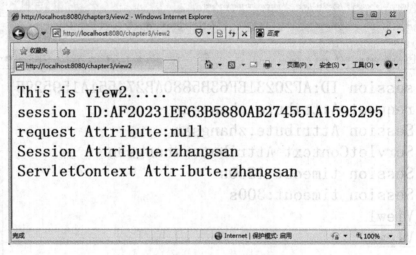

图 3-16　View2Servlet 显示界面

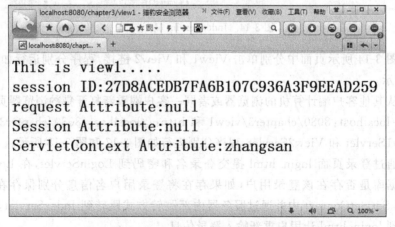

图 3-17　不同会话的 View1Servlet 显示界面

第 3 章 状态管理与作用域对象

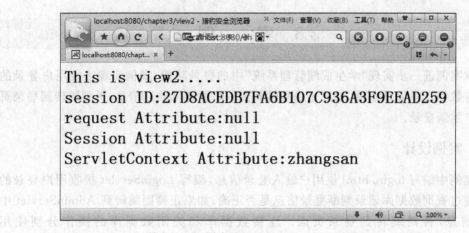

图 3-18 不同会话的 View2Servlet 显示界面

在 IndexServlet 中从 request、Session 和 ServletContext 中读取登录名信息、Session 默认过期时间以及自定义过期时间并显示，从 LoginServlet 到 IndexServlet 的过程属于同一个请求响应过程，属于同一个会话，属于同一个应用，所以在 IndexServlet 结果页面可以正常显示登录用户名信息"zhangsan"，如图 3-14 所示。另外，IndexServlet 中还定义了 View1 和 View2 两个超链接，单击这两个超链接，分别执行 View1Servlet 和 View2Servlet。

View1Servlet 和 View2Servlet 中都实现了从 request、Session 和 ServletContext 中读取登录名信息并显示。从 IndexServlet 到 View1Servlet 和从 IndexServlet 到 View2Servlet 与从 LoginServlet 到 IndexServlet 不属于同一个请求响应过程，所以此时通过 request 对象获得登录用户名信息为 null，从 IndexServlet 到 View1Servlet 和从 IndexServlet 到 View2Servlet 与从 LoginServlet 到 IndexServlet 属于同一个会话，属于同一个应用，所以通过 Session 和 ServletContext 对象正常获得登录用户名信息，如图 3-15 和图 3-16 所示。

当通过其他客户端计算机的浏览器或者同一客户端重新打开其他浏览器页面访问 View1Servlet 和 View2Servlet 时，这和之前的请求不属于同一请求响应过程，也不属于同一个会话，但是属于同一个应用，所以此时通过 request 和 Session 对象获得登录用户名信息为 null，通过 ServletContext 对象可以正常获得登录用户名信息，如图 3-17 和图 3-18 所示。

3.2.4 作用域对象的比较

ServletContext、HttpSession、ServletRequest 对象的作用域依次变小，在能够满足要求的情况下，应选择尽可能小的作用域对象使用。从作用域对象中获得指定参数的信息，如果参数不存在则返回 null。ServletContext、HttpSession、ServletRequest 对象参数独立，没有包含关系，即存放在作用域大的对象中的数据利用作用域小的对象无法获取。例如，存放在 ServletContext 对象中的数据无法通过 ServletRequest 对象获得。

3.3 案 例

本章案例进一步实现"学生成绩管理系统"中的登录功能,案例实现接收用户登录的信息并与数据表中记录比较,如果正确跳转到另一页面显示欢迎信息,否则跳转到登录页面让用户重新登录。

3.3.1 案例设计

在案例中编写 login.html 供用户输入登录信息,编写 LoginServlet 接收用户登录的信息并通过查询数据库记录判断登录信息是否正确,如果正确则跳转到 AdminServlet 中显示该信息,否则跳转到登录页面。连接数据库和关闭数据库的操作分别使用 ConnectionFactory.java 和 ResourceClose.java 类来实现。本章案例使用的主要文件如表 3-1 所示。

表 3-1 本章案例使用的文件

文 件	所在包/路径	功 能
login.html	/	登录界面
LoginServlet.java	com.imut.servlet	接收登录参数、访问数据库、根据结果进行跳转
AdminServlet.java	com.imut.servlet	登录成功后的页面,显示欢迎信息

3.3.2 案例演示

在浏览器地址栏中输入"http://localhost:8080/ch03/login.html",显示登录页面,效果如图 2-16 所示。在该页面中输入用户名和密码,单击"登录"按钮,如果用户名和密码正确则会跳转到显示欢迎信息的页面,运行效果如图 2-17 所示,否则跳转到登录页面,运行效果如图 2-16 所示。

3.3.3 代码实现

创建工程 ch03,根据案例设计描述需要完成各部分的具体代码实现,程序 LoginServlet.java 和 AdminServlet.java 的代码如下所示。

程序 LoginServlet.java 代码如下:

```
package com.imut.servlet;
import java.io.IOException;
import java.io.PrintWriter;
import java.sql.Connection;
import java.sql.PreparedStatement;
import java.sql.ResultSet;
import java.sql.SQLException;
```

```java
import javax.servlet.ServletException;
import javax.servlet.http.HttpServlet;
import javax.servlet.http.HttpServletRequest;
import javax.servlet.http.HttpServletResponse;
import javax.servlet.http.HttpSession;

import com.imut.commmon.ConnectionFactory;
import com.imut.commmon.ResourceClose;
public class LoginServlet extends HttpServlet {
    public void doGet(HttpServletRequest request, HttpServletResponse response)
            throws ServletException, IOException {
        this.doPost(request, response);
    }
    public void doPost(HttpServletRequest request, HttpServletResponse response)
            throws ServletException, IOException {
        //设置响应的MIME类型以及编码格式
        response.setContentType("text/html;charset=UTF-8");
        //设置请求的编码格式
        request.setCharacterEncoding("UTF-8");
        //接收用户登录传递过来的用户名和密码
        String loginName=request.getParameter("loginName");
        String password=request.getParameter("password");
        HttpSession session=request.getSession();
        Connection conn=null;
        PreparedStatement pstmt=null;
        ResultSet rs=null;
        String name=null;
        try{
            conn=ConnectionFactory.getConnection();
            String sql="select * from admin where loginName=? and password=?";
            pstmt=conn.prepareStatement(sql);
            pstmt.setString(1, loginName);
            pstmt.setString(2, password);
            rs=pstmt.executeQuery();
            while(rs.next()){
                name=rs.getString(2);
            }
        }catch(SQLException e){
            e.printStackTrace();
        }finally{
            ResourceClose.close(rs, pstmt, conn);
        }
        if(name!=null){
            session.setAttribute("name", name);
```

```
                response.sendRedirect(request.getContextPath()+"/adminServlet");
            }else{
                response.sendRedirect(request.getContextPath()+"/login.html");
            }
        }
    }
```

程序 LoginServlet.java 在 web.xml 文件中的配置信息如下：

```
<servlet>
    <servlet-name>loginServlet</servlet-name>
    <servlet-class>com.imut.servlet.LoginServlet</servlet-class>
</servlet>
<servlet-mapping>
    <servlet-name>loginServlet</servlet-name>
    <url-pattern>/loginServlet</url-pattern>
</servlet-mapping>
```

程序 AdminServlet.java 代码如下：

```
package com.imut.servlet;
import java.io.IOException;
import java.io.PrintWriter;
import javax.servlet.ServletException;
import javax.servlet.http.HttpServlet;
import javax.servlet.http.HttpServletRequest;
import javax.servlet.http.HttpServletResponse;
import javax.servlet.http.HttpSession;
public class AdminServlet extends HttpServlet {
    public void doGet(HttpServletRequest request, HttpServletResponse response)
            throws ServletException, IOException {
        this.doPost(request, response);
    }
    public void doPost(HttpServletRequest request, HttpServletResponse response)
            throws ServletException, IOException {
        //设置响应的 MIME 类型和编码格式
        response.setContentType("text/html;charset=UTF-8");
        //获取 HttpSession 对象
        HttpSession session=request.getSession();
        //获取 HttpSession 对象作用范围中属性"name"的值,并将其值赋给变量 name
        String name= (String)session.getAttribute("name");
        PrintWriter out=response.getWriter();
        out.println ( " <!DOCTYPE HTML PUBLIC \" -//W3C//DTD HTML 4. 01
        Transitional//EN\">");
        out.println("<HTML>");
        out.println("    <HEAD><TITLE>A Servlet</TITLE></HEAD>");
```

```
//如果变量 name 的值不为空,则表示登录信息正确,显示 name 的值,否则跳转到
//登录页面。
if(name!=null){
    out.println("登录成功,欢迎您:"+name+"<br>");
}else{
    response.sendRedirect(request.getContextPath()+"/login.html");
}
out.println("    </BODY>");
out.println("</HTML>");
out.flush();
out.close();
    }
}
```

程序 AdminServlet.java 在 web.xml 文件中的配置信息如下:

```
<servlet>
    <servlet-name>adminServlet</servlet-name>
    <servlet-class>com.imut.servlet.AdminServlet</servlet-class>
</servlet>
<servlet-mapping>
    <servlet-name>adminServlet</servlet-name>
    <url-pattern>/adminServlet</url-pattern>
</servlet-mapping>
```

【代码分析】程序 LoginServlet.java 使用 HttpServletRequest 对象的 getParameter() 方法接收用户登录的信息,并使用接收到的用户名和密码作为 SQL 查询的条件。如果查询到的结果集 ResultSet 对象不为空,证明用户名和密码正确,则把得到的结果集中的第二列也即真实姓名取出赋给 name 变量。如果 name 不为空,则把 name 的值保存在 HttpSession 作用范围中,并跳转到 AdminServlet 中。如果 name 为空,表明用户名和密码错误,跳转到登录页面。

程序 AdminServlet 首先获取 HttpSession 作用范围属性 name 的值,为了避免用户没有登录就直接访问该页面,需要先对 name 的值进行判断。如果 name 的值不为空,表明登录成功的用户访问该页面,则显示欢迎信息,否则表明未经登录的用户直接访问该页面,则跳转到 login.html 页面供用户登录。

习 题

1. 选择题

(1) 关于 ServletContext,以下哪个说法正确?(　　)

　　A. 由 Servlet 容器负责创建,对于每个客户端请求,Servlet 容器都会创建一个

ServletContext 对象

B. 由 Java Web 应用本身负责为自己创建一个 ServletContext 对象

C. 由 Servlet 容器负责创建，对于每个 Java Web 应用，在启动时，Sevlet 容器都会创建一个 ServletContext 对象

D. 由 Servlet 容器在启动时负责创建，容器内所有 Java Web 应用共享一个 ServletContext 对象

(2) Servlet 中的变量 cookie 表示客户端的一个 Cookie 数据，以下哪个选项中的代码用于删除客户端的相应 Cookie 数据？（　　）

  A. response.deleteCookie(cookie);

  B. cooke.setMaxAge(0);response.addCookie(cookie);

  C. cookie.setMaxAge(－1);response.addCookie(cookie);

  D. request.deleteCookie(cookie);

(3) 下面哪个方法能够读取给定 HttpServletRequest 对象中的所有 Cookie？（　　）

  A. request.getCookies();　　　　B. request.getAttributes();

  C. request.getSession().getCookies();

  D. request.getSession().getAttributes();

(4) 以下哪条语句能够将 session 中的属性删除？（　　）

  A. session.unbind("key");　　　　B. session.remove("key");

  C. session.removeAttribute("key");　　D. session.deleteAttribute("key");

(5) 在 Servlet 里，获取 session 的正确方式是（　　）。

  A. HttpSession session＝request.getSession();

  B. HttpSession session＝request.getHttpSession(true);

  C. HttpSession session＝response.getSession();

  D. HttpSession session＝response.getHttpSessin(true);

(6) 有关 Session 的说法，错误的是（　　）。

  A. Session 在 Servlet 中是 HttpSession 对象

  B. Session 可以通过方法调用设置其存活期

  C. Session 可以用在购物网站中记录之前购买的商品

  D. Session 中设置的属性不可能存在于整个会话期间，并且在其中设置的属性是不能移除的

(7) 下面有关 Cookie 的说法，错误的是（　　）。

  A. Cookie 是客户端技术，它的信息存放在客户端被永久保存

  B. Cookie 是一段小文本信息

  C. Cookie 的存活期可以设置

  D. 只有在浏览器接受 Cookie 的情况下，客户端才可以保存 Cookie

2. 填空题

（1）两种用于保持 HTTP 连接状态的技术是_____和_____。

（2）Session 会话结束有三种方式，这三种方式是_____、_____和_____。

（3）在 Java Web 编程中常见的作用域对象有_____、_____和_____。

3. 简答题

（1）Session 用来在无状态的 HTTP 下，越过多个请求页面来维持状态和识别用户。简述在 Servlet 中使用 Session 的过程。

（2）简述 Java Web 编程中常见的作用域对象有哪些，并且描述它们各自的特性。

4. 程序设计题

（1）实践 Session 管理功能，如图 3-19 所示，实现 index.html 页面，页面中提供用户输入用户名文本框，单击"提交"按钮执行 sessionDemo1.java（一个 Servlet），将用户输入的用户名保存在 Session 对象中，用户在响应中可以添加最喜欢去的地方，再次单击"提交"按钮，跳转到 sessionDemo2.java，在 sessionDemo2.java 响应页面中将用户输入的用户名与最想去的地方显示出来。

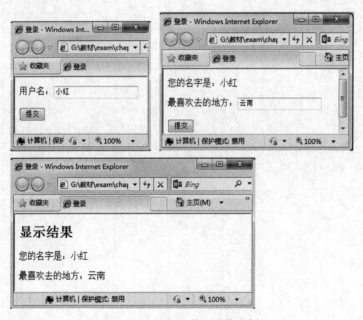

图 3-19　Session 管理功能实例

（2）应用 Cookie 技术编程完成如下功能：制作登录页面，页面如图 3-20(a)所示，验证登录信息，用户名"张三"，密码为"123456"的用户登录，登录成功，页面跳转，跳转后页面显示"张三，欢迎你！"，如图 3-20(b)所示。

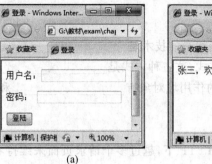 

图 3-20　登录成功

第4章 JSP 语法基础

JSP 是由 Sun Microsystems 公司倡导、许多公司参与共同建立的一种动态 Web 网页技术标准。JSP 技术有点类似 ASP 技术,它是在传统的静态页面 HTML 文件中插入 Java 脚本和 JSP 元素,从而形成 JSP 文件。本章将以理论与实践相结合的方式介绍 JSP 基本概念、JSP 元素、JSP 内置对象、JSP 注释以及 Java Web 程序开发中的路径问题等内容。

4.1 JSP 基本概念

JSP(Java Server Page)的全称是 Java 服务页面,是一种 Web 动态页面技术,事实上 JSP 就是嵌入了 Java 程序段的 HTML 文件。JSP 文件后缀名为.jsp。JSP 由 HTML 要素(静态部分)、JSP 元素(动态部分)和 JSP 注释组成。JSP 元素包括脚本元素、指令元素和动作元素。JSP 运行时文件会被翻译成 Java 类文件(Servlet),并且产生 Java 对象。JSP 和 Servlet 一样是一种服务器端技术,同样运行在 Web 服务器端。JSP 定义在 Java Web 工程中 WebRoot 根路径下或者 WebRoot 根路径的某个目录下,其访问方式和访问 HTML 文件一致。

JSP 在运行时会动态编译成一个 Servlet,其本质上就是一个 Servlet,JSP 是对 Servlet 技术的扩展,但和 Servlet 技术有很大不同。Servlet 和 JSP 在服务器端运行后均可以在客户端产生 Web 动态页面,但是在构建 Web 动态页面方面 JSP 比 Servlet 具有很多优势。Servlet 完全是 Java 程序代码构成,擅长于流程控制和业务逻辑处理,通过 Servlet 来生成动态网页结构不清晰,页面效果不能预览,对编程者要求也高(既要能编程,还要会美工)。JSP 由 HTML 代码和 JSP 元素构成,对页面的静态内容和动态内容进行了有效分离,可以方便地编写 Web 动态页面。

4.2 JSP 元素

JSP 包括脚本元素、指令元素和动作元素,下面分别详细介绍。

4.2.1 脚本元素

1. 声明

声明用来在 JSP 页面中声明变量和定义方法。声明是以"<%!"开头,以"%>"结

束的标签,其中可以包含任意数量的合法的 Java 声明语句。在 JSP 中声明一个变量的代码如下:

```
<%!int count=0; %>
```

上述代码声明了一个名为 count 的变量并将其初始化为 0。声明的变量仅在页面第一次载入时由容器初始化一次,初始化后在后面的请求中一直保持该值。

如下代码声明了一个变量和一个方法:

```
<%!
    String color[]={"red", "green", "blue"};
    String getColor(int i){
        return color[i];
    }
%>
```

2. 脚本

脚本是嵌入在 JSP 页面中的 Java 代码段。脚本是以"<%"开头,以"%>"结束的标签。例如:

```
<%count++; %>
```

脚本在每次访问页面时都被执行,因此 count 变量在每次请求时都增 1。由于脚本可以包含任何 Java 代码,所以它通常被用来实现在 JSP 页面中嵌入的计算逻辑,同时还可以使用脚本输出 HTML 文本。

3. 表达式

表达式是以"<%="开头,以"%>"结束的标签,它作为 Java 语言表达式的占位符。例如:

```
<%=count%>
```

上述代码包含一个符合 Java 语法的表达式。表达式的元素在运行后被自动转化为字符串,然后插入到这个表达式的 JSP 文件的位置显示。因为这个表达式的值已经转化为字符串,所以能在一行文本中插入这个表达式。表达式是一个简化的 out.println("...")语句。

在页面每次被访问时都要计算表达式,然后将其值嵌入到 HTML 的输出中。与变量声明不同,表达式不能以分号结束,因此如下的代码是非法的:

```
<%=count; %>
```

使用表达式可以向输出流输出任何对象或任何基本数据类型的值,也可以打印任何算术表达式、布尔表达式或方法调用返回的值。在 JSP 表达式的百分号和等号之间不能有空格。

例程 4-1：演示 JSP 中声明、脚本和表达式元素的使用。程序为 scriptTest.jsp。
程序 scriptTest.jsp 代码如下：

```jsp
<%@page contentType="text/html;charset=UTF-8"%>
<html>
    <head>
        <title>JSP脚本元素</title>
    </head>
    <body>
    <%!
        // 全局变量,如果不加"!"表示局部变量
        int count=0;
        String color[]={"red", "green", "blue"};
        String getColor(int i){
            return color[i];
        }
    %>
    <%=++count%><br>
    <%=getColor(1)%><br><br>
    <table>
        <tr><td colspan="9">九九乘法表</td></tr>
        <%
        for(int i=0;i<=9;i++){
        %>
        <tr>
            <%
            for(int j=0;j<=i;j++){
            %>
            <td><%=i%> * <%=j%>=<%=i*j%></td>
            <%
            }
            %>
        </tr>
        <%
        }
        %>
    </table>
    </body>
</html>
```

部署 Java Web 项目 chapter4，启动 Web 服务器，通过浏览器访问 scriptTest.jsp，访问方式同访问 HTML 页面文件相同，在浏览器地址栏中输入"http://localhost:8080/chapter4/scriptTest.jsp"，运行结果如图 4-1 所示。

scriptTest.jsp 页面中使用声明定义了一个全局变量 count 和方法 getColor(int i)，

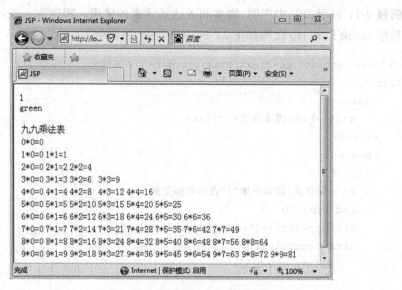

图 4-1　scriptTest.jsp 运行结果

然后在页面中通过表达式元素<%=++count%>和<%=getColor(1)%>来访问变量和方法，第一次运行页面显示结果分别为 1 和 green，如图 4-1 所示。以后每次访问 count 都会增加 1，这一点读者可以验证。scriptTest.jsp 页面利用 JSP 脚本元素和表达式元素在页面中输出九九乘法表，读者可以看出页面出现的 Java 代码均使用标签<%和%>标记，即 JSP 脚本元素。

scriptTest.jsp 页面中的第一行代码<%@ page contentType="text/html;charset=UTF-8"%>为 JSP 的 page 指令元素，用来设置 JSP 页面属性，4.2.2 节将详细介绍。

从 scriptTest.jsp 页面代码可以看出，JSP 就是在 HTML 中嵌入 Java 源代码。当 JSP 执行被"翻译"成 Servlet 时，Java 源代码直接放到 Servlet 的 service 方法中，对于每一个 HTML 元素将被转化为 out.println("HTML 内容")放到 Servlet 的 service 方法中。

4.2.2　指令元素

指令元素不是用来进行逻辑处理或者产生输出代码的，而是通过指令中的属性配置向 JSP 客户端发出一些指示，从而控制 JSP 页面的某些特征。使用 JSP 指令元素的格式一般如下：

<%@指令名 属性名="属性值" 属性名="属性值" …%>

JSP 有三种指令元素：page 指令、include 指令和 taglib 指令。下面分别详细介绍。

1．page 指令

page 指令用于设置 JSP 页面的属性，这些属性将用于和 Web 服务器通信，控制所生成的 Servlet 结构。page 指令作用于整个 JSP 页面，从语法上讲可以将一个指令放在 JSP

文档中的任何地方，不过通常是放在JSP页面的第一行。page指令使用语法格式如下：

<%@page   属性名="属性值" 属性名="属性值"…%>

page指令可以设置的页面属性说明如表4-1所示。

表4-1 JSP页面常用属性

属　　性	描　　述	默　认　值
language	定义JSP页面使用的脚本语言种类	Java
contentType	定义JSP页面字符编码方式和页面响应的MIME类型	MIME：text/html 字符编码：ISO-8859-1
pageEncoding	定义JSP页面字符编码方式，如果contentType对页面字符编码方式也有设置，将以pageEncoding的设置为准	无
import	导入脚本语言中使用的类文件，如果要导入多个类使用逗号","隔开	无
extends	定义JSP页面产生的Servlet时继承的父类，必须指定该类全名，即包名＋类名。请特别谨慎使用这一功能，否则可能会影响JSP文件的编译	HttpJspBase
isErrorPage	可取逻辑值true或者false,逻辑值表示当前JSP页面是否为其他页面的errorPage目标，如果设置为true,则可以使用exception对象显示错误信息；如果设置为false,则不可以使用exception对象，即不能显示错误信息	false
errorPage	设置当前JSP页面发生异常时，负责处理异常事件的目标JSP页面。被指定的目标JSP页面的page指令元素必须指定isErrorPage="true"	默认忽略
session	可取逻辑值true或者false,定义在JSP页面是否可以使用HTTP的session	true
isELIgnored	可取逻辑值true或者false,设置JSP页面是否支持EL表达式。false表示支持，否则不支持	false
buffer	buffer="none|8kb|size kb"。指定内置对象out发送信息到客户端浏览器的信息缓存大小，以kb为单位，可以指定缓存的大小。设置为none即没有缓冲区，所有的输出都不经过缓存而直接输出	8kb
autoFlush	autoFlush="true|false",指定是否当缓存填满时自动刷新输出缓存中的内容。如果为true,则自动刷新，否则当缓存填满后，可能会出现严重的错误。如果把buffer设置为none时，就不能将autoFlush设置为false	true
isThreadSafe	指定JSP页面是否支持多线程访问。设置为true,表示可以同时处理多个客户请求，但应该在JSP页面中添加处理多线程的同步控制代码。设置为false,JSP页面在一个时刻就只能响应一个请求	ture
info	指定任何一段字符串，该字符串被直接加入到翻译好的页面中，可以通过Servlet.getServletInfo()方法得到	无

例程 4-2：演示 JSP 中 page 指令常用属性设置的使用。程序为 pageOrder.jsp。
程序 pageOrder.jsp 代码如下：

```jsp
<%@page language="java" contentType="text/html;charset=UTF-8"
import="java.util.*,java.io.*" %>
<html>
    <head>
        <title>page 指令属性设置</title>
    </head>
    <body>
        <h1>欢迎光临:www.mldn.cn</h1>
        <h1>更多学习资源等待着您……</h1>
        <h1><%=new Date()%></h1>
    </body>
</html>
```

程序 pageOrder.jsp 中使用 page 指令属性设置 language="java"说明页面中只能使用 Java 语言脚本。contentType="text/html;charset=UTF-8"说明 response 对象中的内容类型为 text/html 和响应 JSP 页面的编码格式为 UTF-8，该 JSP 代码段翻译成 Servlet 的代码为 response.setContentType("text/html;charset=UTF-8")。import="java.util.*,java.io.*"说明导入 java.util 包和 java.io 包中的所有类到当前页面。在页面的脚本中可以使用导入的类。程序中就使用了 java.util 包中的 Date 类显示当前日期。

程序运行结果如图 4-2 所示。

图 4-2 pageOrder.jsp 页面运行结果

例程 4-3：演示 page 指令设置 errorPage 和 isErrorPage 属性的使用。程序为 pageOrder2.jsp 和 errorPage.jsp。
程序 pageOrder2.jsp 代码如下：

```jsp
<%@page language="java" import="java.util.*"
errorPage="errorPage.jsp" pageEncoding="UTF-8"%>
<!DOCTYPE HTML PUBLIC "-//W3C//DTD HTML 4.01 Transitional//EN">
<html>
```

```
            <head>
                <title>page 指令属性设置</title>
            </head>
            <body>
                <%Integer.parseInt("aaa"); %>
            </body>
        </html>
```

程序 errorPage.jsp 代码如下：

```
<%@page language="java" isErrorPage="true" import="java.util.*"
pageEncoding="UTF-8"%>
<!DOCTYPE HTML PUBLIC "-//W3C//DTD HTML 4.01 Transitional//EN">
<html>
    <head>
        <title>page 指令属性设置</title>
    </head>
    <body>
        <h1>出现错误<%=exception.getMessage() %></h1>
    </body>
</html>
```

程序 pageOrder2.jsp 中的 Java 脚本语句"Integer.parseInt("aaa");"将产生一个类型转换异常，页面 page 指令元素中设置了属性 errorPage="errorPage.jsp"，即 pageOrder2.jsp 页面产生错误时要定位的目标页面为 errorPage.jsp。errorPage.jsp 中页面 page 指令元素中配置了属性 isErrorPage="true"，即可以作为其他 JSP 页面的 errorPage 目标。在 errorPage.jsp 页面中＜％＝exception.getMessage()％＞使用内置对象 exception（有关内置对象 4.3 节将详细介绍）输出错误简短信息。当访问 pageOrder2.jsp 页面时将产生错误并定位到 errorPage.jsp 页面。访问 pageOrder2.jsp 页面程序运行结果如图 4-3 所示。

图 4-3　pageOrder2.jsp 页面运行结果

2. include 指令

include 指令元素将 file 属性指定页面内容静态包含到当前页面。include 指令使用语法格式如下：

```
<%@include file="xxx.jsp"%>
```

include指令包含指定页面的过程是在翻译阶段完成的,也就是JSP被转化成Servlet的阶段进行的。include指令元素包含页面是在运行之前已经完成包含的动作,所以运行时效率较高。如果被包含的文件内容发生变化,那么当前JSP文件需要重新被翻译。

例程4-4:演示include指令元素的使用。程序为main.jsp和insert.jsp。

程序main.jsp代码如下:

```
<%@ page contentType="text/html; charset=UTF-8" language="java" import="java.util.*" pageEncoding="UTF-8"%>
    <html>
        <head>
            <title>include指令</title>
        </head>
        <body>
            This is my main page.<br><br>
            <%@include file="insert.jsp" %>
        </body>
    </html>
```

程序insert.jsp代码如下:

```
<%@page language="java" import="java.util.*" pageEncoding="UTF-8"%>
<html>
    <head>
        <title>include指令</title>
    </head>
    <body>
        这是被插入的页面.<br>
    </body>
</html>
```

程序main.jsp中显示"This is my main page."文本内容,并且使用include指令<%@include file="insert.jsp" %>将insert.jsp页面包含到main.jsp页面中,在insert.jsp页面显示"这是被插入的页面."文本内容,include指令将main.jsp和insert.jsp两个页面融合在一起。访问main.jsp页面运行效果如图4-4所示。

3. taglib 指令

声明用户在当前JSP页面使用JSTL标签或者用户自定义的标签时使用taglib指令,将

图4-4 main.jsp页面运行结果

标签库描述符文件导入到 JSP 页面。在 JSP 页面中使用 JSTL 标签或者自定义的标签语法格式为：

```
<%@taglib uri="tiglibURL" prefix="tagPrefix" %>
```

uri 属性：定位标签库描述符的位置，唯一标识和前缀相关的标签库描述符，可以使用绝对或相对 URL。

prefix 属性：标签的前缀，区分多个自定义标签。不可以使用保留前缀和空前缀，遵循 XML 命名空间的命名约定。

使用自定义标签或者 JSTL 标签的目的是消除 JSP 文件中出现的 Java 脚本。

例程见第 9 章自定义标签和 JSTL 标签的使用实例。

4.2.3 动作元素

JSP 规范中定义了一系列标准动作。Web 容器也按这个标准实现，可以解析并执行标准动作。所有的标准动作元素都是用"jsp:"作为前缀。与 JSP 指令元素不同的是，JSP 动作元素在请求处理阶段起作用。标准动作元素使用标准的 XML 语法：

```
<jsp:action_name attribute1="value1" attribute2="value2" />
```

或者

```
<jsp:action_name1 attribute1="value1" attribute2="value2" >
    <jsp:action_name2 attribute1="value1" attribute2="value2" />
</jsp:action_name1>
```

使用动作元素时，一定要注意正确结束动作元素，如果＜jsp:action_name1＞是动作元素的开始，一定要有对应的结束标记＜/jsp:action_name1＞。如果动作元素没有动作体，就可以直接结束，如＜jsp:action_name attribute1="value1" attribute2="value2"/＞。

下面介绍常用的动作元素及其用法。

1. ＜jsp:param＞

＜jsp:param＞用于指定参数以及与其对应的值，跳转或者包含的页面可以用 request 对象来读取这些参数的值。＜jsp:param＞的语法格式为：

```
<jsp:param name="参数名" value="参数值"/>
```

2. ＜jsp:forward＞

＜jsp:forward＞用于实现请求的转发，转发的目标组件可以是 JSP 文件、HTML 文件或者 Servlet。＜jsp:forward＞的语法格式为：

```
<jsp:forward page="转发目标组件的绝对或者相对 URL" flush="true|false"/>
```

其中，flush="true|false"用来指定是否使用缓冲区。

例程 4-5：演示<jsp:param>和<jsp:forward>动作元素的使用。程序为 forward.jsp 和 targetPage.jsp。

程序 forward.jsp 的代码如下：

```
<%@page language="java" import="java.util.*" contentType="text/html; charset=UTF-8"%>
<html>
    <head>
        <title>JSP 动作元素</title>
    </head>
    <body>
        <jsp:forward page="targetPage.jsp? name=zhangsan&password=admin123">
            <jsp:param name="age" value="28"/>
            <jsp:param name="phone" value="04716575471"/>
            <jsp:param name="phone" value="13804714214"/>
        </jsp:forward>
    </body>
</html>
```

程序 targetPage.jsp 代码如下：

```
<%@page contentType="text/html; charset=utf-8"%>
<%
    String name=request.getParameter("name");
    String password=request.getParameter("password");
    String age=request.getParameter("age");
    String phones[]=request.getParameterValues("phone");
%>
用户名:<%=name%><br>
密码:<%=password%><br>
年龄:<%=age%><br>
<%for(int i=0;i<phones.length;i++){%>
电话:<%=phones[i]%><br>
<%}%>
```

程序 forward.jsp 中使用<jsp:forward>动作元素将请求转发至 targetPage.jsp 页面，转发请求的同时通过两种方式传递了参数。第一种方式采用<jsp:forward>动作元素在动作体内使用<jsp:param>动作元素传递了参数名为 age 和 phone 的参数，其中名字为 phone 的参数具有两个值；第二种方式采用 URL 传参，传递了参数名为 name 和 password 的两个参数。采用 URL 传参的一般格式为：URL?参数名1=值1&参数名2=值2……

程序 targetPage.jsp 的 Java 脚本中，利用内置对象 request 的 getParameter("参数

名")和 getParameterValues("参数名")方法将 forward.jsp 页面传递的参数获得,然后使用 JSP 表达式将各个参数预先显示。访问 forward.jsp 页面的运行效果如图 4-5 所示。

3. <jsp:include>

<jsp:include>动作元素将 page 属性指定页面内容动态包含到当前页面。<jsp:include>动作元素使用语法格式如下:

图 4-5 forward.jsp 页面运行结果

```
<jsp:include page="relativeURL" flush="true|false">
```

其中,page 代表一个相对路径,即所要包含进来的文件位置,所包含文件可以是静态文件或者动态文件,如果是动态文件可以通过<jsp:forward>动作元素传递参数给所包含的文件。flush 可以接受的值为 boolean 类型,若为 true,缓冲区满时将会被清空,flush 的默认值为 false。

使用<jsp:include>动作元素当前页面和被包含页面的代码分别处理,在页面被请求的时候才编译,被编译成多个 Servlet,页面语法相对独立,处理完成之后再将代码的显示结果(处理结果)组合进来。这一点和 include 指令元素是不同的,include 指令元素的包含动作是在 JSP 被转化成 Servlet 的阶段完成的。

例程 4-6:演示<jsp:param>和<jsp:include>动作元素的使用。程序为 include.jsp 和 innerPage.jsp。

程序 include.jsp 代码如下:

```jsp
<%@page language="java" import="java.util.*" pageEncoding="UTF-8"%>
<html>
    <head>
        <title>JSP 动作元素</title>
    </head>
    <body>
        <jsp:include page="innerPage.jsp">
            <jsp:param name="age" value="19"/>
            <jsp:param name="phone" value="04724587999"/>
            <jsp:param name="phone" value="18751426138"/>
        </jsp:include>
    </body>
</html>
```

程序 innerPage.jsp 代码如下:

```jsp
<%@page contentType="text/html; charset=utf-8"%>
<%
    String age=request.getParameter("age");
    String phones[]=request.getParameterValues("phone");
```

```
       %>
       年龄:<%=age%><br>
       <%for(int i=0;i<phones.length;i++){%>
           电话:<%=phones[i]%><br>
       <%}%>
```

程序 include.jsp 中使用＜jsp:include＞动作元素将 innerPage.jsp 页面动态包含在当前页面 include.jsp 中。在＜jsp:include＞动作元素动作体内使用＜jsp:param＞动作元素向所包含页面 innerPage.jsp 传递了参数名为 age 和 phone 的参数,其中名字为 phone 的参数具有两个值。

程序 innerPage.jsp 的 Java 脚本中,利用内置对象 request 的 getParameter("参数名")和 getParameterValues("参数名")方法获得 include.jsp 页面传递的参数,然后使用 JSP 表达式将各个参数预先显示。访问 include.jsp 页面的运行效果如图 4-6 所示。

图 4-6　include.jsp 页面运行结果

除了上述介绍的＜jsp:param＞、＜jsp:forward＞和＜jsp:include＞三种动作元素外,常用的 JSP 动作元素还有＜jsp:useBean＞、＜jsp:setProperty＞和＜jsp:getProperty＞,有关这三种动作元素将在第 6 章中介绍。

4.3　JSP 内置对象

4.3.1　内置对象概述

内置对象又称隐含对象,内置对象在 JSP 页面初始化时由 Web 容器为用户自动创建,使用 JSP 进行页面编程时可以不加声明和创建,直接在 Java 脚本和表达式中使用这些对象。

JSP 内置对象有 9 个,详细介绍如表 4-2 所示。

表 4-2　JSP 内置对象

内置对象	类　　型	范　围	说　　明
request	javax.servlet.http.HttpServletRequest	request	request 对象用来封装客户端的请求信息,通过该对象可以获得客户端请求信息,然后做出响应。request 对象是 HttpServletRequest 类的实例,具有请求作用域,即完成客户端的请求之前,该对象一直有效
response	javax.servlet.http.HttpServletResponse	page	response 对象封装了响应客户请求的有关信息,它是 HttpServletResponse 类的实例。response 对象具有页面作用域,即访问一个页面时,该页面内的 response 对象只能对这次访问有效,其他页面的 response 对象对当前页面无效

续表

内置对象	类　型	范　围	说　明
session	javax.servlet.http.HttpSession	session	session 对象指的是客户端与服务器的一次会话，从客户端连到服务器的一个 Web 应用开始，直到客户端与服务器断开连接为止。Session 对象是 HttpSession 类的实例，具有会话作用域
application	javax.servlet.jsp.ServletContext	application	application 对象实现了用户间数据的共享，可存放全局变量。它开始于服务器的启动，直到服务器的关闭，在此期间，此对象将一直存在；这样在用户的前后连接或不同用户之间的连接中，可以对此对象的同一属性进行操作；在任何地方对此对象属性的操作，都将影响到其他用户对此的访问。服务器的启动和关闭决定了 application 对象的生命。application 对象是 ServletContext 类的实例
pageContext	javax.servlet.jsp.PageContext	page	pageContext 对象提供了对 JSP 页面内所有的对象及名字空间的访问，该对象使用较少
out	javax.servlet.jsp.JspWriter	page	out 对象是 JspWriter 类的实例，是向客户端输出内容常用的对象
config	javax.servlet.ServletConfig	page	config 对象是在一个 Servlet 初始化时，JSP 引擎向它传递信息用的，此信息包括 Servlet 初始化时所要用到的参数（通过属性名和属性值构成）以及服务器的有关信息（通过传递一个 ServletContext 对象）
page	java.lang.Object	page	page 对象就是指向当前 JSP 页面本身，类似于类中的 this 指针，它是 java.lang.Object 类的实例
exception	java.lang.Throwable	page	exception 对象是一个例外对象，当一个页面在运行过程中发生了例外，就产生这个对象。如果一个 JSP 页面要应用此对象，就必须把 isErrorPage 设为 true，否则无法编译。exception 对象是 java.lang.Throwable 的对象。

4.3.2 内置对象使用

JSP 内置对象在 Web 应用编程中经常用到，下面通过几个实例来说明 JSP 内置对象的使用。

例程 4-7：演示使用内置对象 out 打印九九乘法表。程序为 outTest.jsp。

程序 outTest.jsp 代码如下：

```
<%@page language="java" contentType="text/html;charset=UTF-8"%>
<html>
    <head>
        <title>JSP 内置对象的使用</title>
    </head>
    <body>
```

```
        <%
            //打印九九乘法表
            out.println("<table border=\"1\">");
            for(int i=0;i<=9;i++){
                out.println("<tr>") ;
                for(int j=0;j<=i;j++){
                    out.println("<td>"+i+" * "+j+" = "+i*j+"</td>");
                }
                out.println("</tr>");
            }
            out.println("</table>");
        %>
    </body>
</html>
```

程序 outTest.jsp 中使用 out 对象的 println("输出内容")方法将内容输出到客户端。访问 outTest.jsp 页面运行效果如图 4-7 所示。

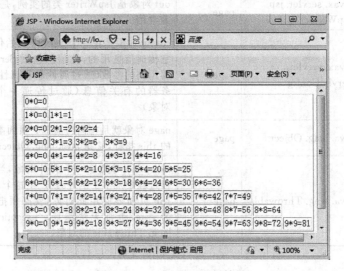

图 4-7　out.jsp 页面运行结果

例程 4-8：演示内置对象 request、response 的使用。程序为 resparam.html 和 resparam.jsp。

程序 resparam.html 代码如下：

```
<!DOCTYPE HTML PUBLIC "-//W3C//DTD HTML 4.01 Transitional//EN">
<html>
    <head>
        <meta http-equiv="content-type" content="text/html; charset=UTF-8">
    </head>
    <body>
        <form action="resparam.jsp" method="post">
```

```
        用户名:<input type="text" name="username">
        <input type="submit" value="提交">
    </form>
    </body>
</html>
```

程序 resparam.jsp 代码如下:

```
<%@page contentType="text/html; charset=UTF-8" language="java" import="java.util.*" pageEncoding="UTF-8"%>
<html>
    <head>
        <title>JSP 内置对象的使用</title>
    </head>
    <body>
        <%
        request.setCharacterEncoding("UTF-8");
        response.setCharacterEncoding("UTF-8");
        // 接收参数
        String name=request.getParameter("username");
        %>
        <h2><%=name%></h2>
        <h2><%=request.getParameter("username")%></h2>
    </body>
</html>
```

程序 resparam.html 中定义了一个＜form＞表单,表单中定义了一个文本框,在文本框中输入数据后单击"提交"按钮,将客户端数据提交到表单的 action 属性中配置的资源,即 resparam.jsp 页面。在 resparam.jsp 中的脚本和表达式中使用内置对象 request 的 getParameter("参数名")方法获得指定参数名的客户端数据,并予以显示。在脚本中还使用内置对象 request 和 response 的 setCharacterEncoding("编码方式")方法设置请求和响应的编码方式。访问 resparam.html 页面运行效果如图 4-8 所示。在如图 4-8 所示界面中输入"张三"后单击提交按钮,将数据"提交"到 resparam.jsp 页面,其运行效果如图 4-9 所示。

图 4-8　resparam.html 页面运行结果　　　图 4-9　resparam.jsp 页面运行结果

例程 4-9:演示内置对象 request、session 和 application 的作用域及其之间的比较。程序为 curPage.jsp、requestPage.jsp、session1Page.jsp 和 session2Page.jsp。

程序 curPage.jsp 代码如下:

```jsp
<%@page language="java" import="java.util.*" pageEncoding="UTF-8"%>
<html>
    <head>
        <title>JSP 内置对象的使用</title>
    </head>
    <body>
        <%
            request.setAttribute("name", "lisi");
            session.setAttribute("name","wangwu");
            application.setAttribute("name","zhaoliu");
            //请求转发
            request.getRequestDispatcher("requestPage.jsp").forward(request,
            response);
        %>
    </body>
</html>
```

程序 requestPage.jsp 代码如下:

```jsp
<%@page language="java" import="java.util.*" pageEncoding="UTF-8"%>
<html>
    <head>
        <title>JSP 内置对象的使用</title>
    </head>
    <body>
        request 中的 name:<%=request.getAttribute("name") %><br>
        session 中的 name:<%=session.getAttribute("name") %><br>
        application 中的 name:<%=application.getAttribute("name") %><br>
        <br>
        <a href="session1Page.jsp">check1</a><br>
        <a href="session2Page.jsp">check2</a><br>
    </body>
</html>
```

程序 session1Page.jsp 和 session2Page.jsp 代码完全相同,代码如下:

```jsp
<%@page language="java" import="java.util.*" pageEncoding="UTF-8"%>
<html>
    <head>
        <title>JSP 内置对象的使用</title>
    </head>
    <body>
```

```
            request 中的 name:<%=request.getAttribute("name") %><br>
            session 中的 name:<%=session.getAttribute("name") %><br>
            application 中的 name:<%=application.getAttribute("name") %><br>
        </body>
    </html>
```

　　程序 curPage.jsp 中使用内置对象 request、session 和 application 的 setAttribute ("参数名","参数值")方法将参数值保存在对应的对象之中,然后将请求转发至 requestPage.jsp。requestPage.jsp 中使用内置对象 request、session 和 application 的 getAttribute("参数名")来获得对象中保存的指定参数的值,由于 curPage.jsp 页面和 requestPage.jsp 页面共享同一个请求,那么显然也共享同一个会话和应用,所以访问 curPage.jsp 页面跳转至 requestPage.jsp 页面的显示效果如图 4-10 所示。

　　requestPage.jsp 页面中定义了两个链接 check1 和 check2,分别用来访问 session1Page.jsp 和 session2Page.jsp 页面,由于 curPage.jsp 页面与 session1Page.jsp 和 session2Page.jsp 均不共享同一请求,但是共享同一个会话,显然也共享同一个应用,所以分别单击链接 check1 和 check2,其页面运行效果如图 4-11 和图 4-12 所示。

图 4-10　requestPage.jsp 页面运行结果

图 4-11　session1Page.jsp 页面运行结果

　　在其他客户端直接访问 session1Page.jsp 或者 session2Page.jsp,不同客户端发起的请求既不共享同一个请求,也不共享同一个会话,但是共享同一个应用,所以运行效果如图 4-13 所示。

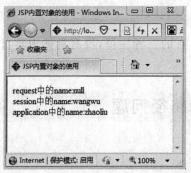
图 4-12　session2Page.jsp 页面运行结果　　图 4-13　不同客户端 session1Page.jsp 页面运行结果

4.4　JSP 注释

JSP 是嵌入了 Java 程序段的 HTML 文件,所以 JSP 注释包含三种：HTML 注释,Java 注释,JSP 注释。

HTML 注释：<!-- HTML 注释内容 -->,又称显式注释,注释内容在页面运行时发送到客户端。

Java 注释：在 JSP 页面的 Java 脚本中可以使用 Java 注释。包括单行注释和多行注释,两者又称为隐式注释,编译时被忽略。例如："//Java 单行注释内容"或者"/*Java 多行注释内容*/"。

JSP 注释：<%--JSP 注释内容--%>。

例程 4-10：演示 JSP 中注释的使用。程序为 comment.jsp。

程序 comment.jsp 代码如下：

```
<%@page contentType="text/html;charset=UTF-8" %>
<html>
    <head>
        <title>JSP 注释</title>
    </head>
    <body>
        <!--HTML 注释,此注释内容会发送到客户端-->
        <%
            // 可以使用 Java 的单行注释
            /*
             *    多行注释
             */
        %>
        <%--
            JSP 中使用的注释
        --%>
    </body>
</html>
```

程序 comment.jsp 中的代码只是编写了 JSP 具有的三种注释,没有具体内容,访问 comment.jsp 页面将得到一个空白页面。

4.5　Java Web 中的路径问题

4.5.1　路径的基本概念

在 Java Web 开发中,路径分为绝对路径和相对路径。

绝对路径为某一资源(HTML、JSP 或 Servlet)在服务器上的真正的路径,即物理路径。例如,"http://localhost:8080/chapter4/index.jsp"是一个 URL 绝对路径。该路径是访问 Web 应用 chapter4 中 WebRoot 下 index.jsp 页面的访问路径。

相对路径为相对于某个基准目录的路径。在 Servlet 中,"/"代表 Web 应用的根目录,"./"代表当前目录,"../"代表当前目录的上一级目录。例如,"../main.jsp"为一个相对路径,即访问当前目录的上一级目录下的 main.jsp 文件。

在 Java Web 开发中,有如下几个基本的目录。

(1) Web 服务器的根目录:http://localhost:8080/。

(2) Web 应用的根目录:Web 应用 chapter4 的根目录为 http://localhost:8080/chapter4/。

(3) 同级目录:http://localhost:8080/chapter4/userManager/addUser.jsp 和 http://localhost:8080/chapter4/userManager/updateUser.jsp 说明 JSP 页面 addUser.jsp 和 updateUser.jsp 处于同级目录。

4.5.2 路径相关函数

通过 JSP 的内置对象 request 和 application 可以使用和 Java Web 中路径有关的几个函数。

例程 4-11:演示路径相关函数的使用。程序为 pathTest.jsp。

pathTest.jsp 程序代码如下:

```
<%@page language="java" import="java.util.*" pageEncoding="UTF-8"%>
<!DOCTYPE HTML PUBLIC "-//W3C//DTD HTML 4.01 Transitional//EN">
<html>
    <head>
        <title>路径相关函数</title>
    </head>
    <body>
        getContextPath():<%=request.getContextPath() %><br>
        getServletPath():<%=request.getServletPath() %><br>
        getRequestURI():<%=request.getRequestURI() %><br>
        getRequestURL():<%=request.getRequestURL() %><br>
        getRealPath("/"):<%=application.getRealPath("/") %><br>
    </body>
</html>
```

程序运行结果如图 4-14 所示。

从图 4-14 中 pathTest.jsp 的运行结果可以看出,路径相关函数的意义如下。

getContextPath():返回 Web 应用的相对路径。

getServletPath():返回 request 请求相对于 Web 应用的相对路径。

getRequestURI():返回 request 请求相对于 Web 服务器的相对路径。

getRequestURL():返回 request 请求的绝对路径。

getRealPath("/"):返回"/"对应于本地磁盘的绝对路径,即当前应用的物理地址。

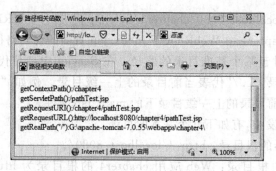

图 4-14　pathTest.jsp 运行结果

4.5.3　Java Web 开发中经常涉及的路径问题

1. web.xml 中的路径

在配置文件 web.xml 中，配置 Servlet 映射＜url-pattern＞/xxxx＜/url-pattern＞中的"/"代表当前 Web 应用的根目录。

2. 转发和重定向中的路径

（1）如果路径中不包含"/"，如 request.getRequestDispatcher("index.jsp").forward(request,response) 和 response.sendRedirect("index.jsp") 就表示在同级目录中寻找 index.jsp 文件。

（2）如果路径中包含"/"（"/"是指路径中的第一个"/"）：

① 转发，例如 request.getRequestDispatcher("/index.jsp").forward(request,response)代表到"Web 应用（例如："http://localhost:8080/chapter4/"）的根"目录下寻找 index.jsp 文件。

② 重定向，例如 response.sendRedirect("/index.jsp")代表到"Web 服务器（例如："http://localhost:8080/"）的根"目录下寻找 index.jsp 文件。

3. 表单和链接中的路径

＜form action="/xxxx"＞或＜a href="/xxxx"＞＜/a＞中的路径，"/"代表的是 Web 服务器的根目录。

4. page 指令元素中 errorPage 属性的路径

errorPage 属性值如果以"/"开头，表示相对于当前 Web 应用的根目录，如果没有"/"开头的路径表示当前目录。

4.6　案　　例

本章案例主要实现"基本信息管理"模块中的"教师信息维护"子模块功能。主要实现的功能包括用户登录、在系统首页上显示用户登录状态及登录者信息、分页显示所有教师信息、增加教师信息、根据条件查询教师信息、删除教师信息、查看和修改教师信息等。

4.6.1 案例设计

在登录功能中,用户在登录页面输入用户名和密码,如果输入正确则跳转到系统首页并显示用户登录状态和登录者信息,否则跳转到登录页面弹出警告框并让用户重新登录。本模块涉及的主要源文件如表 4-3 所示。

表 4-3 用户登录模块使用文件

文件	所在包/路径	功能
login.jsp	/	用户登录界面
LoginServlet.java	com.imut.servlet	接收登录信息,与数据库交互判断登录是否成功,根据结果进行跳转
index.jsp	/	系统首页,显示用户登录状态及登录信息

"教师信息维护"模块实现分页显示所有教师信息、增加教师信息、删除教师信息、修改教师信息以及查看教师信息等功能。本模块各功能涉及的主要源文件如表 4-4 所示。

表 4-4 教师信息维护模块使用文件

文件	所在包/路径	功能
Teacher.java	com.imut.javabean	封装教师信息的类
TeacherDBAccess.java	com.imut.javabean	操作数据库表的类
AddTeacherServlet.java	com.imut.servlet/base	增加教师信息,跳转到教师列表页面
CheckTeacherServlet.java	com.imut.servlet/base	查询教师信息,跳转到教师列表页面
DelTeacherServlet.java	com.imut.servlet/base	删除教师信息,跳转到教师列表页面
ListAllTeacherServlet.java	com.imut.servlet/base	列出所有教师信息
ShowTeacherServlet.java	com.imut.servlet/base	显示某一教师信息
UpdateTeacherServlet.java	com.imut.servlet/base	更新教师信息,跳转到教师列表页面
addTeacher.jsp	/base	增加教师信息页面
teacherList.jsp	/base	教师列表页面
teacherShow.jsp	/base	显示某一教师信息的页面
Page.java	com.imut.servlet/commmon	分页信息类

4.6.2 案例演示

在浏览器地址栏中输入"http://localhost:8080/ch04/login.jsp",显示登录页面。在该页面中输入用户名和密码,单击"登录"按钮,如果用户名和密码正确则会跳转到 index.jsp 页面,同时显示登录状态及登录者信息,运行效果如图 4-15 所示,否则跳转到 login.jsp 页面弹出警告框提示重新登录,效果如图 4-16 所示。

图 4-15 登录成功首页

图 4-16 登录错误警告框

在 index.jsp 页面中单击"基本数据管理",进入"教师信息维护"模块的教师列表页面,效果如图 4-17 所示。在教师列表页面中单击"添加教师"超级链接,出现添加教师信息的页面,效果如图 4-18 所示。如果没有输入信息,会弹出警告框提示信息不能为空,输入完毕教师信息后,单击"提交"按钮,提示添加成功并显示教师列表页面。在教师列表页

图 4-17 教师列表页面

面中输入查询条件,查询结果将显示在"查询"按钮下方的教师列表处,如果没有输入查询条件则查询所有教师信息。在"职称"下拉框中选中"副教授"进行查询,其效果如图 4-19 所示。在教师列表页面中单击"删除"超链接,将会删除该教师信息,删除完成后显示教师列表页面。在教师列表页面中单击"查看/修改"超链接,或者单击教师姓名,将显示教师信息查看/修改页面,效果如图 4-20 所示。修改教师信息完成后单击"提交"按钮,提示修改成功后显示教师列表页面。

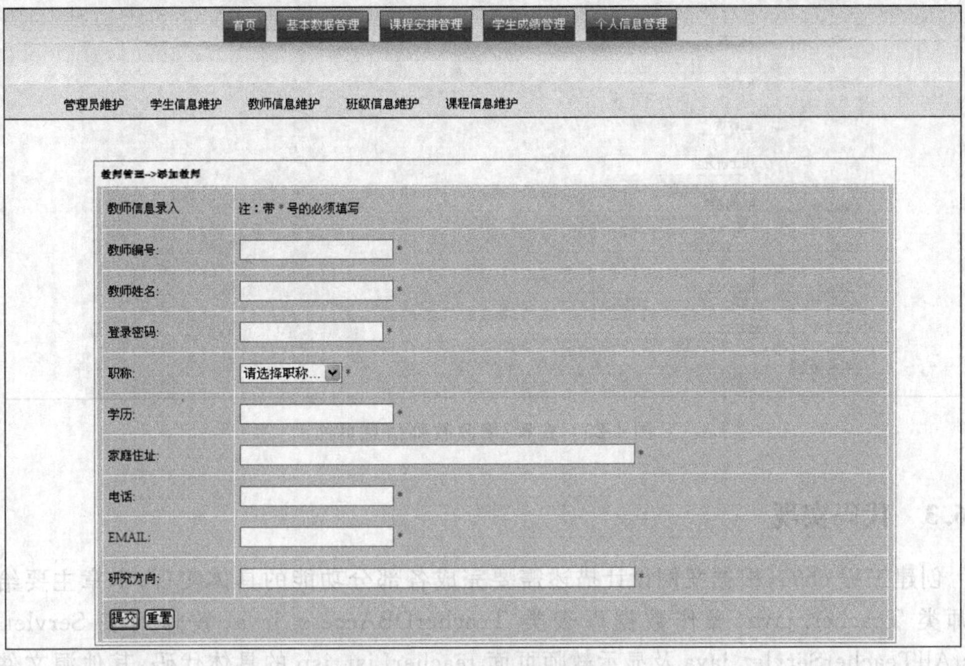

图 4-18 添加教师信息页面

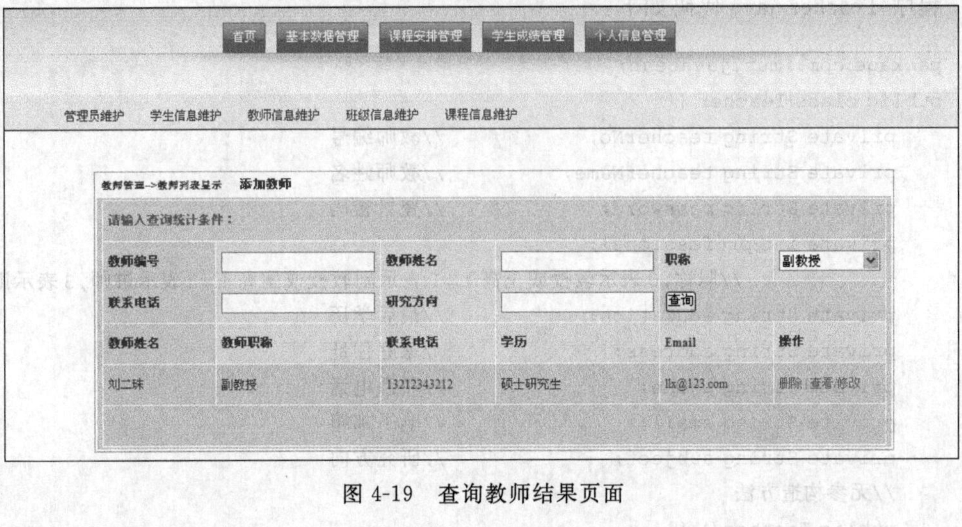

图 4-19 查询教师结果页面

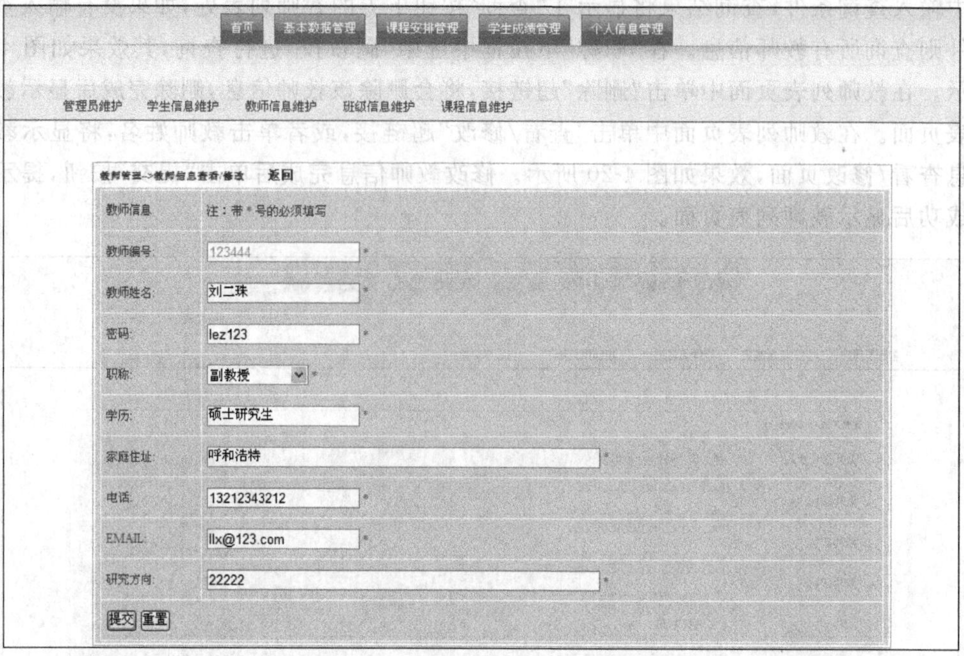

图 4-20 查看/修改教师信息页面

4.6.3 代码实现

创建工程 ch04，根据案例设计描述需要完成各部分功能的具体实现，本章主要给出教师类 Teacher.java、操作数据库表类 TeacherDBAccess.java、教师列表 Servlet 类 ListAllTeacherServlet.java 及显示教师页面 teacherList.jsp 的具体代码，其他源文件代码请参见随书电子资源。

程序 Teacher.java 代码如下：

```java
package com.imut.javabean;
public class Teacher {
    private String teacherNo;           //教师编号
    private String teacherName;         //教师姓名
    private String password;            //登录密码
    private int professional;
            //职称，0 表示教授或正高工，1 表示副教授或副高工，2 表示讲师，3 表示助教
    private String education;           //最后学历
    private String address;             //家庭住址
    private String phone;               //联系电话
    private String email;               //电子邮箱
    private String subject;             //研究方向
    //无参构造方法
    public Teacher() {
    }
```

```java
//有参构造方法
public Teacher(String teacherNo, String teacherName, String password,
        int professional, String education, String address,
        String phone, String email, String subject) {
    super();
    this.teacherNo=teacherNo;
    this.teacherName=teacherName;
    this.password=password;
    this.professional=professional;
    this.education=education;
    this.address=address;
    this.phone=phone;
    this.email=email;
    this.subject=subject;
}
//getter、setter方法
public String getTeacherNo() {
    return teacherNo;
}
public void setTeacherNo(String teacherNo) {
    this.teacherNo=teacherNo;
}
public String getTeacherName() {
    return teacherName;
}
public void setTeacherName(String teacherName) {
    this.teacherName=teacherName;
}
public String getPassword() {
    return password;
}
public void setPassword(String password) {
    this.password=password;
}
public int getProfessional() {
    return professional;
}
public void setProfessional(int professional) {
    this.professional=professional;
}
public String getEducation() {
    return education;
}
public void setEducation(String education) {
```

```java
        this.education=education;
    }
    public String getAddress() {
        return address;
    }
    public void setAddress(String address) {
        this.address=address;
    }
    public String getPhone() {
        return phone;
    }
    public void setPhone(String phone) {
        this.phone=phone;
    }
    public String getEmail() {
        return email;
    }
    public void setEmail(String email) {
        this.email=email;
    }
    public String getSubject() {
        return subject;
    }
    public void setSubject(String subject) {
        this.subject=subject;
    }
}
```

程序 TeacherDBAccess.java 代码如下：

```java
package com.imut.javabean;
import java.sql.Connection;
import java.sql.PreparedStatement;
import java.sql.ResultSet;
import java.sql.SQLException;
import java.sql.Statement;
import java.util.ArrayList;
import java.util.HashMap;
import java.util.List;
import java.util.Map;
import com.imut.commmon.ConnectionFactory;
import com.imut.commmon.Page;
import com.imut.commmon.ResourceClose;
public class TeacherDBAccess {
    //添加教师方法
```

```java
public void addTeacher(Teacher teacher){
    Connection conn=null;
    PreparedStatement pstmt=null;
    ResultSet rs=null;
    try{
        conn=ConnectionFactory.getConnection();
        String sql="insert into Teacher values(?,?,?,?,?,?,?,?,?)";
        pstmt=conn.prepareStatement(sql);
        pstmt.setString(1, teacher.getTeacherNo());
        pstmt.setString(2, teacher.getTeacherName());
        pstmt.setString(3, teacher.getPassword());
        pstmt.setInt(4, teacher.getProfessional());
        pstmt.setString(5, teacher.getEducation());
        pstmt.setString(6, teacher.getAddress());
        pstmt.setString(7, teacher.getPhone());
        pstmt.setString(8, teacher.getEmail());
        pstmt.setString(9, teacher.getSubject());
        pstmt.executeUpdate();
    }catch(Exception e){
        e.printStackTrace();
    }finally{
        ResourceClose.close(rs, pstmt, conn);
    }
}
//删除教师方法
public void delTeacher(String teacherNo){
    Connection conn=null;
    PreparedStatement pstmt=null;
    ResultSet rs=null;
    try{
        conn=ConnectionFactory.getConnection();
        String sql="delete from Teacher where teacherNo=?";
        pstmt=conn.prepareStatement(sql);
        pstmt.setString(1, teacherNo);
        pstmt.executeUpdate();
    }catch (SQLException e) {
        e.printStackTrace();
    }finally{
        ResourceClose.close(rs, pstmt, conn);
    }
}
//修改教师信息
public void updateTeacher(Teacher teacher){
    Connection conn=null;
```

```java
            PreparedStatement pstmt=null;
            ResultSet rs=null;
            try{
                conn=ConnectionFactory.getConnection();
                String sql="update Teacher set teacherNo=?,teacherName=?,password=?,professional=?,"+
                    "education =?, address =?, phone =?, email =?, subject =? where teacherNo=?";
                pstmt=conn.prepareStatement(sql);
                pstmt.setString(1, teacher.getTeacherNo());
                pstmt.setString(2, teacher.getTeacherName());
                pstmt.setString(3, teacher.getPassword());
                pstmt.setInt(4, teacher.getProfessional());
                pstmt.setString(5, teacher.getEducation());
                pstmt.setString(6, teacher.getAddress());
                pstmt.setString(7, teacher.getPhone());
                pstmt.setString(8, teacher.getEmail());
                pstmt.setString(9, teacher.getSubject());
                pstmt.setString(10, teacher.getTeacherNo());
                pstmt.executeUpdate();
            }catch (SQLException e) {
                e.printStackTrace();
            }finally{
                ResourceClose.close(rs, pstmt, conn);
            }
        }
        //根据教师编号查找教师
        public Teacher findTeacherByTeacherNo(String teacherNo){
            Teacher teacher=null;
            Connection conn=null;
            PreparedStatement pstmt=null;
            ResultSet rs=null;
            try{
                conn=ConnectionFactory.getConnection();
                String sql="select * from Teacher where teacherNo=?";
                pstmt=conn.prepareStatement(sql);
                pstmt.setString(1, teacherNo);
                rs=pstmt.executeQuery();
                while(rs.next()){
                    teacher=new Teacher();
                    teacher.setTeacherNo(rs.getString(1));
                    teacher.setTeacherName(rs.getString(2));
                    teacher.setPassword(rs.getString(3));
                    teacher.setProfessional(rs.getInt(4));
```

```java
                teacher.setEducation(rs.getString(5));
                teacher.setAddress(rs.getString(6));
                teacher.setPhone(rs.getString(7));
                teacher.setEmail(rs.getString(8));
                teacher.setSubject(rs.getString(9));
            }
        }catch (SQLException e) {
            e.printStackTrace();
        }finally{
            ResourceClose.close(rs, pstmt, conn);
        }
        return teacher;
    }
    //列表显示所有教师列表--分页
    public Map findAllTeacher(int curPage){
        Teacher teacher=null;
        ArrayList list=new ArrayList();
        Connection conn=null;
        Statement pstmt=null;
        ResultSet rs=null;
        ResultSet r=null;
        Map map=null;
        Page pa=null;
        try{
            conn=ConnectionFactory.getConnection();
            String sql="select * from Teacher order by teacherNo";
            pstmt= conn. createStatement (ResultSet. TYPE _ SCROLL _ INSENSITIVE,
            ResultSet.CONCUR_READ_ONLY);
            rs=pstmt.executeQuery(sql);
            pa=new Page();//声明分页类对象
            pa.setPageSize(5);
            pa.setPageCount(rs);
            pa.setCurPage(curPage);
            r=pa.setRs(rs);
            r.previous();
            for(int i=0;i<pa.getPageSize();i++){
                if(r.next()){
                    teacher=new Teacher();
                    teacher.setTeacherNo(r.getString(1));
                    teacher.setTeacherName(r.getString(2));
                    teacher.setPassword(rs.getString(3));
                    teacher.setProfessional(r.getInt(4));
                    teacher.setEducation(r.getString(5));
                    teacher.setAddress(r.getString(6));
```

```java
                    teacher.setPhone(r.getString(7));
                    teacher.setEmail(r.getString(8));
                    teacher.setSubject(r.getString(9));
                    list.add(teacher);
                }else{
                    break;
                }
            }
            map=new HashMap();
            map.put("list",list);
            map.put("pa",pa);
        }catch (SQLException e) {
            e.printStackTrace();
        }finally{
            ResourceClose.close(rs, pstmt, conn);
            ResourceClose.close(r, null, null);
        }
        return map;
    }
    //多条件查询教师
    public List findAllTeacherByMostCon(String teacherNo,String teacherName,
            Integer professional,String phone,String subject){
        Teacher teacher=null;
        ArrayList list=new ArrayList();
        Connection conn=null;
        PreparedStatement pstmt=null;
        ResultSet rs=null;
        
        //构造多条件查询的SQL语句
        String sql="select * from Teacher where 1=1 ";
        //模糊查询
        if(teacherNo!=null&&!teacherNo.equals("")){
            sql+=" and teacherNo like '%"+teacherNo+"%'";
        }
        if(teacherName!=null&&!teacherName.equals("")){
            sql+=" and teacherName like '%"+teacherName+"%'";
        }
        if(phone!=null&&!phone.equals("")){
            sql+=" and phone like '%"+phone+"%'";
        }
        if(subject!=null&&!subject.equals("")){
            sql+=" and subject like '%"+subject+"%'";
        }
        if(professional!=null&&!professional.equals("")){
```

```
            sql+=" and professional="+professional;
        sql+=" order by teacherNo";
        try{
            conn=ConnectionFactory.getConnection();
            pstmt=conn.prepareStatement(sql);
            rs=pstmt.executeQuery();
            while(rs.next()){
                teacher=new Teacher();
                teacher.setTeacherNo(rs.getString(1));
                teacher.setTeacherName(rs.getString(2));
                teacher.setPassword(rs.getString(3));
                teacher.setProfessional(rs.getInt(4));
                teacher.setEducation(rs.getString(5));
                teacher.setAddress(rs.getString(6));
                teacher.setPhone(rs.getString(7));
                teacher.setEmail(rs.getString(8));
                teacher.setSubject(rs.getString(9));
                list.add(teacher);
            }
        }catch (SQLException e) {
            e.printStackTrace();
        }finally{
            ResourceClose.close(rs, pstmt, conn);
        }
        return list;
    }
}
```

程序 ListAllTeacherServlet.java 代码如下：

```
package com.imut.servlet.base;
import java.io.IOException;
import java.sql.ResultSet;
import java.sql.SQLException;
import java.util.ArrayList;
import java.util.HashMap;
import java.util.List;
import java.util.Map;
import javax.servlet.ServletException;
import javax.servlet.http.HttpServlet;
import javax.servlet.http.HttpServletRequest;
import javax.servlet.http.HttpServletResponse;
import com.imut.commmon.Page;
import com.imut.javabean.TeacherDBAccess;
```

```java
public class ListAllTeacherServlet extends HttpServlet {
    public void doGet(HttpServletRequest request, HttpServletResponse response)
            throws ServletException, IOException {
        doPost(request, response);
    }
    public void doPost(HttpServletRequest request, HttpServletResponse response)
            throws ServletException, IOException {
        response.setContentType("text/html;charset=UTF-8");
        int curPage=1;
        String temp=request.getParameter("curPage");
        if(temp!=null){
            curPage=Integer.parseInt(request.getParameter("curPage"));
        }
        TeacherDBAccess dbAccess=new TeacherDBAccess();
        Map map=dbAccess.findAllTeacher(curPage);
        ArrayList list=(ArrayList) map.get("list");
        Page pa= (Page) map.get("pa");
        request.setAttribute("curPage", pa.getCurPage());
                                            //向显示页传递当前页页码
        request.setAttribute("pageCount",pa.getPageCount());
                                            //向显示页传递总页数
        request.setAttribute("list", list);    //向显示页传递结果集
        request. getRequestDispatcher ( "/base/teacherList. jsp "). forward
        (request, response);
    }
}
```

程序 ListAllTeacherServlet.java 在 web.xml 文件中的配置信息如下：

```xml
<servlet>
    <servlet-name>ListAllTeacherServlet</servlet-name>
    <servlet-class>com.imut.servlet.base.ListAllTeacherServlet</servlet-class>
</servlet>
<servlet-mapping>
    <servlet-name>ListAllTeacherServlet</servlet-name>
    <url-pattern>/base/listAllTeacherServlet</url-pattern>
</servlet-mapping>
```

程序 teacherList.jsp 代码如下：

```
<%@page language="java" contentType="text/html; charset=UTF-8" pageEncoding="UTF-8"%>
<%@page import="java.util.*"%>
<%@page import="com.imut.javabean.*"%>
<!DOCTYPE html PUBLIC "-//W3C//DTD XHTML 1.0 Transitional//EN" "http://www.w3.
```

```
org/TR/xhtml1/DTD/xhtml1-transitional.dtd">
<html xmlns="http://www.w3.org/1999/xhtml">
<head>
<title>教师管理</title>
<link rel="stylesheet" type="text/css" id="css" href="../style/main.css" />
<link rel="stylesheet" type="text/css" id="css" href="../style/style1.css" />
<link rel="stylesheet" type="text/css" id="css" href="../style/style.css" />
<style type="text/css">

<!--
table{border-spacing:1px; border:1px solid #A2C0DA;}
td, th{padding:2px 5px; border-collapse: collapse; text-align: left; font-weight:normal;}
thead tr th{height:50px;background:#B0D1FC;border:1px solid white;}
thead tr th.line1{background:#D3E5FD;}
thead tr th.line4{background:#C6C6C6;}
tbody tr td{height:35px;background:#CBE2FB;border:1px solid white; vertical-align:middle;}
tbody tr td.line4{background:#D5D6D8;}
tbody tr th{height:35px;background: #DFEDFF;border:1px solid white; vertical-align:middle;}
tfoot tr td{height:35px;background: #FFFFFF;border:1px solid white; vertical-align:middle;}
-->
</style>
<script type="text/javascript" src="../js/common.js" ></script>
</head>
<%
    String message= (String)session.getAttribute("message");
    if(message!=null){
%>
    <script type="text/javascript">
        alert("<%=message %>");
    </script>
<%
        session.removeAttribute("message");
    }
%>
<body>
<div id="btm">
<div id="main">
    <div id="header">
        <div id="top"></div>
        <div id="logo">
```

```html
            <h1>教师管理</h1></div>
        <div id="mainnav">
            <ul>
                <li><a href="../index.jsp">首页</a></li>
                <li><a href="listAllTeacherServlet">基本数据管理</a></li>
                <li><a href="#">课程安排管理</a></li>
                <li><a href="#">学生成绩管理</a></li>
                <li><a href="#">个人信息管理</a></li>
            </ul>
            <span>
            </span>
        </div>
    </div>
    <div id="tabsJ">
        <ul>
            <li><a href="#" title="管理员维护"><span>管理员维护</span></a></li>
            <li><a href="#" title="学生信息维护"><span>学生信息维护</span></a></li>
            <li><a href="listAllTeacherServlet" title="教师信息维护"><span>教师信息维护</span></a></li>
            <li><a href="#" title="班级信息维护"><span>班级信息维护</span></a></li>
            <li><a href="#" title="课程信息维护"><span>课程信息维护</span></a></li>
        </ul>
    </div>
    <div id="content" align="center">
        <div id="center">
<BR /><BR />
<form method="post" action="checkTeacherServlet" >
<table width="800" align="center" cellpadding="0" cellspacing="0">
    <thead>
    <tr>
        <td width="70%"><h5>教师管理-->教师列表显示       <a href="addTeacher.jsp">添加教师</a></h5></td>
    </tr>
    </thead>
    <tr>
        <td colspan="2" width="100%">
        <table width="100%">
        <thead>
            <tr>
                <th align="center" class="line1" scope="col" colspan="12">
                    <b><font color="red">请输入查询统计条件:</font></b>
                </th>
            </tr>
            <tr>
```

```
<th align="center" class="line1" scope="col" colspan="3">
    <b>教师编号</b>
</th>
<%
    String teacherNo = (String) request.getAttribute
    ("teacherNo");
%>
<th align="center" scope="col" colspan="1">
    <input name="teacherNo" value="<%=teacherNo==null?"
    ":teacherNo%>"/>
</th>
<th align="center" scope="col" colspan="3">
    <b>教师姓名</b>
</th>
<%
    String teacherName = (String) request.getAttribute
    ("teacherName");
%>
<th align="center" scope="col" colspan="1">
    <input name="teacherName" value="<%=teacherName==
    null?"":teacherName%>"/>
</th>
    <th align="center" scope="col" colspan="3">
    <b>职称</b>
</th>
<th align="center" scope="col" colspan="1">
<select name="professional">
<%
    Integer professional= (Integer)request.getAttribute
    ("professional");
    if(professional!=null&&professional==0){
%>
        <option value="">请选择职称...</option>
        <option value="0" selected>教授</option>
        <option value="1">副教授</option>
        <option value="2">讲师</option>
        <option value="3">助教</option>
        <option value="4">其他</option>
<%
    }else if(professional!=null&&professional==1){
%>
        <option value="">请选择职称...</option>
        <option value="0">教授</option>
        <option value="1" selected>副教授</option>
```

```
                <option value="2">讲师</option>
                <option value="3">助教</option>
                <option value="4">其他</option>
            <%
            }else if(professional!=null&&professional==2){
            %>
                <option value="">请选择职称...</option>
                <option value="0">教授</option>
                <option value="1">副教授</option>
                <option value="2" selected>讲师</option>
                <option value="3">助教</option>
                <option value="4">其他</option>
            <%
            }else if(professional!=null&&professional==3){
            %>
                <option value="">请选择职称...</option>
                <option value="0">教授</option>
                <option value="1">副教授</option>
                <option value="2">讲师</option>
                <option value="3" selected>助教</option>
                <option value="4">其他</option>
            <%
            }else if(professional!=null&&professional==4){
            %>
                <option value="">请选择职称...</option>
                <option value="0">教授</option>
                <option value="1">副教授</option>
                <option value="2">讲师</option>
                <option value="3">助教</option>
                <option value="4" selected>其他</option>
            <%
            }else if(professional==null){
            %>
                <option value="" selected>请选择职称...</option>
                <option value="0">教授</option>
                <option value="1">副教授</option>
                <option value="2">讲师</option>
                <option value="3">助教</option>
                <option value="4">其他</option>
            <%
            }
            %>
            </select>
        </th>
```

```html
        </tr>
        <tr>
            <th align="center" class="line1" scope="col" colspan="3">
                <b>联系电话</b>
            </th>
            <%
                String phone = (String) request.getAttribute
                ("phone");
            %>
            <th align="center" scope="col" colspan="1">
                <input name=" phone" value="<%=phone==null?"
                ":phone%>"/>
            </th>
            <th align="center" scope="col" colspan="3">
                <b>研究方向</b>
            </th>
            <%
                String subject = (String) request.getAttribute
                ("subject");
            %>
            <th align="center" scope="col" colspan="1">
                <input name="subject" value="<%=subject==null?":
                subject%>"/>
            </th>
            <th align="center" class="line1" scope="col" colspan="4">
                <input type="submit" value="查询"/>
            </th>
        </tr>
        <tr>
            <th width="15%" align="center" class="line1" scope=
            "col" colspan="2">
                <b>教师姓名</b>
            </th>
            <th width="15%" align="center" scope="col" colspan="2">
                <b>教师职称</b>
            </th>
            <th width="15%" align="center" scope="col" colspan="2">
                <b>联系电话</b>
            </th>
            <th width="15%" align="center" scope="col" colspan="2">
                <b>学历</b>
            </th>
            <th width="15%" align="center" scope="col" colspan="2">
                <b>Email</b>
```

```html
                </th>
                <th width="15%" align="center" colspan="2">
                    <b>操作</b>
                </th>
            </tr>
        </thead>
        <tbody>
            <%
                ArrayList list=(ArrayList)request.getAttribute("list");
                if(list!=null){
                    Iterator e=list.iterator();
                    while(e.hasNext()){
                        Teacher teacher=(Teacher)e.next();
            %>
            <tr>
                <td width="15%" align="center" colspan="2">
                    <a href=" showTeacherServlet? teacherNo = <%=
                    teacher.getTeacherNo() %>" > <% = teacher.
                    getTeacherName() %></a>
                </td>
                <td width="15%" align="center" colspan="2">
                    <%  if(teacher.getProfessional()==0)%>教授
                    <%  if(teacher.getProfessional()==1)%>副教授
                    <%  if(teacher.getProfessional()==2)%>讲师
                    <%  if(teacher.getProfessional()==3)%>助教
                    <%  if(teacher.getProfessional()==4)%>其他
                </td>
                <td width="15%" align="center" colspan="2">
                    <%=teacher.getPhone() %>
                </td>
                <td width="15%" align="center" colspan="2">
                    <%=teacher.getEducation() %>
                </td>
                <td width="15%" align="center" colspan="2">
                    <%=teacher.getEmail() %>
                </td>
                <td width="15%" align="center" colspan="2">
                    <a href="deleteTeacherServlet? teacherNo = <%=
                    teacher.getTeacherNo()%>"><font color=red>删除
                    </font></a>|
                    <a href=" showTeacherServlet? teacherNo = <%=
                    teacher.getTeacherNo()%>"><font color=red>查看/
                    修改</font></a>
```

```html
                </td>
            </tr>
<%
            }
        }
%>
</tbody>
<tbody>
    <tr>
        <td align="center" colspan="12">
<%
            Integer pageCount = (Integer) request.getAttribute
            ("pageCount");
            Integer curPage = (Integer) request.getAttribute
            ("curPage");
            if(pageCount!=null&&curPage!=null){
                if(pageCount!=0&&curPage!=1){
%>
                    <a href="listAllTeacherServlet? curPage=1"/>首
                    页</a>
                    <a href=" listAllTeacherServlet? curPage = <% =
                    curPage-1 %>"/>前一页</a>
<%
                }
%>
<%
                if(pageCount==0||curPage==1){
%>
                    首页    前一页
<%
                }
%>
<%
                if(pageCount!=0&&curPage!=pageCount){
%>
                    <a href=" listAllTeacherServlet? curPage = <% =
                    curPage+1 %>">下一页</a>
                    <a href=" listAllTeacherServlet? curPage = <% =
                    pageCount%>"/>尾页</a>
<%
                }
%>
<%
                if(pageCount==0||curPage==pageCount){
```

```
                %>
                            下一页 尾页
                <%
                        }
                %>
                         
                        第<%=curPage%>页/共<%=pageCount%>页
                         
                <%
                        }
                %>
                    </td>
                </tr>
            </tbody>
        </table>
                    </td>
                </tr>
            </table>
        </form>
        </div>
    </div>
    <div id="footer">
        <div id="copyright">
            <div id="copy">
            <p align="center">CopyRight&copy;2010</p>
            <p>内蒙古工业大学信息工程学院软件工程系</p>
                </div>
            </div>
            <div id="bgbottom"></div>
        </div>
    </div>
</div>
</body>
</html>
```

【代码分析】程序 Teacher.java 中定义了教师的属性及设置和访问属性的方法。

程序 TeacherDBAccess.java 定义了操作数据库方法：增加教师信息、删除教师信息、修改教师信息、根据教师号查找教师信息、分页显示所有教师信息以及多条件查询教师信息。

程序 ListAllTeacherServlet.java 首先从 request 作用范围获取当前页码，根据当前页码查找教师信息。把当前页码、总页数和查询到的教师集合保存到 request 作用范围，然后跳转到 teacher.jsp 页面进行显示。

程序 teacherList.jsp 提供了"添加教师"的超链接，并将链接地址设置为

"addTeacher.jsp"。在查询功能中，首先从 request 作用范围中获取各属性的值，如果为 null 则将其设置为空字符串，然后将各属性的值赋给相应表单组件的 value 属性。接着从 request 作用范围中获取 list 集合对象，如果不为空，则使用 Iterator 迭代输出集合中 Teacher 对象的各个属性值。在每位教师信息的最后通过超链接提供"删除"和"查看/修改"的操作，将教师编号通过 URL 重写的方式附加在超链接地址后面来确定具体要操作哪一位教师信息。最后从 request 作用范围中取得分页需要的各属性值来实现具体分页操作。

习　题

1. 选择题

(1) 对于声明＜％！……％＞的说法错误的是(　　)。
　　A. 一次可声明多个变量和方法
　　B. 一个声明仅在一个页面中有效
　　C. 声明的变量将作为局部变量
　　D. 在预定义中声明的变量将在页面加载时被初始化

(2) 可以在以下哪个标记之间插入 Java 程序段？(　　)
　　A. ＜％ 和 ％＞　　　B. ＜％ 和 /＞　　　C. ＜/ 和 ％＞　　　D. ＜％ 和 ！＞

(3) JSP 中的内置对象 application 相当于下列哪个类型的对象？(　　)
　　A. HttpSession　　　　　　　　　　B. HttpServletReponse
　　C. HTTPServletRequest　　　　　　D. ServletContext

(4) 下面对象不属于 JSP 内置对象的是(　　)。
　　A. jspWriter 对象　　　　　　　　　B. response 对象
　　C. application 对象　　　　　　　　D. page 对象

(5) include 指令用于在 JSP 页面静态插入一个文件，插入文件可以是 JSP 页面、HTML 网页、文本文件或一段 Java 代码，但必须保证插入后形成的文件是(　　)。
　　A. 一个完整的 HTML 文件　　　　B. 一个完整的 JSP 文件
　　C. 一个完整的 TXT 文件　　　　　D. 一个完整的 Java 源文件

(6) JSP 页面中 request.getParameter(String)得到的数据，其类型是(　　)。
　　A. double　　　　B. int　　　　C. String　　　　D. integer

(7) aa.jsp 把请求转发给 bb.jsp。aa.jsp 在请求范围内存放了一个 String 类型的 username 属性，bb.jsp 如何获取该属性？(　　)
　　A. ＜％
　　　　String username=request.getAttribute("username");
　　％＞
　　B. ＜％
　　　　String username=(String)request.getAttribute("username");

C. <%
   String username=request.getParameter("username")
   %>

D. <%
   String username=(String)application.getAttribute("username")
   %>

(8) 某应用中的 test.jsp 文件的源代码如下：

```
<!%int a=0; %>
<%
    int b=0;
    a++;
    b++;
%>
a:<%=a %><br>
b:<%=b %><br>
```

当浏览器第二次访问该 test.jsp 时得到的返回结果是（　　）。

　　A. a=0 b=0　　　　　　　　　　B. a=1 b=1
　　C. a=2 b=1　　　　　　　　　　D. a=1 b=0

(9) 在 JSP 中为内置对象定义了 4 种作用范围，即 Application Scope、Session Scope、Page Scope 和（　　）。

　　A. Request Scope　　　　　　　B. Response Scope
　　C. Out Scope　　　　　　　　　D. Writer Scope

(10) 以下哪个部署符元素可用来指定异常类型以便映射到错误页面上？（　　）

　　A. <error>　　　　　　　　　　B. <error-page>
　　C. <exception>　　　　　　　　D. <exception-type>

(11) 可以利用 request 对象的哪个方法获取客户端的表单信息？（　　）

　　A. request.getParameter()　　　　B. request.outParameter()
　　C. request.writeParameter()　　　　D. request.handlerParameter()

2. 填空题

(1) JSP 的全称是_____，是一种 Web 动态页面技术，事实上 JSP 就是嵌入了_____的 HTML 文件，JSP 文件后缀名为_____。JSP 由_____、_____和_____组成。

(2) JSP 有三种指令元素，它们分别是：_____、_____和_____。_____指令元素将 file 属性指定页面内容静态包含到当前页面。_____动作元素将 page 属性指定页面内容动态包含到当前页面。

(3) 内置对象又称_____，内置对象在 JSP 页面初始化时由 Web 容器为用户自动创建,编程者使用 JSP 进行页面编程时可以不加声明和创建,直接在_____和表达式中使用这些对象。

(4) JSP 页面中,输出型注释的内容写在_____和_____之间。

(5) Page 指令的属性 Language 的默认值是_____。

3. 简答题

(1) 简述 JSP 的指令元素使用方法。

(2) 简述 JSP 的动作元素使用方法。

(3) JSP 中静态 include 和动态 include 有什么区别?

(4) JSP 内置对象有哪些?它们的作用是什么?

4. 程序设计题

(1) 请编写 JSP 程序实现如图 4-21 所示的简易加法器。要求:输入完"加数"和"被加数"后,单击"提交计算"按钮,结果将显示在"答案"文本框中。

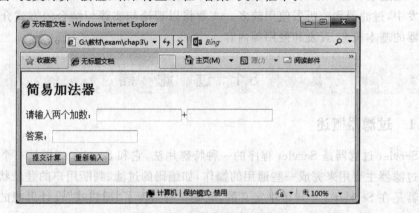

图 4-21 简易加法器界面

(2) 编制一个网站主页的访问计数器,计数器初始值用配置初始化参数的方法定为 200。每增加一个访问者,计数器加 1。

(3) 用 JSP 来实现第 3 章习题 4 中程序设计题(1)的功能,在 index.jsp 页面中,提供用户输入用户名文本框,单击"提交"按钮,将用户输入的用户名保存在 Session 对象中,进入 session.jsp,在 session.jsp 页面中可以添加最喜欢去的地方,再次单击"提交"按钮,跳转到 result.jsp 页面,页面中将用户输入的用户名与最喜欢去的地方显示出来。

过滤器和监听器

过滤器是在 Servlet 2.3 规范中引入的新功能,并在 Servlet 2.4 规范中得到增强。过滤器是在服务端运行的 Web 组件程序,可以过滤客户端向服务器发送的请求,也可以过滤服务器向客户端做出的响应。Servlet 过滤器是 Servlet 的一种特殊的用法,主要用来完成一些通用的操作,例如用户认证等操作。监听器可以监听客户端的请求、服务端的操作等。通过监听器,可以自动激发一些操作,比如监听在线用户的数量。在 Java Web 应用开发中,过滤器和监听器使用较多。本章将以理论与实践相结合的方式介绍过滤器和监听器的基本概念、开发和使用等内容。

5.1 过滤器

5.1.1 过滤器概述

Servlet 过滤器是 Servlet 程序的一种特殊用法,它和 Servlet 一样是一个特殊的 Java 类。过滤器主要用来完成一些通用的操作,如编码的过滤、判断用户的登录状态。Servlet 过滤器是在 Servlet 2.3 规范中定义的,它是一种采用了"插拔式"设计思想的 Web 组件,能够对 Web 服务器接收到的客户端请求和向客户端发出的响应进行拦截过滤,过滤器支持对 Servlet 程序和 JSP 页面的基本请求处理功能。

Servlet 过滤器本身不产生请求和响应,它只提供过滤作用。当用户发起 Web 请求时,Web 服务器首先判断是否存在过滤器和这个请求的目标资源相关,如果存在关联,Web 服务器将把请求交给过滤器去处理,在过滤器中可以对请求的内容做出改变,然后再将请求转交给被请求的目标资源。当被请求的资源做出响应时,Web 服务器同样会将响应先转发给过滤器,在过滤器中可以对响应做出处理,然后再将响应发送给客户端。在上述整个过程中客户端和目标资源是不知道过滤器存在的。

一个过滤器对请求做了两次(分别对 request 和 response 过滤)过滤,其实过滤器是对请求中的 Request 和 Response 进行了拦截。拦截到了将进行处理,处理完后再返回到其原来的调用流程上去,这一点体现了责任链模式。

在一个 Web 应用程序中可以配置多个过滤器,从而形成过滤器链。在请求目标资源时,过滤器链中的过滤器依次对请求做出处理,在接收到响应时再按照相反的顺序对响应做出处理。Servlet 过滤器链的工作流程如图 5-1 所示。

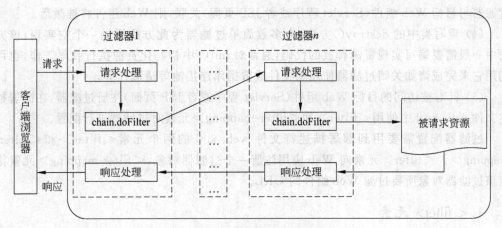

图 5-1　过滤链的工作流程

5.1.2　Filter 接口

所有的 Servlet 过滤器类都必须实现 javax.servlet.Filter 接口,该接口中定义了以下三个过滤器必须实现的方法。

（1）public void init(FilterConfig filterConfig)：过滤器的初始化方法,通常用来做资源的初始化工作。Servlet 容器在创建过滤器实例时调用这个方法,在这个方法中可以利用 FilterConfig 对象读出在 web.xml 文件中为该过滤器配置的初始化参数。

（2）public void doFilter（ServletRequest request, ServletResponse response, FilterChain chain）：用于完成实际的过滤操作,当客户请求访问与过滤器相关联的 URL 时,Web 服务器将先调用过滤器的这个方法。FilterChain 对象负责将请求和响应向后传递。

（3）public void destroy()：过滤器在即将被销毁时执行这个方法,释放过滤器申请的资源。

5.1.3　过滤器开发步骤

开发一个 Servlet 过滤器的步骤如下。

（1）创建一个实现了 javax.servlet.Filter 接口的类。

（2）重写类中的 init(FilterConfig filterConfig)方法,该方法中一般编写初始化 Filter 需要执行的代码。例如,读入为过滤器配置的初始化参数,申请过滤器需要的资源。该方法只在此过滤器第一次初始化时执行,不是每次调用过滤器都执行它。对于简单的过滤器,可提供此方法的一个空实现。

（3）重写类中的方法 doFilter(ServletRequest request, ServletResponse response, FilterChain chain),该方法中编写实现过滤操作的代码。在代码处理过程中,可以从 ServletRequest 参数中得到全部的请求信息,对于简单的过滤器,大多数过滤逻辑是基于这个对象的,从 ServletResponse 参数中得到全部的响应信息。在方法代码的最后需要调用 FilterChain 对象的 doFilter()方法,来激活下一个相关的过滤器。如果不存在另一个过滤器与目标Web组件(Servlet程序或者JSP页面)关联,则Web组件将被激活。

(4) 重写类中的destroy()方法,大多数简单过滤器为此方法提供一个空实现,该方法中一般需要编写实现资源释放的代码,通常对init()中初始化资源执行收尾工作,也可利用它来完成诸如关闭过滤器使用的文件或数据库连接池等清除任务。

(5) 针对被访问的目标Web组件(Servlet程序或者JSP页面)注册过滤器,在部署描述文件web.xml中使用＜filter＞和＜filter-mapping＞元素对过滤器进行配置。

过滤器配置需要用到部署描述符文件web.xml的两个元素＜filter＞和＜filter-mapping＞。＜filter＞元素向Web应用注册一个过滤器对象,＜filter-mapping＞元素指定该过滤器对象所要过滤Web组件的URL。

1. ＜filter＞元素

＜filter＞元素位于部署描述符文件web.xml的前部,所有＜filter-mapping＞、＜servlet＞或＜servlet-mapping＞元素之前。＜filter＞元素常用的子元素如下。

＜filter-name＞:这是一个必需的元素,它给过滤器分配一个选定的名字。

＜filter-class＞:这是一个必需的元素,它指定过滤器实现类的完全限定名即类的全名。

＜init-param＞:这是一个可选的元素,它定义可利用FilterConfig的getInitParameter方法读取的初始化参数。单个过滤器元素可包含多个＜init-param＞元素。

2. ＜filter-mapping＞元素

＜filter-mapping＞元素位于web.xml文件中＜filter＞元素之后,＜serlvet＞元素之前。＜filter-mapping＞元素常用的子元素如下。

＜filter-name＞:这是一个必需的元素,该名称必须与用＜filter＞元素中声明过的过滤器名称相匹配。

＜url-pattern＞:此元素声明一个以斜杠/开始的模式,它指定过滤器应用的URL。所有＜filter-mapping＞元素中必须提供＜url-pattern＞,但不能对单个＜filter-mapping＞元素提供多个＜url-pattern＞元素项。如果希望过滤器适用于多个模式,可重复整个＜filter-mapping＞元素。

5.1.4 过滤器应用举例

例程 5-1:演示使用过滤器过滤未登录用户或者登录失败用户使其重新返回到登录页面登录。程序为login.jsp、LoginServlet.java、LoginFilter.java和index.jsp。

程序LoginServlet.java代码如下:

```
package com.ch05.filter;
import java.io.IOException;
import java.sql.Connection;
import java.sql.PreparedStatement;
```

```java
import java.sql.ResultSet;
import javax.servlet.ServletException;
import javax.servlet.http.HttpServlet;
import javax.servlet.http.HttpServletRequest;
import javax.servlet.http.HttpServletResponse;
import javax.servlet.http.HttpSession;
public class LoginServlet extends HttpServlet{
    /*
    *验证登录状态,如果登录成功,将用户名信息和登录信息保存到
    *HttpSession级的attribute中
    **/
    public void doPost (HttpServletRequest request,HttpServletResponse response)
        throws ServletException,IOException{
        response.setContentType("text/html");
        request.setCharacterEncoding("UTF-8");
        response.setCharacterEncoding("UTF-8");
        String name=request.getParameter("name");
        String password=request.getParameter("password");
        HttpSession session=request.getSession();
        PreparedStatement pstmt=null;
        ResultSet rs=null;
        Connection conn=null;
        try{
            String sql="select * from usertbl where name=? and password=?";
            conn=ConnectionFactory.getConnection();
            pstmt=conn.prepareStatement(sql);
            pstmt.setString(1, name);
            pstmt.setString(2, password);
            rs=pstmt.executeQuery();
            if(rs.next()){
                session.setAttribute("name", name);
                session.setAttribute("message", "恭喜您,登录成功!");
                response.sendRedirect ( "http://localhost:8080/chapter5/index.
                jsp");
            }else{
                session.setAttribute ("message", "对不起,登录信息有误,请重新
                登录!");
                request.getRequestDispatcher ("/login.jsp").forward (request,
                response);
            }
        }catch(Exception e){
            e.printStackTrace();
        }finally{
            ResourceClose.close(rs, pstmt, conn);
```

        }
    }
}

程序 LoginFilter.java 代码如下:

```java
package com.ch05.filter;
import java.io.IOException;
import javax.servlet.Filter;
import javax.servlet.FilterChain;
import javax.servlet.FilterConfig;
import javax.servlet.ServletException;
import javax.servlet.ServletRequest;
import javax.servlet.ServletResponse;
import javax.servlet.http.HttpServletRequest;
import javax.servlet.http.HttpServletResponse;
import javax.servlet.http.HttpSession;
public class LoginFilter implements Filter{
    public void init(FilterConfig arg0)
        throws ServletException {
        /*
        *包含初始化 Filter 时需要执行的代码,该代码执行一次
        **/
    }
    public void doFilter(ServletRequest request, ServletResponse response,
            FilterChain chain) throws IOException, ServletException {
        HttpServletRequest req= (HttpServletRequest)request;
        HttpServletResponse res= (HttpServletResponse)response;
        HttpSession session=req.getSession();
        String name= (String)session.getAttribute("name");
        if (name!=null) {
            chain.doFilter(req, res);
        } else {
            res.sendRedirect(req.getContextPath()+"/login.jsp");
        }
    }
    public void destroy() {
        /*
        *包含资源释放的代码,通常对 init()中的初始化的资源执行收尾工作;
        **/
    }
}
```

程序 login.jsp 代码如下:

```jsp
<%@ page language="java" import="java.util.*" contentType="text/html;
```

```
    charset=utf-8"%>
<html>
    <head>
        <title>登录页面</title>
    </head>
    <body>
    <%
        String message=(String)session.getAttribute("message");
        if(message!=null){
            out.println(message);
            session.removeAttribute("message");
        }
    %>
    <form action="login" method="post">
        用户名:<input type="text" name="name"/><br>
        密  码:<input type="password" name="password"/><br>
        <input type="submit" value="登录/>
        <input type="reset" value="取消"/>
    </form>
    </body>
</html>
```

程序 index.jsp 代码如下:

```
<%@page language="java" import="java.util.*" contentType="text/html; charset=utf-8" %>
<html>
    <head>
        <title>系统首页</title>
    </head>
    <body>
        <br>
        <br>
        <font size=5 color=red><b>
        <%=session.getAttribute("name") %>,<% = session.getAttribute("message") %>
        </b></font>
        <br>
        <font size=5 color=red><b>
        欢迎您使用 xxx 系统
        </b></font>
    </body>
</html>
```

LoginFilter.java 配置代码如下:
```
<filter>
```

```xml
    <filter-name>loginFilter</filter-name>
    <filter-class>com.ch05.filter.LoginFilter</filter-class>
</filter>
<filter-mapping>
    <filter-name>loginFilter</filter-name>
    <url-pattern>/index.jsp</url-pattern><!--每当用户访问 index.jsp 时,执行 LoginFilter 过滤器-->
</filter-mapping>
```

LoginServlet.java 配置代码如下：

```xml
<servlet>
    <servlet-name>login</servlet-name>
    <servlet-class>com.ch05.filter.LoginServlet</servlet-class>
</servlet>
<servlet-mapping>
    <servlet-name>login</servlet-name>
    <url-pattern>/login</url-pattern>
</servlet-mapping>
```

程序 LoginServlet.java 实现从客户端获得登录用户名和密码信息,利用用户名和密码访问数据库(本例涉及的 UserTbl 表,访问数据库用到 ConnectionFactory.java 和 ResourceClose.java 类参见例程 2-6)验证是否存在该用户,如果用户存在,则将用户名信息和登录成功信息分别存放在 Session 对象的 name 和 message 属性中,并将请求跳转至 index.jsp 页面;如果用户不存在,则将登录失败信息存放在 session 对象的 message 属性中,并将请求转发至 login.jsp 页面,供用户重新登录。

过滤器 LoginFilter.java 程序 doFilter 方法首先将方法参数 request 和 response 对象由原来的 ServletRequest 类型和 ServletResponse 类型转化为 HttpServletRequest 类型和 HttpServletResponse 类型。然后从 Session 获得 name 属性值,如果 name 不为 null (登录成功),调用 FilterChain 对象的 doFilter 方法,由于本例中只存在 LoginFilter 一个过滤器,不存在下一个过滤器,直接将请求发送至 LoginFilter 过滤器所关联的 Web 组件,即 index.jsp。过滤器关联的 Web 组件在过滤器配置信息中<filter-mapping>元素的子元素<url-pattern>体现,本例中 LoginFilter 过滤器配置信息中<url-pattern>/index.jsp</url-pattern>说明该过滤器关联的 Web 组件为 index.jsp,即每当用户发送访问 index.jsp 的请求必须首先满足过滤器 LoginFilter 的过滤条件。如果 name 为 null (登录失败),请求跳转至 login.jsp 页面,提示用户登录失败,需要重新登录。

程序 login.jsp 实现了登录信息输入的页面,在 Java 脚本中从 session 对象获得保存登录信息的 message 属性值并显示在页面上。用户输入完成后通过单击登录按钮将用户信息数据提交至 LoginServlet 处理。

程序 index.jsp 为过滤器关联的 Web 组件,该页面 Java 脚本中从 session 对象获得保存用户名信息和登录信息的 name 属性和 message 属性值并显示在页面上。

运行例程 5-1,首先访问 login.jsp 页面,运行效果如图 5-2 所示。在登录页面上输入不正确的用户名和密码,程序直接跳转到登录页面并在页面显示上登录失败信息,运行结果显示如图 5-3 所示。在登录页面上输入正确的用户名"zhangsan"和密码"zs123",程序将访问 index.jsp,LoginFilter 过滤器关联 index.jsp,所以要运行过滤器 LoginFilter。用户名、密码正确,在 session 对象中存在用户名信息,能够满足过滤器过滤的条件,index.jsp 页面被激活,并在页面上显示 session 对象中的用户名 name 属性和登录成功信息 message 属性的值。运行效果如图 5-4 所示。

图 5-2 登录页面

图 5-3 登录错误页面

图 5-4 index.jsp 运行页面

如果在浏览器地址栏中输入"http://localhost:8080/chapter5/index.jsp"直接访问 index.jsp,此时 session 对象中不存在用户名 name 属性信息,无法通过过滤器过滤,直接跳转至登录页面,如图 5-2 所示。

例程 5-2:利用过滤器实现对整个 Web 应用的 HTTP 请求响应做相应的编码处理。程序为 EncodeFilter.java。

程序 EncodeFilter.java 代码如下:

```
package com.ch05.filter;
import java.io.IOException;
import javax.servlet.Filter;
import javax.servlet.FilterChain;
import javax.servlet.FilterConfig;
import javax.servlet.ServletException;
import javax.servlet.ServletRequest;
import javax.servlet.ServletResponse;
import javax.servlet.http.HttpServletRequest;
import javax.servlet.http.HttpServletResponse;
/**
 * 编码过滤器:如果在初始化参数中配置了 encode,
```

```
 * 那么将所有的过滤的请求和响应均设置为指定的编码；
 * 否则,直接通过过滤器不做任何处理。
 */
public class EncodeFilter implements Filter {
    private FilterConfig config;
    //初始化配置参数
    private static final String INIT_PARAM_ENCODE="encode";
    //初始化方法
    public void init(FilterConfig config) throws ServletException {
        this.config=config;
    }
    public void doFilter(ServletRequest request,ServletResponse response,
            FilterChain chain) throws IOException, ServletException {
        HttpServletRequest req= (HttpServletRequest) request;
        HttpServletResponse res= (HttpServletResponse) response;
        String encode=config.getInitParameter(INIT_PARAM_ENCODE);
        if(encode!=null && !encode.isEmpty()){
            req.setCharacterEncoding(encode);
        }
        chain.doFilter(request,response);
        if(encode!=null && !encode.isEmpty()){
            res.setCharacterEncoding(encode);
        }
    }
    public void destroy() {
    }
}
```

在web.xml中配置过滤器EncodeFilter.java,代码如下：

```
<filter>
    <filter-name>EncodeFilter</filter-name>
    <filter-class>com.ch05.filter.EncodeFilter</filter-class>
    <init-param>
        <param-name>encode</param-name>
        <param-value>UTF-8</param-value>
    </init-param>
</filter>
<filter-mapping>
    <filter-name>EncodeFilter</filter-name>
    <url-pattern>/*</url-pattern>
</filter-mapping>
```

在上述配置代码中,通过＜filter＞的子元素 ＜init-param＞配置初始化参数 encode,值为UTF-8。＜filter-mapping＞的子元素＜url-pattern＞配置值为"/*",即过滤器过滤的目标资源为整个 Web 应用下的所有 Web 组件,这意味着用户访问该 Web 应用的任何 Web 资源都将执行 EncodeFilter 过滤器。

EncodeFilter 过滤器开发配置完成并启动 Web 应用后,将 Web 应用所有的请求和响应均设置为初始化参数 encode 指定的编码。

通过例程 5-1 和例程 5-2 过滤器程序的编写、配置、运行以及运行结果分析,可以看出过滤器的执行流程如图 5-5 所示。

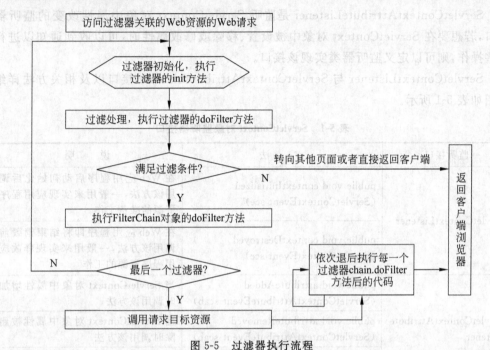

图 5-5 过滤器执行流程

5.2 监 听 器

5.2.1 监听器简介

监听器是一个实现了某一特定接口的普通 Java 类,该类专门用于监听某一特定 Java 对象的方法调用或属性改变事件,当被监听对象发生上述事件后,监听器某个方法将立即被执行而做出相应的动作或者反应。

在 Servlet 规范中定义了多种类型的监听器,它们用于监听的事件源分别为 ServletContext、HttpSession 和 ServletRequest 这三个作用域对象。当这三个对象域对象创建、销毁或者其中属性增加、删除等事件发生时,将执行对应监听器类的特定方法。

5.2.2 监听器接口

Servlet 监听器用来监听 ServletContext、HttpSession 和 ServletRequest 作用域对象相关事件,针对这三个作用域对象 Servlet API 提供了对应的监听接口。

1. ServletContext 对象监听器接口

与 ServletContext 对象相关的监听器接口有两个,分别为 ServletContextListener 与 ServletContextAttributeListener。

ServletContextListener 是用来监听 Web 应用程序生命周期的监听器接口,若想要知道何时 Web 应用程序已经初始化或即将结束销毁,可以定义监听器类实现该接口。

ServletContextAttributeListener 是监听 ServletContext 对象中属性改变的监听器接口,若想要在 ServletContext 对象中被设置、移除或修改属性时,可以收到通知以进行一些操作,则可以定义监听器类实现该接口。

ServletContextListener 与 ServletContextAttributeListener 接口以及相关方法详细介绍如表 5-1 所示。

表 5-1 ServletContext 对象监听器接口

监听接口	方法	说明
ServletContextListener	public void contextInitialized (ServletContextEvent sce)	在 Web 应用程序启动初始化后调用该方法,一般用来实现应用程序资源的准备工作
	public void contextDestroyed (ServletContextEvent sce)	在 Web 应用程序即将结束销毁前调用该方法,一般用来实现释放应用程序资源的工作
ServletContextAttributeListener	public void attributeAdded (ServletContextAttributeEvent scab)	当 ServletContext 对象中属性增加时调用该方法
	public void attributeRemoved (ServletContextAttributeEvent scab)	当 ServletContext 对象中属性被删除时调用该方法
	public void attributeReplaced (ServletContextAttributeEvent scab)	当 ServletContext 对象中属性值被改变时调用该方法

2. HttpSession 对象监听器接口

与 HttpSession 对象相关的监听器接口有 4 个,分别为 HttpSessionListener、HttpSessionAttributeListener、HttpSessionBindingListener 和 HttpSessionActivationListener。

HttpSessionListener 是用来监听会话对象生命周期的监听器接口,若想要在 HttpSession 对象创建或结束时,做些相对应动作,则可以定义监听器类实现该接口。

HttpSessionAttributeListener 是用来监听会话对象属性改变的监听器接口,当在会话对象中加入属性、移除属性或修改属性时,做些相对应的动作,则可以定义监听器类实现该接口。

HttpSessionBindingListener 是用来监听会话对象的对象绑定监听器接口,若有个即将加入 HttpSession 的属性对象,希望在设置给 HttpSession 成为属性或从 HttpSession 中移除时,可以收到 HttpSession 的通知,则可以用属性对象所属类型实现该接口。

HttpSessionActivationListener 是用来监听会话对象的对象迁移监听器接口,对象迁移是 Web 服务器为了更好地利用资源或进行服务器负载平衡等而对特定对象采取的措施。在使用分布式环境时，Web 应用程序的对象可能分散在多个服务器中，当希望服务器之间存在会话对象的转移时，做些相对应动作，则可以定义监听器类实现该接口。当会话从一个服务器转移到另一个服务器时，首先在原来的服务器上将其中的属性对象进行序列化（会话对象中属性对象必须实现 java.io.Serializable），转移至另一服务器后，对其中属性对象反序列化。

HttpSessionListener、HttpSessionAttributeListener、HttpSessionBindingListener 和 HttpSessionActivationListener 接口以及相关方法详细介绍如表 5-2 所示。

表 5-2 HttpSession 对象监听器接口

监听接口	方　法	说　明
HttpSessionListener	public void sessionCreated（HttpSessionEvent se）	当 HttpSession 对象创建时执行该方法
	public void sessionDestroyed（HttpSessionEvent se）	当 HttpSession 对象销毁时执行该方法
HttpSessionAttributeListener	public void attributeAdded（HttpSessionBindingEvent se）	当 HttpSession 对象中属性增加时调用该方法
	public void attributeRemoved（HttpSessionBindingEvent se）	当 HttpSession 对象中属性被删除时调用该方法
	public void attributeReplaced（HttpSessionBindingEvent se）	当 HttpSession 对象中属性值被改变时调用该方法
HttpSessionBindingListener	public void valueBound（HttpSessionBindingEvent event）	当实现 HttpSessionBindingListener 接口的属性对象被加入 HttpSession 时调用该方法
	public void valueUnbound（HttpSessionBindingEvent event）	当实现 HttpSessionBindingListener 接口的属性对象从 HttpSession 对象中移除时调用该方法
HttpSessionActivationListener	public void sessionDidActivate（HttpSessionEvent se）	当 HttpSession 对象要从一个服务器迁移至另一个服务器后，就会对所有属性对象做反序列化，此时会调用该方法
	public void sessionWillPassivate（HttpSessionEvent se）	当 HttpSession 对象要从一个服务器迁移至另一个服务器时，必须先在原来的服务器上序列化 HttpSession 对象中所有的属性对象，此时会调用该方法

3. ServletRequest 对象监听器接口

与 ServletRequest 对象相关的监听器接口有两个，分别为 ServletRequestListener 与 ServletRequestAttributeListener。

ServletRequestListener 是用来监听 HttpServletRequest 生命周期的监听器接口，若想要知道何时 HttpServletRequest 对象生成或即将结束销毁，可以定义监听器类实现该接口。

ServletRequestAttributeListener 是用来监听 ServletRequest 对象中属性改变的监听器接口，若想要在 ServletRequest 对象中被设置、移除或修改属性，可以收到通知以进行一些操作，则可以定义监听器类实现该接口。

ServletRequestListener 与 ServletRequestAttributeListener 接口以及相关方法详细介绍如表 5-3 所示。

表 5-3　ServletRequest 对象监听器接口

监听接口	方　　法	说　　明
ServletRequestListener	public void requestInitialized(ServletRequestEvent sce)	在 ServletRequest 对象生成时调用该方法
	public void requestDestroyed(ServletRequesttEvent sce)	在 ServletRequest 对象即将结束销毁前调用该方法
ServletRequestAttributeListener	public void attributeAdded(ServletRequestAttributeEvent scab)	当 ServletRequest 对象中属性增加时调用该方法
	public void attributeRemoved(ServletRequestAttributeEvent scab)	当 ServletRequest 对象中属性被删除时调用该方法
	public void attributeReplaced(ServletRequestAttributeEvent scab)	当 ServletRequest 对象中属性值被改变时调用该方法

5.2.3　监听器开发

Servlet 监听器开发分为两步，分别如下。

（1）创建相应的监听器类，根据实际需要实现相应的接口并覆盖其中相应的抽象方法。例如，定义一个监听器类 XxxxxListenner 用来监听 HttpSession 对象的创建与销毁，其结构代码如下所示：

```
public class XxxxxListenner implements HttpSessionListener{
    public void sessionCreated(HttpSessionEvent arg0) {
        …;
    }
    public void sessionDestroyed(HttpSessionEvent arg0) {
        …;
    }
}
```

（2）在部署描述文件 web.xml 中使用＜listener＞元素配置监听器（实现接口 HttpSessionBindingListener 接口的监听器例外，不需要在 web.xml 文件中配置）。例如，在 web.xml 文件中配置步骤（1）中定义的监听器类 XxxxxListenner，其配置代码如下：

```
<listener>
    <listener-class>
        xxxPackage.xxxPackage.XxxxxListenner
```

```
        </listener-class>
    </listener>
```

下面通过几个例程来演示监听器的开发以及监听器在实际项目中的应用。

例程 5-3：演示监听器的开发步骤，监听 Web 应用的初始化和销毁。程序为 MyServletContextListenner.java。

首先，定义监听器类 MyServletContextListenner，该类实现接口 ServletContextListener，代码如下：

```
package com.ch05.listener;
import javax.servlet.ServletContextEvent;
import javax.servlet.ServletContextListener;
public class MyServletContextListenner implements ServletContextListener {
    public void contextDestroyed(ServletContextEvent arg0) {
        /* Web 应用停止时执行 */
        System.out.println("web Application destoryed!");
    }
    public void contextInitialized(ServletContextEvent arg0) {
        /* Web 应用启动时执行 */
        System.out.println("web Application initialiezed!");
    }
}
```

其次，在 web.xml 文件中配置监听器 MyServletContextListenner，配置代码如下：

```
<listener>
    <listener-class>
        com.ch05.listener.MyServletContextListenner
    </listener-class>
</listener>
```

程序 MyServletContextListenner 实现了 ServletContextListener 接口并覆盖了其中的 contextInitialized 和 contextDestroyed 方法，该监听器用来监听 Web 应用的初始化和销毁，当 Web 应用初始化后或者即将销毁时分别执行 contextInitialized 方法或者 contextDestroyed 方法。contextInitialized 方法和 contextDestroyed 方法的实现逻辑比较简单，分别打印"web Application initialiezed!"字符串和"web Application destoryed!"字符串。

当 Web 应用启动或者重新启动时，将进行 Web 应用的初始化，此时将执行监听器的 contextInitialized 方法，程序在控制台打印出"web Application initialiezed!"，如图 5-6 所示。

当通过 MyEclipse 正常结束 Web 应用运行时，将进行 Web 应用的销毁，在 Web 应用即将销毁时执行监听器的 contextDestroyed 方法，程序在控制台打印出"web Application destoryed!"，如图 5-7 所示。

例程 5-4：演示利用实现了 HttpSessionListener 接口的监听器实现统计系统的在线

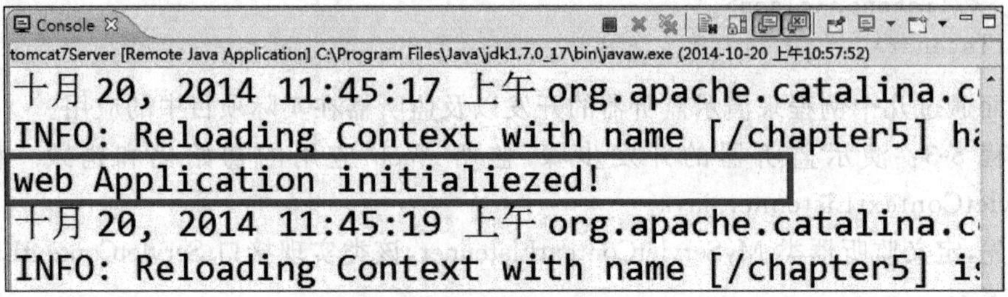
Console
tomcat7Server [Remote Java Application] C:\Program Files\Java\jdk1.7.0_17\bin\javaw.exe (2014-10-20 上午10:57:52)
十月 20, 2014 11:45:17 上午 org.apache.catalina.c
INFO: Reloading Context with name [/chapter5] ha
web Application initialiezed!
十月 20, 2014 11:45:19 上午 org.apache.catalina.c
INFO: Reloading Context with name [/chapter5] is

图 5-6　Web 应用初始化运行效果

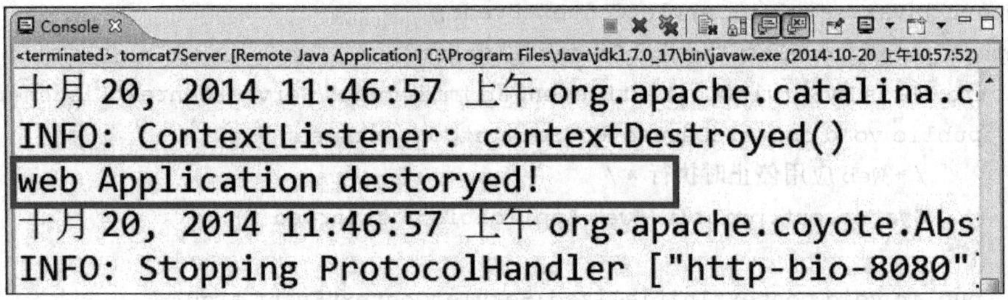
Console
<terminated> tomcat7Server [Remote Java Application] C:\Program Files\Java\jdk1.7.0_17\bin\javaw.exe (2014-10-20 上午10:57:52)
十月 20, 2014 11:46:57 上午 org.apache.catalina.c
INFO: ContextListener: contextDestroyed()
web Application destoryed!
十月 20, 2014 11:46:57 上午 org.apache.coyote.Abs
INFO: Stopping ProtocolHandler ["http-bio-8080"

图 5-7　Web 应用销毁运行效果

人数功能。程序为 CountOnLineListenner.java 和 countOnLine.jsp。

监听器类 CountOnLineListenner 的代码如下：

```
package com.ch05.listener;
import java.util.HashSet;
import javax.servlet.ServletContext;
import javax.servlet.http.HttpSession;
import javax.servlet.http.HttpSessionEvent;
import javax.servlet.http.HttpSessionListener;
public class CountOnLineListenner implements HttpSessionListener {
  public void sessionCreated(HttpSessionEvent event) {
  HttpSession session=event.getSession();
  ServletContext application=session.getServletContext();
  //在 application 范围由一个 HashSet 集保存所有的 session
  HashSet sessions=(HashSet) application.
  getAttribute("sessions");
      if (sessions==null) {
         sessions=new HashSet();
         application.setAttribute("sessions", sessions);
      }
      //新创建的 session 均添加到 HashSet 集中
      sessions.add(session);
      application.setAttribute("sessions", sessions);
  }
```

```
        public void sessionDestroyed(HttpSessionEvent event){
            HttpSession session=event.getSession();
            ServletContext application=session.getServletContext();
            HashSet sessions=(HashSet)application.getAttribute("sessions");
            //销毁的 session 均从 HashSet 集合中移除
            if(session!=null){
                sessions.remove(session);
            }
            application.setAttribute("sessions", sessions);
        }
    }
```

系统访问页面 countOnLine.jsp 代码如下：

```
<%@page language="java" import="java.util.*" pageEncoding="UTF-8"
import="java.util.*,com.ch05.listener.*"%>
<html>
    <head>
        <title>统计系统在线人数</title>
    </head>
    <body>
        <%
            HashSet hs=(HashSet)application.getAttribute("sessions");
        %>
        <center><font color=red size=4>系统当前的在线人数为：
        <%=hs.size()%></font></center>
        <br>
        <br>
        <br>
        <center><font color=red size=8><b>XXXXX 信息系统主页</b></font>
        </center>
    </body>
</html>
```

监听器类 CountOnLineListenner 的配置代码如下：

```
<listener>
    <listener-class>
        com.ch05.listener.CountOnLineListenner
    </listener-class>
</listener>
```

例程 5-4 中监听器类 CountOnLineListenner 实现了监听接口 HttpSessionListener，并覆盖了其中的 sessionCreated 方法和 sessionDestroyed 方法，该监听器可以用来监听 HttpSession 对象的创建和销毁。当有新的 HttpSession 对象创建时，将执行

sessionCreated 方法。当有 HttpSession 对象销毁时将执行 sessionDestroyed 方法。sessionCreated 方法实现了有新的 HttpSession 对象创建时，就把该对象加入到 HashSet 集合（其中不能存放重复元素对象）中，并把该集合对象存放在 ServletContext 作用域对象中。sessionDestroyed 方法实现了有 HttpSession 对象销毁时，将该对象从 HashSet 集合中移除，并将 HashSet 集合对象放在 ServletContext 作用域对象中（替换作用域对象中原有的 HashSet 集合对象）。注：应用程序启动后，用户访问系统将开启一个新的会话，即创建了一个新的 HttpSession 对象。在 Web 页面获得 ServletContext 作用域对象中的 HashSet 集合对象，该集合对象的 size()方法值即为系统在线人数。

页面 countOnLine.jsp 首先在 Java 脚本中获得 ServletContext 作用域对象中的 HashSet 集合对象，然后利用表达式<%=hs.size()%>显示集合对象中的 HttpSession 对象个数，即系统在线人数。

通过浏览器访问 countOnLine.jsp 页面程序运行效果如图 5-8 所示。页面将显示当前系统开启的会话数即系统在线人数。

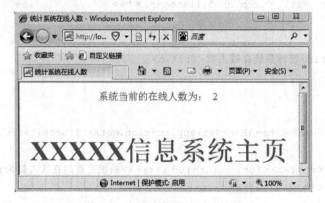

图 5-8　在线人数统计页面效果

例程 5-5：演示监听接口 HttpSessionBindingListener 的使用。程序为 User.java 和 user.jsp。

监听器类 User.java 的代码如下：

```
package com.ch05.listener;
import javax.servlet.http.HttpSessionBindingEvent;
import javax.servlet.http.HttpSessionBindingListener;
public class User implements HttpSessionBindingListener{
    private String name;
    private String data;
    public User(String name){
        this.name=name;
    }
    public void valueBound(HttpSessionBindingEvent arg0) {
        this.data=name+"被绑定在 HttpSession 对象中";
    }
```

```
        public void valueUnbound(HttpSessionBindingEvent arg0) {
            this.data=name+"被移除出 HttpSession 对象";
        }
        public String getData(){
            return this.data;
        }
}
```

系统访问页面 user.jsp 代码如下:

```
<%@page language="java" import="java.util.*" pageEncoding="UTF-8"
 import="com.ch05.listener.User"%>
<html>
    <head>
        <title>对象绑定监听</title>
    </head>
    <body>
        <%
            User user=new User("张三");
            session.setAttribute("user", user);
        %>
        <center><font color=red size=4>
        <%=user.getData()%></font></center>
        <br>
        <%
            session.removeAttribute("user");
        %>
        <center><font color=red size=4>
        <%=user.getData()%></font></center>
    </body>
</html>
```

例程 5-5 中监听器 User.java 实现了 HttpSessionBindingListener 接口并覆盖了其中的 valueBound 和 valueUnbound 方法,当向 HttpSession 对象保存 User 对象或者从 HttpSession 对象移除 User 对象时,分别执行 valueBound 或者 valueUnbound 方法。监听器 User.java 和上述监听器有所不同,该监听器不需要在部署文件 web.xml 中进行配置就可以直接使用。页面 user.jsp 中先使用 Java 脚本创建 User 对象 user,然后将 user 存放在 HttpSession 对象中,此时<%=user.getData()%>显示"xx 被绑定在 HttpSession 对象中";当使用 Java 脚本从 HttpSession 对象中移除对象 user 时,此时<%=user.getData()%>将显示"xx 被移除出 HttpSession 对象"。访问 user.jsp 页面,其运行效果如图 5-9 所示。

图 5-9 user.jsp 页面运行效果

5.3 案 例

本章案例在第 4 章案例的基础上增加实现"个人信息管理"模块中的个人信息修改功能。主要使用过滤器技术实现登录判断，登录成功后根据不同的用户角色提供相应模块的访问权限，使用监听器技术实现用户在线人数统计的功能。

5.3.1 案例设计

本章将用户角色分为两类，一类是管理员用户，一类是教师用户。管理员用户可以访问首页、进行基本信息管理以及个人信息管理。教师用户可以访问首页和进行个人信息管理，不能进行基本信息管理。

首先使用过滤器技术过滤未登录用户，如果没有登录则提示登录并转到登录页面，如果已登录则继续使用过滤器技术判断用户角色，根据配置信息决定用户可以访问的页面。

定义一个用户类来封装用户角色信息，当登录用户是管理员时，将用户类中的用户类型属性设置为"0"，当登录用户是教师时，将用户类中的用户类型属性设置为"1"。

使用监听器技术监听登录用户人数，并将其显示在首页上。

本章案例使用的主要源文件如表 5-4 所示。

表 5-4 本章案例使用的文件

文　件	所在包/路径	功　能
login.jsp	/	用户登录界面
LoginServlet.java	com.imut.servlet	接收登录信息，与数据库表交互判断登录是否成功，登录成功判断用户类型，跳转至系统首页，否则跳转至登录页面重新登录
index.jsp	/	首页，显示登录状态、登录者类别、登录者信息及在线人数
User.java	com.imut.javabean	封装用户信息的类
Admin.java	com.imut.javabean	管理员类
AdminDBAccess.java	com.imut.javabean	管理员操作数据类
ShowPersonServlet.java	com.imut.servlet.person	根据用户类别显示不同用户的查看/修改页面的 Servlet
LoginFilter.java	com.imut.filter	过滤未登录用户
AdminFilter.java	com.imut.filter	过滤未登录用户及非管理员用户
TeacherFilter.java	com.imut.filter	过滤未登录用户及非教师用户
CountListener.java	com.imut.listenter	监听在线用户人数
adminShow.jsp	/person	显示管理员用户的个人查看/修改页面
teacherShow.jsp	/person	显示教师用户的个人查看/修改页面

5.3.2 案例演示

用户没有登录,直接通过浏览器地址 http://localhost:8080/ch05/index.jsp 访问首页,单击菜单"基本数据管理"或"个人数据管理"会弹出警告框提示登录,效果如图 5-10 所示,之后自动跳转到登录页面供用户重新登录。

管理员用户登录系统,在首页中会显示用户类别为管理员及目前在线人数,效果如图 5-11 所示。单击菜单中的"基本数据管理"显示"教师信息维护模块",运行效果如图 4-17 所示,可对教师信息进行增删改查等操作。单击菜单中的"个人信息管理",显示"管理员个人信息查看/修改"页面,运行效果如图 5-12 所示。

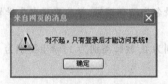
图 5-10 未登录访问系统提示

图 5-11 管理员用户登录首页

图 5-12 管理员个人信息查看/修改

教师用户登录系统,在首页中显示的用户类别为教师,同时显示在线人数,效果如图 5-13 所示。单击菜单中的"基本数据管理"会弹出警告对话框提示只有管理员才可访问,效果如图 5-14 所示,然后返回首页。单击菜单中的"个人信息管理",显示当前登录教师的"教师信息查看/修改"页面,效果如图 4-20 所示。

图 5-13 教师用户登录首页

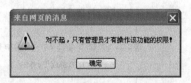
图 5-14 教师用户越权访问提示

5.3.3 代码实现

创建工程 ch05,根据案例设计描述需要完成各部分功能的具体实现。本章主要给出用户类 User.java、登录 Servlet 类 LoginServlet.java、登录过滤器 LoginFilter.java、管理员过滤器 AdminFilter.java 以及在线人数统计监听器 CountListener.java 的具体代码,其他源文件代码请参考随书电子资源。

程序 User.java 代码如下:

```java
package com.imut.javabean;
public class User {
    private String loginName;    //用户登录名,教师登录名为教师编号,学生登录名为学号
    private String name;         //用户真实姓名
    private int userType;        //用户类别,管理员类别为 0,教师类别为 1,学生类别为 2
    //无参构造方法
    public User() {
        super();
    }
    //有参构造方法
    public User(String loginName, String name, int userType) {
        super();
        this.loginName=loginName;
        this.name=name;
        this.userType=userType;
    }
    //getter、setter 方法
    public String getLoginName() {
        return loginName;
    }
    public void setLoginName(String loginName) {
        this.loginName=loginName;
    }
    public String getName() {
        return name;
    }
    public void setName(String name) {
        this.name=name;
    }
    public int getUserType() {
        return userType;
    }
    public void setUserType(int userType) {
        this.userType=userType;
    }
```

}

程序 LoginServlet.java 代码如下：

```java
package com.imut.servlet;
import java.io.IOException;
import javax.servlet.ServletException;
import javax.servlet.http.HttpServlet;
import javax.servlet.http.HttpServletRequest;
import javax.servlet.http.HttpServletResponse;
import javax.servlet.http.HttpSession;
import com.imut.javabean.Admin;
import com.imut.javabean.AdminDBAccess;
import com.imut.javabean.Teacher;
import com.imut.javabean.TeacherDBAccess;
import com.imut.javabean.User;
public class LoginServlet extends HttpServlet {
    public void doGet(HttpServletRequest request, HttpServletResponse response)
            throws ServletException, IOException {
        doPost(request, response);
    }
    public void doPost(HttpServletRequest request, HttpServletResponse response)
            throws ServletException, IOException {
        HttpSession session=request.getSession();
                      //获得一个 Session 对象,用于存放一些提示信息返回到前台
        request.setCharacterEncoding("UTF-8");   //设置请求的编码方式
        response.setCharacterEncoding("UTF-8");  //设置响应的编码方式
        //通过 request 对象从前台登录页面表单获得登录名和密码数据
        String loginName=request.getParameter("loginName");
        String password=request.getParameter("password");
        //声明 AdminDBAccess、TeacherDBAccess 的对象,调用其中的登录方法
        AdminDBAccess db1=new AdminDBAccess();
        TeacherDBAccess db2=new TeacherDBAccess();
        Admin admin=db1.login(loginName,password);
        Teacher teacher=db2.login(loginName, password);
        User user=null;
        if(admin!=null||teacher!=null){
            if(admin!=null){
                user=new User();
                user.setLoginName(admin.getLoginName());
                user.setName(admin.getName());
                user.setUserType(0);
            }else if(teacher!=null){
                user=new User();
                user.setLoginName(teacher.getTeacherNo());
```

```
                user.setName(teacher.getTeacherName());
                user.setUserType(1);
            }
            //如果登录成功将获得的User对象存在session对象中(十分重要,
            //程序以后要使用)
            session.setAttribute("user",user);
            //如果登录成功跳转到系统主页面 index.jsp
            response.sendRedirect(request.getContextPath()+"/index.jsp");
        }else{
            //如果登录失败将登录失败提示信息放入session对象
            session.setAttribute("message","登录信息有误,请重新登录!!!");
            //如果登录失败跳转到系统登录页面login.jsp,要求用户重新登录
            response.sendRedirect(request.getContextPath()+"/login.jsp");
        }
    }
}
```

程序 LoginServlet.java 在 web.xml 文件中配置信息如下：

```
<servlet>
    <servlet-name>login</servlet-name>
    <servlet-class>com.imut.LoginServlet</servlet-class>
</servlet>
<servlet-mapping>
    <servlet-name>login</servlet-name>
    <url-pattern>/login</url-pattern>
</servlet-mapping>
```

程序 LoginFilter.java 代码如下：

```java
package com.imut.filter;
import java.io.IOException;
import javax.servlet.Filter;
import javax.servlet.FilterChain;
import javax.servlet.FilterConfig;
import javax.servlet.ServletException;
import javax.servlet.ServletRequest;
import javax.servlet.ServletResponse;
import javax.servlet.http.HttpServletRequest;
import javax.servlet.http.HttpServletResponse;
import javax.servlet.http.HttpSession;
import com.imut.javabean.User;
public class LoginFilter    implements Filter{
    public void init(FilterConfig arg0) throws ServletException {
        // 包含初始化Filter时需要执行的代码,该代码执行一次
```

```java
    }
    public void doFilter(ServletRequest request, ServletResponse response,
            FilterChain chain) throws IOException, ServletException {
        HttpServletRequest req=(HttpServletRequest) request;
                                //将 request 强制转换为 HttpServletRequest 类型
        HttpServletResponse res=(HttpServletResponse) response;
                                //将 response 强制转换为 HttpServletResponse 类型
        HttpSession session=req.getSession();
                                //获得一个 Session 对象,用于存放一些提示信息返回
                                到前
                                //台或者获取 Session/对象中的信息
        //从 session 中取出 user 对象(该对象在登录成功后放入 session)
        User user=(User)session.getAttribute("user");
        //如果用户登录成功,将继续执行用户请求操作,否则返回用户未登录提示信息
        //并跳转到系统登录页面
        if (user!=null) {
            chain.doFilter(req, res);          //继续执行用户请求的操作
        } else {
            //未登录成功将返回用户没有登录提示信息并跳转到系统登录页面
            session.setAttribute("message","对不起,只有登录后才能访问系统!");
            res.sendRedirect(req.getContextPath()+"/login.jsp");
        }
    }
    public void destroy() {
        //包含资源释放的代码,通常对 init()中的初始化的资源执行收尾工作;
    }
}
```

程序 LoginFilter.java 在 web.xml 文件中的配置信息如下：

```xml
<filter>
    <filter-name>login</filter-name>
    <filter-class>com.imut.filter.LoginFilter</filter-class>
</filter>
<filter-mapping>
    <filter-name>login</filter-name>
    <url-pattern>/person/*</url-pattern>
</filter-mapping>
```

程序 AdminFilter.java 代码如下：

```java
package com.imut.filter;
import java.io.IOException;
import javax.servlet.Filter;
import javax.servlet.FilterChain;
```

```java
import javax.servlet.FilterConfig;
import javax.servlet.ServletException;
import javax.servlet.ServletRequest;
import javax.servlet.ServletResponse;
import javax.servlet.http.HttpServletRequest;
import javax.servlet.http.HttpServletResponse;
import javax.servlet.http.HttpSession;
import com.imut.javabean.User;
public class AdminFilter   implements Filter{
    public void init(FilterConfig arg0) throws ServletException {
        // 包含初始化Filter时需要执行的代码,该代码执行一次
    }
    public void doFilter(ServletRequest request, ServletResponse response,
         FilterChain chain) throws IOException, ServletException {
        HttpServletRequest req= (HttpServletRequest) request;
                        //将request强制转换为HttpServletRequest类型
        HttpServletResponse res= (HttpServletResponse) response;
                        //将response强制转换为HttpServletResponse类型
        HttpSession session=req.getSession();
        //获得一个Session对象,用于存放一些提示信息返回到前台或者获取Session对
        //象中的信息从session中取出user对象(该对象在登录成功后放入session)
        User user= (User)session.getAttribute("user");
        //如果用户登录成功并且登录用户为管理员类型,将继续执行用户请求操作,否则返回
        //越权操作或者未登录提示信息并跳转到系统首页或者登录页面
        if (user!=null&&user.getUserType()==0){
            chain.doFilter(req, res);           //继续执行用户请求的操作
        }else if(user!=null&&(user.getUserType()==1)){
            //登录成功但是登录用户不是管理员类型将返回越权操作提示信息并跳转到
            //系统首页
             session.setAttribute("message","对不起,只有管理员才有操作该功能的
             权限!");
            res.sendRedirect(req.getContextPath()+"/index.jsp");
                                        //跳转到系统首页
        }else{
            //未登录成功将返回用户没有登录提示信息并跳转到系统登录页面
            session.setAttribute("message","对不起,只有登录后才能访问系统!");
            res.sendRedirect(req.getContextPath()+"/login.jsp");
        }
    }
    public void destroy() {
        //包含资源释放的代码,通常对init()中的初始化的资源执行收尾工作;
    }
}
```

AdminFilter.java在web.xml文件中的配置信息如下:

```xml
<filter>
    <filter-name>admin</filter-name>
    <filter-class>com.imut.filter.AdminFilter</filter-class>
</filter>
<filter-mapping>
    <filter-name>admin</filter-name>
    <url-pattern>/base/*</url-pattern>
</filter-mapping>
```

程序 CountListener.java 代码如下:

```java
package com.imut.listener;
import javax.servlet.http.HttpSessionEvent;
import javax.servlet.http.HttpSessionListener;
public class CountListener implements HttpSessionListener{
    private static long linedNumber=0;         //初始化在线人数变量
    //监听 session 创建的方法
    public void sessionCreated(HttpSessionEvent arg0) {
        linedNumber++;
    }
    ////监听 session 销毁的方法
    public void sessionDestroyed(HttpSessionEvent arg0) {
        linedNumber--;
    }
    //返回在线人数
    public static long getLinedNumber(){
        return linedNumber;
    }
}
```

程序 CountListener.java 在 web.xml 文件中的配置信息如下:

```xml
<listener>
    <listener-class>com.imut.listener.CountListener</listener-class>
</listener>
```

【代码分析】程序 User.java 中定义了 loginName、name 和 userType 属性,其中 userType 用于标识用户类别,同时提供了设置和访问属性的方法。

程序 LoginServlet.java 中,首先接收用户登录的信息,然后声明 AdminDBAccess、TeacherDBAccess 的对象,调用其中的登录方法。如果登录成功判断用户类型,并将该类用户对象转化为 User 对象,同时设置对应的登录名、真实姓名、用户类型等属性值,最后跳转到系统首页,如果登录失败跳转至登录页面。

程序 LoginFilter.java 中,只判断用户是否登录,如果已经登录可以继续执行请求操作,并且能够访问 web.xml 中配置信息指定的"/person"路径下的任何资源,否则返回用户没有登录的提示信息并跳转到系统登录页面。

程序 AdminFilter.java 中，首先判断用户是否登录，如果已经登录并且用户类别是管理员用户则可以继续执行请求，并且能够访问 web.xml 中配置信息指定的"/base"路径和"/person"路径下的任何资源。如果已经登录并且用户类别是教师用户，会提示越权操作，不能访问"/base"路径下的资源同时跳转到 index.jsp 页面。此时，教师用户作为已经登录的用户只能访问"/person"路径下的资源。如果用户没有登录将返回用户没有登录提示信息并跳转到系统登录页面。

习　题

1. 选择题

(1) 以下哪一项为过滤器的主要作用？（　　）
　　A. 用来过滤错误信息
　　B. 用来对数据进行拦截，拦截后这些数据就不起作用了
　　C. 在运行时由 Servlet 容器来拦截处理请求和响应
　　D. 过滤器的作用十分有限，而且会破坏工程的文件

(2) 如果要取得 ServletContext 初始参数，则可以执行的方法是（　　）。
　　A. getContextParameter()　　　　B. getParameter()
　　C. getInitParameter()　　　　　　D. getAttribute()

(3) 关于过滤器的描述，正确的是（　　）。
　　A. Filter 接口定义了 init()、service()与 destroy()方法
　　B. 会传入 ServletRequest 与 ServletResponse 至 Filter
　　C. 要执行下一个过滤器，必须执行 FilterChain 的 next()方法
　　D. 如果要取得初始参数，要使用 FilterConfig 对象

(4) 在开发过滤器时，以下说法正确的是（　　）。
　　A. 必须考虑前后过滤器之间的关系
　　B. 挂上过滤器后不改变应用程序原有的功能
　　C. 设计 Servlet 时必须考虑到未来加装过滤器的需求
　　D. 每个过滤器要设计为独立相互不影响的组件

(5) 关于 Filter 接口上的 doFilter()方法的说明，有误的是（　　）。
　　A. 会传入两个参数 ServletRequest、ServletResponse
　　B. 会传入三个参数 ServletRequest、ServletResponse、FilterChain
　　C. 前一个过滤器调用 FilterChain 的 doFilter()后，会执行目前过滤器的 doFilter()方法
　　D. 前一个过滤器的 doFilter 执行后，会执行目前过滤器的 doFilter()方法

(6) 以下一段代码
```
HttpSession session=request.getSession();
User user=new User();
```

session.setAttribute("user",user);

以下哪个监听器可以实现统计在线人数？（　　）

　　A. 实现 HttpSessionBindingListener

　　B. 实现 HttpSessionListener

　　C. 实现 HttpSessionActivationListener

　　D. 以上皆非

(7) 过滤器的结构是（　　）。

　　A. 实质是一个 Servlet，所以它的定义和 Servlet 完全一样

　　B. 过滤器需要实现 java.servlet.Filter

　　C. 过滤器必须实现三个方法：init、doFilter、destroy

　　D. 过滤器必须要在 web.xml 中配置，配置方式和 Servlet 完全一样

2. 填空题

(1) 过滤器是在_____运行的 Web 组件程序，可以过滤客户端给服务器发的请求，也可以过滤服务器给客户端的响应。

(2) 所有的 Servlet 过滤器类都必须实现_____接口，该接口中定义了三个过滤器必须实现的方法：_____、_____和_____。

(3) 过滤器配置需要用到部署文件 web.xml 中的两个元素_____和_____。

(4) 监听器可以完成对_____的监听，主要监听_____、_____和_____的操作。

(5) _____是用来监听 HttpServletRequest 生命周期的监听器接口，若想要知道何时 HttpServletRequest 对象生成或即将结束销毁，可以定义监听器类实现该接口。

(6) _____是用来监听会话对象的生命周期的监听器接口，若想要在 HttpSession 对象创建或结束时，做些相对应动作，则可以定义监听器类实现该接口。

3. 简答题

(1) 简述 Servlet 过滤器的作用及过滤器开发的一般步骤。

(2) 简述 Servlet 监听器的作用及其开发方法。

4. 程序设计题

(1) 利用 HttpSessionListener 监听器实现在线人数统计，及在页面上显示目前登录用户的名称和最后活动时间。

(2) 实现一个简单的性能测评过滤器，可用来记录请求与响应间的时间差，记录 Servlet 处理请求到响应所需花费的时间。

JavaBean 组件

在 Java Web 应用中,JavaBean 被用来封装数据和业务逻辑,以实现业务逻辑和显示逻辑的分离。JavaBean 是一个可重用的组件,通过使用它,还可以减少在 JSP 中脚本代码的使用,使得应用程序易于维护、系统具有更强的健壮性和灵活性。本章将以理论与实践相结合的方式介绍 JavaBean 的概念、JavaBean 的使用、JavaBean 的作用范围等内容。

6.1 JavaBean 的概念

JavaBean 是用 Java 语言描述的软件组件模型,实际上就是一个 Java 类。在编写 JSP 页面时,对于一些常用的功能,可以将它们的共同功能抽取出来,组织为 JavaBean。当需要在某个页面中使用该功能时,只要调用该 JavaBean 中相应的方法,而不必在每个页面中都编写实现该功能的详细代码,这样就实现了代码的重用。当需要进行修改时,只要修改 JavaBean 即可,没必要再去修改每一个调用该 JavaBean 的页面,这样就实现了良好的可维护性。JavaBean 需要遵循 Sun 制定的 JavaBean 规范文档中描述的有关约定。任何遵循以下规范的 Java 类都可以在 JSP 页面中作为 JavaBean 来使用。

(1) JavaBean 必须是 public 类。

(2) JavaBean 必须具有一个无参的 public 构造方法,通过定义不带参数的构造方法或使用默认的构造方法均可满足这个要求。

(3) JavaBean 属性的访问权限一般定义为 private 或 protected,而不是定义为 public 的。

(4) 对每个属性一般定义两个 public 方法,一个是访问属性的方法,方法名应该定义为 getXxx() 格式;另一个是设置属性的方法,方法应定义为 setXxx(形式参数) 格式。

例程 6-1:演示 JavaBean 的定义。程序为 BookBean.Java。

程序 BookBean.Java 代码如下:

```
package com.ch06.bean;
public class BookBean {
    //声明属性
    private String isbn;
```

```
    private String name;
//无参构造方法
public BookBean(){
}
//定义访问属性
public String getIsbn() {
    return isbn;
}
//定义设置属性
public void setIsbn(String isbn) {
    this.isbn=isbn;
}
public String getName() {
    return name;
}
public void setName(String name) {
    this.name=name;
}
}
```

程序 BookBean.java 使用两个 private 属性封装了图书信息，并提供了访问和修改这些信息的方法。

当属性是布尔类型的值时，访问方法应该定义为 isXxxx()。例如，假设 JavaBean 中有一个 boolean 类型的属性 valid，则访问方法应该定义为：

```
public boolean isValid(){
    return valid;
}
```

除了访问方法和修改方法外，JavaBean 类中还可以定义其他的方法来实现某种业务逻辑。

6.2 JavaBean 的使用

要想在 JSP 中使用 JavaBean 组件，必须应用<jsp:useBean>、<jsp:setProperty>、<jsp:getProperty>等 JSP 的操作指令。

6.2.1 <jsp:useBean>

<jsp:useBean>动作用来在指定作用域中查找或创建一个 bean 实例。

1. < jsp:useBean> 使用语法

```
<jsp:useBean id="beanInstanceName"
    scope="page | request | session | application"
```

```
{class="package.class" |type="package.class" |
class="package.class" type="package.class" |
beanName="{package.class |<%=expression %>" type="package.class"
}
{/>| other elements</jsp:useBean>
}
```

2. <jsp:useBean> 属性解析

1) id 属性

id 属性用来唯一标识一个 bean 实例,该属性是必需的。注意,不能在一个转换单元中的多个<jsp:useBean>中使用相同的 id 属性值。

2) scope 属性

scope 属性用于指定 bean 实例的作用域。JavaBean 在 JSP 页面中的访问范围由 4 个作用域决定,分别是:page、request、session、application。该属性是可选的,默认值是 page 作用域,只能在当前页面使用。关于 JavaBean 的作用范围,将在 6.3 节中详细讨论。

3) class 属性

class 属性指定创建 bean 实例的 Java 类,如果容器在指定的作用域中不能找到一个现存的 bean 实例,它将使用 class 属性指定类的无参构造方法创建一个 bean 实例。如果该类属于某个包,则必须指定类的全名,如 com.ch06.Book。

4) type 属性

type 属性指定由 id 属性声明的 bean 对象的数据类型。

5) beanName 属性

如果 JSP 页面使用的 JavaBean 实例是由序列化的 bean 创建的,则应该使用 beanName 属性指定序列化的 bean 名称。

3. <jsp:useBean> 使用示例

```
<jsp:useBean id="book" scope="page" class="com.ch06.Book" />
```

上述代码的作用是在 page 作用范围也即当前页面查找指定名称为 book 的对象,如果没有找到,则使用 com.ch06.Book 类的无参构造方法创建 book 实例。

4. 与<jsp:useBean>等价的 Java 脚本代码

以上代码等价于以下脚本代码:

```
<%
com.ch06.Book book=null;
book=new com.ch06.Book();
if(pageCotext.getAttribute("book")==null){
    book=new Book();
    pageContext.setAttribute("book",book);
```

```
    }else{
        book=(Book)pageContext.getAttribute("book");
    }
%>
```

6.2.2 <jsp:setProperty>

<jsp:setProperty>动作用来指定JavaBean实例某个属性的值。

1. <jsp:setProperty> 使用语法

```
<jsp:setProperty name="beanInstanceName"
    {property="propertyName" value="{String |<%=expression %>" |
    property="propertyName" [param="paramName"] |
    property=" * "
} />
```

2. <jsp:setProperty> 属性解析

1) name 属性

name 属性用来标识一个 bean 实例，name 属性值必须与前面用 JSP 动作＜jsp:useBean＞声明的 id 属性值相同。该属性是必需的。

2) peroperty 属性

peroperty 属性用来设置指定的 bean 的属性，容器根据指定的 bean 的属性调用相应的 setXxx()方法。如果 property 的值是"＊"，表示所有名字和 bean 属性名字匹配的请求参数都将被传递给相应属性的 setXxx()方法。该属性是必需的。

3) value 属性

value 属性为 bean 所指定的属性设置值。该属性是可选的。

4) param 属性

param 属性指定某个请求参数名，并用该参数值设置 bean 的属性值。该属性可选。value 属性和 param 属性不能同时使用。

3. <jsp:setProperty> 使用示例

代码"＜jsp:setProperty name="book" property="name" value="Web开发技术"/＞"的作用是将 book 对象的 name 属性值设置为"Web 开发技术"，等价于以下代码的作用：

```
<%
    book.setName("Web开发技术");
%>
```

代码"＜jsp:setProperty name="book" property="name" param="bookName"/＞"的作用是用请求参数 bookName 的值为 book 对象的 name 属性赋值，等价于以下代码的

作用:

```
<%
    book.setName(request.getParameter("bookName"));
%>
```

代码"<jsp:setProperty name="book" property="*"/>"的作用是使用请求参数的每个值为 book 对象的每个与请求参数同名的属性赋值,等价于以下脚本代码:

```
<%
    book.setName(request.getParameter("name"));
    book.setPrice(request.getParameter("price"));
%>
```

6.2.3 <jsp:getProperty>

<jsp:getProperty>动作用于获取 JavaBean 对象的属性值,并显示在页面上。

1. JSP 动作<jsp:getProperty> 语法

```
<jsp:getProperty name="beanInstanceName" property="propertyName" />
```

2. JSP 动作<jsp:getProperty> 属性解析

1) name 属性

name 属性用来标识一个 bean 实例,name 属性值必须与前面用 JSP 动作<jsp:useBean>声明的 id 属性值相同。该属性是必需的。

2) property 属性

property 属性指定要输出的属性名。该属性是必需的。

3. JSP 动作<jsp:getProperty> 使用示例

```
<jsp:getProperty name="book" property="name" />
```

该代码的作用是输出 book 对象中 name 属性的值。

```
<jsp:getProperty name="book" property="price" />
```

该代码的作用是输出 book 对象中 price 属性的值。

4. 与 JSP 动作<jsp:getProperty> 等价的代码

```
<%
    out.print(book.getName());
    out.print(book.getPrice());
%>
```

例程 6-2:演示 JavaBean 的使用。程序为 BookBean.java、book.html、book.jsp。

BookBean.java 同例程 6-1 中的 BookBean.java。

程序 book.html 代码如下：

```html
<!DOCTYPE html>
<html>
    <head>
        <title>book.html</title>
        <meta http-equiv="content-type" content="text/html; charset=UTF-8">
    </head>
    <body>
        <form action="book.jsp" method="post">
            ISBN:<input type="text" name="isbn"><br>
            书名:<input type="text" name="name"><br>
            <input type="submit" value="提交">
            <input type="reset" value="取消">
        </form>
    </body>
</html>
```

程序 book.jsp 代码如下：

```jsp
<%@page language="java" import="java.util.*" pageEncoding="UTF-8"%>
<!DOCTYPE HTML PUBLIC "-//W3C//DTD HTML 4.01 Transitional//EN">
<html>
    <head>
        <title>JavaBean 的使用</title>
    </head>
    <body>
        <%
            request.setCharacterEncoding("UTF-8");
        %>
        <jsp:useBean id="book" scope="page" class="com.ch06.bean.BookBean"/>
        <jsp:setProperty property="*" name="book"/>
        ISBN:<jsp:getProperty property="isbn" name="book"/><br>
        书名:<jsp:getProperty property="name" name="book"/>
    </body>
</html>
```

程序 book.html 提供了两个文本框供用户输入图书信息。在程序 book.jsp 中通过 <jsp:useBean id="book" scope="page" class="com.ch06.bean.BookBean"/> 指定在 page 作用域中创建一个 com.ch06.bean.BookBean 类型的对象 book；然后通过 <jsp:setProperty property="*" name="book"/> 指定为 book 对象的属性设置值，设置的属性是与 book.html 表单组件 name 属性值相同的属性，设置的属性值就是文本框中输入并传递过来的值；最后通过 <jsp:getProperty property="isbn" name="book"/> 和 <jsp:getProperty property="name" name="book"/> 分别将 book 对象的 isbn 属性值

和 name 属性值输出。

运行程序 book.html,输入 ISBN 号和书名信息,效果如图 6-1 所示。单击"提交"按钮转到 book.jsp 页面,将表单传递过来的参数值显示出来,效果如图 6-2 所示。

图 6-1 输入信息页面

图 6-2 显示信息页面

6.3 JavaBean 的作用范围

之前章节介绍了作用域对象和 JSP 内置对象的作用域,JavaBean 也存在作用范围的问题。JavaBean 的作用范围共有 4 种,分别是 page 作用范围、request 作用范围、session 作用范围和 application 作用范围。

1. page 作用范围

当<jsp:useBean>中属性 scope 的值设置为"page"时,表示 JavaBean 对象只在创建它的当前页面有效,当返回到客户端或跳转到其他资源时,则该 JavaBean 对象会被释放掉。page 范围的对象存储在 pageContext 中。

例程 6-3:演示 page 使用范围的计数器。程序为 CounterBean.java、pagecounter.jsp。

程序 CounterBean.java 代码如下:

```
package com.ch06.bean;
public class CounterBean {
    int count=0;
    public CounterBean() {
```

```
        }
        public int getCount() {
            count++;
            return count;
        }
        public void setCount(int count) {
            this.count=count;
        }
}
```

程序 pagecounter.jsp 代码如下:

```
<%@page language="java" import="java.util.*" pageEncoding="UTF-8"%>
<!DOCTYPE HTML PUBLIC "-//W3C//DTD HTML 4.01 Transitional//EN">
<html>
    <head>
        <title>page 使用范围</title>
    </head>
    <body>
        <%--使用<jsp:useBean--%>
        <jsp:useBean id="counter" class="com.ch06.bean.CounterBean" scope=
        "page" />
        <%--使用<jsp:getProperty>获取 counter 属性的值--%>
        计数器的值:<jsp:getProperty property="count" name="counter"/>
    </body>
</html>
```

程序 CounterBean.java 定义了一个名为 count 的属性值并将其初始化为 0,在 getCount()方法中将 count 加 1 并返回。在 pagecounter.jsp 中实例化 CounterBean 对象并获取该对象的 count 的值。用浏览器打开 pagecounter.jsp 页面,不论如何刷新页面,计数值始终保持为 1 不变。因为 scope 的值设置为"page",此时 JavaBean 的有效期为从访问该页开始到执行完该页为止,当刷新页面时,前一个 JavaBean 已经被销毁了。运行效果如图 6-3 所示。

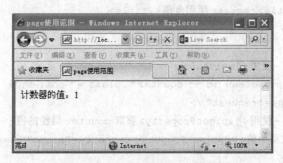

图 6-3 page 范围的计数器

2. request 作用范围

当<jsp:useBean>中属性 scope 的值设置为"request"时,表示 JavaBean 对象除了在当前页面中有效之外,还可以在通过 forward 方法跳转的目标页面中有效。

例程 6-4:演示 request 使用范围计数器。程序为 CounterBean.java、requestcounter.jsp、getrequestcounter.jsp。

CounterBean.java 代码同例程 6-3 中 CounterBean.java。

requestcounter.jsp 代码如下:

```jsp
<%@page language="java" import="java.util.*" pageEncoding="UTF-8"%>
<!DOCTYPE HTML PUBLIC "-//W3C//DTD HTML 4.01 Transitional//EN">
<html>
    <head>
        <title>reqeust 使用范围</title>
    </head>
    <body>
        <%--使用<jsp:useBean--%>
        <jsp:useBean id="counter" class="com.ch06.bean.CounterBean" scope="request" />
        <%--使用<jsp:getProperty>获取 counter 属性的值--%>
        计数器的值:<jsp:getProperty property="count" name="counter"/>
        <%--将请求转发到 getrequestcounter.jsp 页面执行--%>
        <jsp:forward page="getrequestcounter.jsp"></jsp:forward>
    </body>
</html>
```

getrequestcounter.jsp 代码如下:

```jsp
<%@page language="java" import="java.util.*" pageEncoding="UTF-8"%>
<!DOCTYPE HTML PUBLIC "-//W3C//DTD HTML 4.01 Transitional//EN">
<html>
    <head>
        <title>request 使用范围</title>
    </head>
    <body>
        <%--使用<jsp:useBean--%>
        <jsp:useBean id="counter" class="com.ch06.bean.CounterBean" scope="request" />
        <%--使用<jsp:getProperty>获取 counter 属性的值--%>
        计数器的值:<jsp:getProperty property="count" name="counter"/>
    </body>
</html>
```

当把 scope 的值设置为"request"时,此时 JavaBean 在一次请求范围内有效。在程序

requestcounter.jsp 中实例化 CounterBean 对象并且获取该对象属性 count 的值,此时值为 1,然后通过<jsp:forward>跳转动作将该请求转到 getrequestcounter.jsp 页面中执行。在该页中首先查找 CounterBean 对象,因为在该请求范围中 CounterBean 类的对象已经存在,所以不用重新实例化直接使用即可。之后通过<jsp:getProperty property="count"name="counter"/>再获取对象属性 count 的值,相当于调用 counter 对象的 getCount()方法,将 count 的值加 1 后的结果也即 2 输出,执行效果如图 6-4 所示。

图 6-4 request 范围的计数器

(3) session 作用范围

当<jsp:useBean>中属性 scope 的值设置为"session"时,表示 JavaBean 对象可以存在 session 中,该对象可以被同一个用户的所有页面共享。

例程 6-5:演示 session 使用范围的计数器。程序为 CounterBean.java、sessioncounter.jsp。CounterBean.java 同例程 6-3 中的 CounterBean.java。

sessioncounter.jsp 代码如下:

```
<%@page language="java" import="java.util.*" pageEncoding="UTF-8"%>
<!DOCTYPE HTML PUBLIC "-//W3C//DTD HTML 4.01 Transitional//EN">
<html>
    <head>
        <title>session 使用范围</title>
    </head>
    <body>
        <%--使用<jsp:useBean--%>
        <jsp:useBean id="counter" class="com.ch06.bean.CounterBean" scope="session" />
        <%--使用<jsp:getProperty>获取 counter 属性的值--%>
        计数器的值:<jsp:getProperty property="count" name="counter"/>
    </body>
</html>
```

当把 scope 的值设置为"session"时,此时 JavaBean 在同一个客户端用户发出的所有请求中有效。在程序 sessioncounter.jsp 中实例化 CounterBean 对象并且获取该对象属性 count 的值,此时值为 1,然后每刷新一次页面,count 的值总会在前一次的结果上再加 1。刷新 4 次页面后的执行效果如图 6-5 所示。

图 6-5 session 范围的计数器

4. application 作用范围

当<jsp:useBean>中属性 scope 的值设置为"application"时,表示 JavaBean 对象可以存在 application 中,该对象可以被所有用户的所有页面共享到。

例程 6-6:演示 application 使用范围计数器。程序为:CounterBean.java、applicationcounter.jap。

CounterBean.javat 同例程 6-3 中的 CounterBean.java。

applicationcounter.jsp 代码如下:

```
<%@page language="java" import="java.util.*" pageEncoding="UTF-8"%>
<!DOCTYPE HTML PUBLIC "-//W3C//DTD HTML 4.01 Transitional//EN">
<html>
    <head>
        <title>application 使用范围</title>
    </head>
    <body>
        <%--使用<jsp:useBean--%>
        <jsp:useBean id="counter" class="com.ch06.bean.CounterBean" scope="application" />
        <%--使用<jsp:getProperty>获取 counter 属性的值--%>
        计数器的值:<jsp:getProperty property="count" name="counter"/>
    </body>
</html>
```

在程序 applicationcounter.jsp 中将 scope 的值设置为"application",表示所有访问服务器的用户都会共享这个 Bean 实例。刷新页面时,count 的值会增加,如果只有一个用户访问页面,则会观察到 count 的值每次会增加 1,如果有多个用户访问页面,则观察到的值可能会是增加了大于 1 的值。图 6-6 是一个用户刷新 5 次页面的显示效果。

图 6-6　application 范围的计数器

6.4 案　　例

本章案例在第 5 章案例的基础上增加实现"基本信息管理"模块中"班级信息维护"子模块功能。案例主要使用 JavaBean 封装各实体信息,通过操作 JavaBean 实现操作数据库表记录的功能。

6.4.1 案例设计

为了便于编写代码及后期维护,将班级信息定义成 JavaBean,它的属性对应数据库表 classTbl 的各字段,通过操作 JavaBean 来操作 classTbl 表。具体实现列表显示所有班级信息、添加班级信息、查询班级信息、删除班级信息以及查看/修改某一班级信息等功能。

本章案例使用的主要文件如表 6-1 所示。

表 6-1　本章案例使用的文件

文　件	所在包/路径	功　　能
ClassTbl.java	com.imut.javabean	封装班级信息的 JavaBean
ClassDBAccess.java	com.imut.javabean	操作数据库表的类
AddClassServlet.java	com.imut.servlet.base	增加班级信息的 Servlet
CheckClassServlet.java	com.imut.servlet.base	查询班级信息的 Servlet
DelClassServlet.java	com.imut.servlet.base	删除班级信息的 Servlet
ListClassServlet.java	com.imut.servlet.base	列出所有班级信息的 Servlet
ShowClassServlet.java	com.imut.servlet.base	查看班级信息的 Servlet
UpdateClassServlet.java	com.imut.servlet.base	更新班级信息的 Servlet
addClass.jsp	/base	添加班级信息的界面
classList.jsp	/base	显示所有班级信息的界面
classShow.jsp	/base	查看/修改某一班级信息的界面

6.4.2 案例演示

在浏览器地址栏中输入"http://localhost:8080/ch06/login.jsp",以管理员用户登录系统,单击"基本数据管理"菜单中的"班级信息维护",显示班级列表页面,效果如图 6-7 所示。单击"添加班级"超链接,出现添加班级页面,效果如图 6-8 所示,提示添加班级成功之后跳转到班级列表页面。在班级列表页面中输入查询条件,查询结果将显示在"查询"按钮下方的班级列表处,如果没有输入查询条件则查询所有班级信息。根据"班级名称"查询结果的效果图如图 6-9 所示。在班级列表页面中单击"删除"超链接,将会删除该班级信息,删除完成后显示班级列表页面。在班级列表页面单击班级编号或"查看/修改"超链接,将显示班级信息查看/修改页面,效果如图 6-10 所示。修改班级信息完成后单击"提交"按钮,提示修改成功后显示班级列表页面。

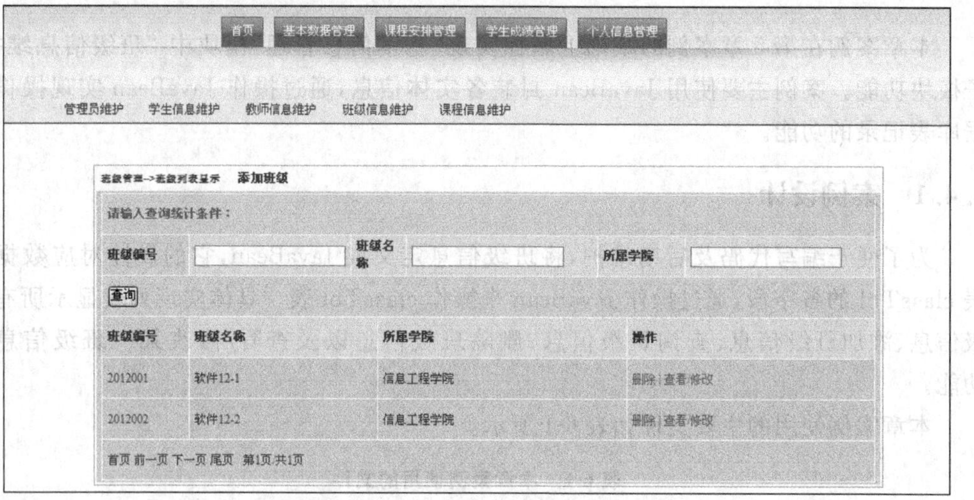

图 6-7 班级列表页面

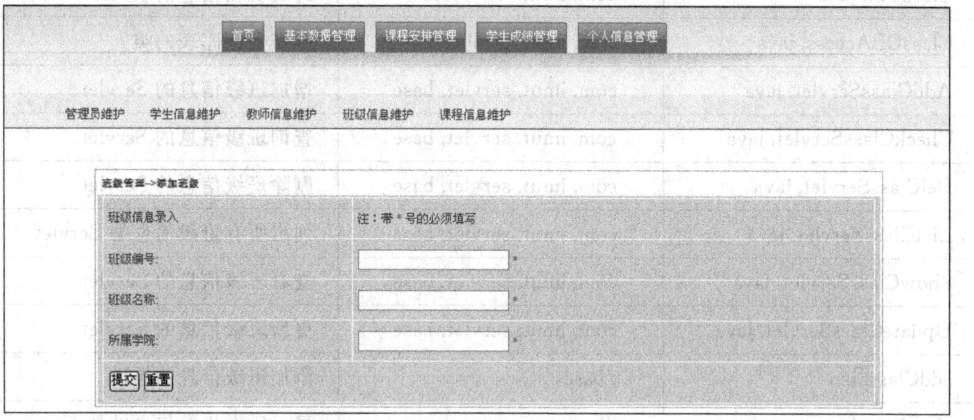

图 6-8 添加班级信息页面

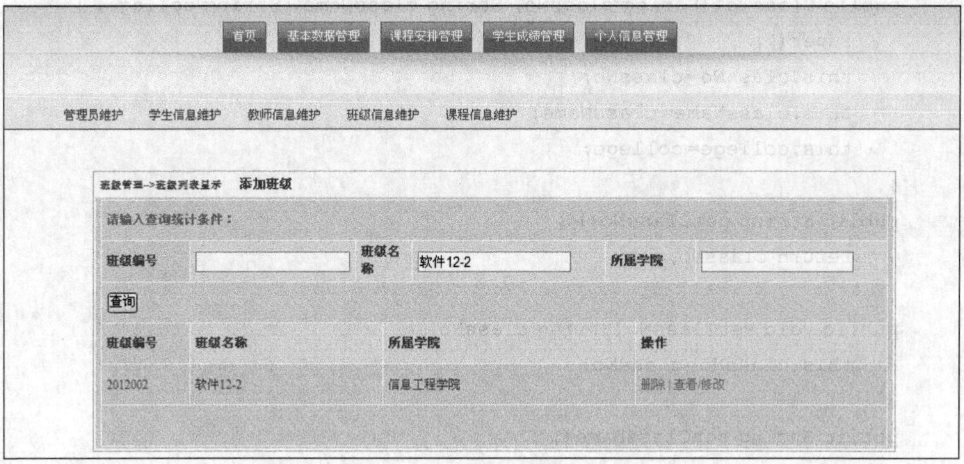

图 6-9　查询班级结果页面

图 6-10　查看/修改班级页面

6.4.3　代码实现

创建工程 ch06，根据案例设计描述需要完成各部分功能的具体实现，本章主要给出班级类 ClassTbl.java、操作数据库表类 ClassDBAccess.java 的具体代码，其他源文件代码请参考随书电子资源。

程序 ClassTbl.java 代码如下：

```
package com.imut.javabean;
public class ClassTbl {
    private String classNo;
    private String className;
    private String college;
    public ClassTbl() {
    }
```

```java
    public ClassTbl(String classNo, String className, String college) {
        super();
        this.classNo=classNo;
        this.className=className;
        this.college=college;
    }
    public String getClassNo() {
        return classNo;
    }
    public void setClassNo(String classNo) {
        this.classNo=classNo;
    }
    public String getClassName() {
        return className;
    }
    public void setClassName(String className) {
        this.className=className;
    }
    public String getCollege() {
        return college;
    }
    public void setCollege(String college) {
        this.college=college;
    }
}
```

程序 ClassDBAccess.java 代码如下：

```java
package com.imut.javabean;
import java.sql.Connection;
import java.sql.PreparedStatement;
import java.sql.ResultSet;
import java.sql.SQLException;
import java.sql.Statement;
import java.util.ArrayList;
import java.util.HashMap;
import java.util.List;
import java.util.Map;
import com.imut.commmon.ConnectionFactory;
import com.imut.commmon.Page;
import com.imut.commmon.ResourceClose;
public class ClassDBAccess {
    //添加班级方法
    public void addClass(ClassTbl classTbl){
        Connection conn=null;
```

```java
        PreparedStatement pstmt=null;
        ResultSet rs=null;
        try{
            conn=ConnectionFactory.getConnection();
            String sql="insert into ClassTbl values(?,?,?)";
            pstmt=conn.prepareStatement(sql);
            pstmt.setString(1, classTbl.getClassNo());
            pstmt.setString(2, classTbl.getClassName());
            pstmt.setString(3, classTbl.getCollege());
            pstmt.executeUpdate();
        }catch(Exception e){
            e.printStackTrace();
        }finally{
            ResourceClose.close(rs, pstmt, conn);
        }
    }
}
//删除班级方法
public void delClassTbl(String classNo){
    Connection conn=null;
    PreparedStatement pstmt=null;
    ResultSet rs=null;
    try{
        conn=ConnectionFactory.getConnection();
        String sql="delete from ClassTbl where classNo=?";
        pstmt=conn.prepareStatement(sql);
        pstmt.setString(1, classNo);
        pstmt.executeUpdate();
    }catch (SQLException e) {
        e.printStackTrace();
    }finally{
        ResourceClose.close(rs, pstmt, conn);
    }
}
//修改班级信息
public void updateClassTbl(ClassTbl classTbl){
    Connection conn=null;
    PreparedStatement pstmt=null;
    ResultSet rs=null;
    try{
        conn=ConnectionFactory.getConnection();
        String sql="update ClassTbl set classNo=?,className=?,college=?
        where classNo=?";
        pstmt=conn.prepareStatement(sql);
        pstmt.setString(1, classTbl.getClassNo());
```

```java
            pstmt.setString(2, classTbl.getClassName());
            pstmt.setString(3, classTbl.getCollege());
            pstmt.setString(4, classTbl.getClassNo());
            pstmt.executeUpdate();
        }catch (SQLException e) {
            e.printStackTrace();
        }finally{
            ResourceClose.close(rs, pstmt, conn);
        }
    }
    //根据班级编号查找班级
    public ClassTbl findClassTblByClassNo(String classNo){
        ClassTbl classTbl=null;
        Connection conn=null;
        PreparedStatement pstmt=null;
        ResultSet rs=null;
        try{
            conn=ConnectionFactory.getConnection();
            String sql="select * from ClassTbl where classNo=?";
            pstmt=conn.prepareStatement(sql);
            pstmt.setString(1, classNo);
            rs=pstmt.executeQuery();
            while(rs.next()){
                classTbl=new ClassTbl();
                classTbl.setClassNo(rs.getString(1));
                classTbl.setClassName(rs.getString(2));
                classTbl.setCollege(rs.getString(3));
            }
        }catch (SQLException e) {
            e.printStackTrace();
        }finally{
            ResourceClose.close(rs, pstmt, conn);
        }
        return classTbl;
    }
    //列表显示所有班级列表-分页
    public Map findAllClassTbl(int curPage){
        ClassTbl classTbl=null;
        ArrayList list=new ArrayList();
        Connection conn=null;
        Statement pstmt=null;
        ResultSet rs=null;
        ResultSet r=null;
        Map map=null;
```

```java
        Page pa=null;
        try{
            conn=ConnectionFactory.getConnection();
            String sql="select * from ClassTbl order by classNo";
            pstmt= conn. createStatement (ResultSet. TYPE _ SCROLL _ INSENSITIVE,
            ResultSet.CONCUR_READ_ONLY);
            rs=pstmt.executeQuery(sql);
            pa=new Page();//声明分页类对象
            pa.setPageSize(5);
            pa.setPageCount(rs);
            pa.setCurPage(curPage);
            r=pa.setRs(rs);
            r.previous();
            for(int i=0;i<pa.getPageSize();i++){
                if(r.next()){
                    classTbl=new ClassTbl();
                    classTbl.setClassNo(r.getString(1));
                    classTbl.setClassName(r.getString(2));
                    classTbl.setCollege(r.getString(3));
                    list.add(classTbl);
                }else{
                    break;
                }
            }
            map=new HashMap();
            map.put("list",list);
            map.put("pa",pa);
        }catch (SQLException e) {
            e.printStackTrace();
        }finally{
            ResourceClose.close(rs, pstmt, conn);
            ResourceClose.close(r, null, null);
        }
        return map;
    }
    public List findAllClassTbl(){
        ClassTbl classTbl=null;
        ArrayList list=new ArrayList();
        Connection conn=null;
        Statement pstmt=null;
        ResultSet rs=null;
        try{
            conn=ConnectionFactory.getConnection();
            String sql="select * from ClassTbl order by classNo";
```

```java
        pstmt=conn.createStatement();
        rs=pstmt.executeQuery(sql);
        while(rs.next()){
            classTbl=new ClassTbl();
            classTbl.setClassNo(rs.getString(1));
            classTbl.setClassName(rs.getString(2));
            classTbl.setCollege(rs.getString(3));
            list.add(classTbl);
        }
    }catch (SQLException e) {
        e.printStackTrace();
    }finally{
        ResourceClose.close(rs, pstmt, conn);
    }
    return list;
}
//多条件查询班级
public List findAllClassTblByMostCon (String classNo, String className,
String college){
    ClassTbl classTbl=null;
    ArrayList list=new ArrayList();
    Connection conn=null;
    PreparedStatement pstmt=null;
    ResultSet rs=null;
    //构造多条件查询的SQL语句
    String sql="select * from ClassTbl where 1=1 ";
    //模糊查询
    if(classNo!=null&&!classNo.equals("")){
        sql+=" and classNo like '%"+classNo+"%'";
    }
    if(className!=null&&!className.equals("")){
        sql+=" and className like '%"+className+"%'";
    }
    if(college!=null&&!college.equals("")){
        sql+=" and college like '%"+college+"%'";
    }
    sql+=" order by classNo";
    try{
        conn=ConnectionFactory.getConnection();
        pstmt=conn.prepareStatement(sql);
        rs=pstmt.executeQuery();
        while(rs.next()){
            classTbl=new ClassTbl();
            classTbl.setClassNo(rs.getString(1));
```

```
                classTbl.setClassName(rs.getString(2));
                classTbl.setCollege(rs.getString(3));
                list.add(classTbl);
            }
        }catch (SQLException e) {
            e.printStackTrace();
        }finally{
            ResourceClose.close(rs, pstmt, conn);
        }
        return list;
    }
}
```

【代码分析】程序 ClassTbl.java 中定义了班级的各个属性、无参构造方法、三参构造方法以及设置属性的方法 setXXX()和获取属性的方法 getXxx()。需要注意的是,属性名和数据库中 classTbl 表中的字段名一致。

程序 ClassDBAccess.java 中定义了操作数据库 classTbl 表的方法,包括添加班级、删除班级、修改班级信息、根据班级编号查找班级、分页显示所有班级列表以及根据多个条件查询班级信息等方法。增删改等方法使用 JavaBean ClassTbl 对象的属性值更新 classTbl 表中的记录,而查询方法使用从 classTbl 表中查询到的结果集为 JavaBean ClassTbl 对象赋值从而进行列表显示。

习　　题

1. 选择题

(1) 指定 useBean 动作标记以便包括名为 TestBean 的 JavaBean 的选项是(　　)。
　　A. <jsp:useBean id="test" name="TestBean">
　　B. <jsp:useBean id="test" name="TestBean"/>
　　C. <jsp:useBean id="test">
　　D. <%jsp:useBean id="test" name="TeatBean"%>

(2) JavaBean 的属性必须声明为 private,方法必须声明为以下哪种访问类型?(　　)
　　A. private　　　　B. static　　　　C. protect　　　　D. public

(3) 如果要编写一个 Bean,并将该 Bean 存放在 WEB-INF/classes/jsp/example/mybean 目录下,则包(package)名称为(　　)。
　　A. package mybean;
　　B. package classes.jsp.example.mybean;
　　C. package jsp.example;
　　D. package jsp.example.mybean;

(4) 对于以下<jsp:useBean>标签:

```
<jsp:useBean id="myBean" class="mypack.CounterBean" scope="request"/>
```
它与哪个选项中的 Java 程序段等价？（   ）

A.
```
<%
    CounterBean myBean=(CounterBean) request.getAttribute("myBean");
    if(myBean==null)myBean=new CounterBean();
%>
```

B.
```
<%
    CounterBean myBean=(CounterBean) request.getAttribute("myBean");
    if(myBean==null){
        myBean=new CounterBean();
        request.setAttribute("CounterBean",myBean);
    }
%>
```

C.
```
<%
    CounterBean myBean=(CounterBean) request.getAttribute("myBean");
    if(myBean==null){
        myBean=new CounterBean();
        request.setAttribute("myBean",myBean);
    }
%>
```

D.
```
<%
    myBean=new CounterBean();
    request.setAttribute("myBean",myBean);
%>
```

(5) 有关 JavaBean 的描述中正确的是(　　)。

    A. JavaBean 是一个 Java 类

    B. 如果属性为 xxx,则与属性有关的方法为 getXXX()和 setXXX()

    C. 对于布尔类型的属性,可以用 isXxx()方法

    D. 类中可以有不带参数的构造方法

(6) JavaBean 可以通过相关 jsp 动作指令进行调用,下面哪个不是 JavaBean 可以使用的 jsp 动作指令？（　　）

    A. <jsp:useBean>　　　　　　　　B. <jsp:setProperty>

    C. <jsp:getProperty>　　　　　　　D. <jsp:setParameter>

(7) 在 JSP 页面中使用<jsp:setPropety name="Bean 的名字" property="Bean 属性名" param="表单参数名"/>格式,用表单参数为 Bean 属性赋值,要求 Bean 的属性名字(　　)。

    A. 必须和表单参数类型一致

    B. 必须和表单参数名称一一对应

    C. 必须和表单参数数量一致

    D. 名称不一定对应

(8) <jsp:useBean id="bean 的名称" scope="bean 的有效范围" class="包名.类名"/>动作指令中,scope 的值不可以是（　　）。

  A. page   B. request   C. session   D. response

2. 填空题

(1) 在 JSP 中使用 JavaBean 的标签是_____,其中 id 的用途是_____。

(2) 使用 Bean 首先要在 JSP 页面中使用_____指令将 Bean 引入。

(3) 在 test.jsp 文件中包含如下代码

```
<%@page import="mypack.CounterBean" %>
<jsp:useBean id="myRequestBean" scope="request" class="mypack.CounterBean"/>
<jsp:useBean id="myRequestBean1" scope="request" class="mypack.CounterBean"/>
<jsp:setProperty name="myRequestBean" property="count"
     Value="<%=myRequestBean1.getCount()+1 %>"/>
<jsp:getProperty name="myRequestBean1" property="count"/>
```

通过浏览器第一次访问 test.jsp,得到的结果是_____。

(4) 要使 JavaBean 在整个应用程序的生命周期内,被该应用程序中的任何 JSP 文件所使用,则该 JavaBean 的 Scope 属性必须设置为_____。

(5) JavaBean 中用一组 set 方法设置 Bean 的私有属性值,get 方法获得 Bean 的私有属性值。set 和 get 方法名称与属性名称之间必须对应,也就是:如果属性名称为 xxx,那么 set 和 get 方法的名称必须为_____和_____。

3. 简答题

试说明什么是 JavaBean,如何在 JSP 网页中载入 JavaBean。

4. 程序设计题

(1) 编写 MyBean 的源代码,其中有两个属性 name 和 password。实现简单的<form>表单,提交 name,password 参数给 get.jsp,要求 get.jsp 使用<jsp:useBean>指令获取并显示 name,password 参数。

(2) 创建 JavaBean,在 JavaBean 中含有学生信息:姓名、性别、年龄、学号、成绩等属性;创建 JSP 页面,通过表单提交输入的学生信息。并要求在 Servlet 或者 JSP 页面中显示学生信息。

MVC 设计模式

设计模式(Design Pattern)是一套被反复使用、多数人知晓的、经过分类编目的、代码设计经验的总结。编写程序使用设计模式是为了重用代码、让代码更容易被他人理解、保证代码可靠性。毫无疑问,设计模式使代码编制真正工程化,是软件工程的基石脉络,如同大厦的结构一样。MVC 设计模式是用一种将程序流程控制、业务逻辑和数据显示分离的方法来编写程序代码,使得程序代码可复用、易维护。本章将以理论结合实践的方式介绍 JSP Model 1、JSP Model 2 以及 MVC 设计模式。

7.1 JSP 开发模型

利用 JSP 进行 Java Web 应用开发有两种开发模型:JSP Model 1 和 JSP Model 2。

7.1.1 JSP Model 1

1. 传统的 JSP Model 1

在早期的 Java Web 应用开发中,JSP 文件既负责处理业务逻辑和控制程序的运行流程,还负责数据的显示。即用 JSP 文件来独立自主地完成系统功能的所有任务。传统的 JSP Model 1 如图 7-1 所示。

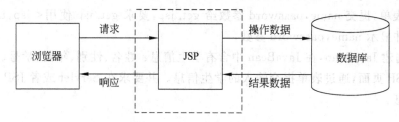

图 7-1 传统的 JSP Model 1

例程 7-1:利用传统的 JSP Model 1 实现系统登录功能。程序为 login.jsp、loginCheck.jsp 和 success.jsp。

程序 login.jsp 代码如下:

```
<%@ page language="java" import="java.util.*" pageEncoding="UTF-8"%>
<%@ taglib uri="http://java.sun.com/jsp/jstl/core" prefix="c"%>
```

```
<c:if test="$ {!(empty sessionScope.message)}">
    <script type="text/javascript">
        alert('<c:out value="$ {sessionScope.message}"/>');
    </script>
    <c:remove var="message" scope="session"/>
</c:if>
<html>
    <head>
            <title>系统登录页面</title>
    </head>
    <body>
            <form action="loginCheck.jsp" method="post">
            <table>
                <tr>
                    <td colspan="2" align="center"><b>用户登录</b></td>
                </tr>
                <tr>
                    <td>用户名</td>
                    <td><input type=text name=name></td>
                </tr>
                <tr>
                    <td>密码</td>
                    <td><input type=password name=password></td>
                </tr>
                <tr>
                    <td colspan="2" align=center>
                        <input type=submit value=提交>
                        <input type=reset value=取消>
                    </td>
                </tr>
            </table>
            </form>
    </body>
</html>
```

程序 loginCheck.jsp 代码如下：

```
<%@page language="java" import="java.sql.*,com.ch07.model.*"
pageEncoding="UTF-8"%>
<html>
    <head>
        <title>登录处理页面</title>
    </head>
    <body>
        <%
```

```java
//从表单获取数据
request.setCharacterEncoding("UTF-8");
response.setCharacterEncoding("UTF-8");
String name=request.getParameter("name");
String passwd=request.getParameter("password");
//JDBC 操作
String driver="oracle.jdbc.driver.OracleDriver";
String url="jdbc:oracle:thin:@127.0.0.1:1521:XE";
String userName="webbook";
String password="webbook";
//定义资源变量
Connection conn=null;
PreparedStatement stm=null;
ResultSet rs=null;
User user=null;
try {
    //1、注册驱动
    Class.forName(driver);
    //2、获得连接
    conn=DriverManager.getConnection(url, userName, password);
    //3、获得 statement
    String sql="select * from usertbl where name=? and password=?";
    stm=conn.prepareStatement(sql);
    stm.setString(1,name);
    stm.setString(2,passwd);
    //4、执行 SQL
    rs=stm.executeQuery();
    //5、处理结果集
    if(rs.next()){
        user=new User();
        user.setName(rs.getString(1));
        user.setPassword(rs.getString(2));
        user.setSex(rs.getString(3));
        user.setPrivence(rs.getString(4));
        user.setAuthor(rs.getString(5));
    }
}catch (ClassNotFoundException e) {
    e.printStackTrace();
} catch (SQLException e) {
    e.printStackTrace();
}finally{
    try {
        //6、关闭资源
        if(rs!=null)rs.close();
```

```
                if(stm!=null)stm.close();
                if(conn!=null)conn.close();
            }catch (SQLException e) {
                e.printStackTrace();
            }
        }
        //根据是否登录成功,跳转到不同的页面
        if(user!=null){
            session.setAttribute("user",user);
            response.sendRedirect("success.jsp");
        }else{
            session.setAttribute("message","登录信息有误,请重新登录!!!");
            response.sendRedirect("login.jsp");
        }
    %>
    </body>
</html>
```

程序 success.jsp 代码如下:

```
<%@page language="java" import="com.ch07.model.*" pageEncoding="UTF-8"%>
<html>
    <head>
        <title>系统主页面</title>
    </head>
    <body>
        <h4>
        用户
        <font color=red><%=((User)session.getAttribute("user")).getName() %>
        </font>
        您好!欢迎您使用本系统!
        </h4>
        <br>
        <h4>系统操作主页面</h4>
    </body>
</html>
```

程序 login.jsp 为登录页面。页面代码

```
<c:if test="$ {!(empty sessionScope.message)}">
    ...
</c:if>
```

利用 JSTL 标签(有关 JSTL 标签内容将在第 9 章讲解)和 JavaScript 脚本实现通过页面弹出框的方式显示 Session 对象中保存的用户登录信息的 message 值。页面提供了表单用来提交用户输入的用户名和密码信息,表单属性 action="loginCheck.jsp"定义了页面

提交目标资源为 loginCheck.jsp 页面。

程序 loginCheck.jsp 为登录功能处理页面。loginCheck.jsp 首先接收登录页面提交的用户名和密码信息数据,其次进行数据库访问操作,查看是否存在该用户,然后根据查询结果跳转到不同的页面。如果存在该用户,将用户对象保存在 Session 对象中,跳转至登录成功页面,否则将登录错误信息 message 对象保存在 Session 对象中,跳转至登录页面供用户重新登录。可以看出程序 loginCheck.jsp 实现了登录功能的业务处理逻辑、数据库访问操作和程序流程控制。

程序 success.jsp 为登录成功页面。success.jsp 从 Session 对象中取出用户对象显示用户名信息以及欢迎使用系统信息。

访问登录 login.jsp 页面,页面运行效果如图 7-2 所示。在登录页面上输入正确的用户名 zhangsan 和密码 zs123,系统运行至登录成功页面,如图 7-3 所示。如果在登录页面上输入错误的用户名和密码,程序将弹出登录错误提示框,如图 7-4 所示。

图 7-2 登录页面

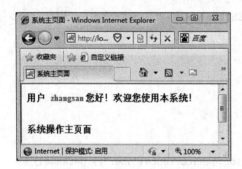

图 7-3 登录成功页面

图 7-4 登录错误提示框

从例程 7-1 可以看出,使用传统的 JSP Model 1 进行 Java Web 应用开发,将业务处理代码、程序流程控制代码以及数据显示逻辑代码全部混杂在一起,很难分离出单独的业务模型,造成程序很难维护和扩展,几乎没有任何可重用的价值。

2. 改进的 JSP Model 1

改进的 JSP Model 1 利用 JSP 页面与 JavaBean 组件共同协作来完成系统功能的所有任务,JSP 文件负责程序的流程控制逻辑和数据显示逻辑任务,JavaBean 负责处理业务逻辑任务。改进的 JSP Model 1 如图 7-5 所示。

例程 7-2:利用改进的 JSP Model 1 实现系统登录功能。程序为 login1.jsp、loginCheck1.jsp、success.jsp、DBAccess.java 和 User.java。

程序 login1.jsp 为登录页面,除了表单属性 action="loginCheck1.jsp"定义了页面提交目标资源为 loginCheck1.jsp 页面外,其余代码与例程 7-1 中 login.jsp 代码完全一致。

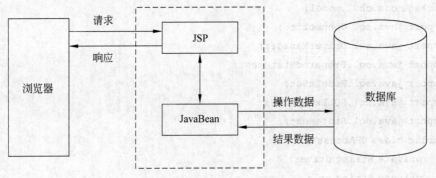

图 7-5　改进的 JSP Model 1

程序 success.jsp 为登录成功页面,代码与例程 7-1 中 success.jsp 代码完全一致。loginCheck1.jsp 程序代码如下:

```jsp
<%@ page language="java" import="java.sql.*,com.ch07.model.*" pageEncoding="UTF-8"%>
<jsp:useBean id="dbAccess" class="com.ch07.model.DBAccess" scope="page">
</jsp:useBean>
<html>
    <head>
        <title>登录处理页面</title>
    </head>
    <body>
    <%
        //从表单获取数据
        request.setCharacterEncoding("UTF-8");
        response.setCharacterEncoding("UTF-8");
        String name=request.getParameter("name");
        String passwd=request.getParameter("password");
        User user=dbAccess.login(name,passwd);
        //根据是否登录成功,跳转到不同的页面
        if(user!=null){
            session.setAttribute("user",user);
            response.sendRedirect("success.jsp");
        }else{
            session.setAttribute("message","登录信息有误,请重新登录!!!");
            response.sendRedirect("login1.jsp");
        }
    %>
    </body>
</html>
```

程序 DBAccess.java 代码如下:

```java
package com.ch07.model;
import java.sql.Connection;
import java.sql.DriverManager;
import java.sql.PreparedStatement;
import java.sql.ResultSet;
import java.sql.SQLException;
import java.sql.Statement;
public class DBAccess {
    private String driver;
    private String url;
    private String userName;
    private String password;
    //无参构造方法
    public DBAccess() {
        this.driver="oracle.jdbc.driver.OracleDriver";
        this.url="jdbc:oracle:thin:@127.0.0.1:1521:XE";
        this.userName="webbook";
        this.password="webbook";
    }
    //获得数据库连接的方法
    public Connection getConnection(){
        try{
            Class.forName(driver);
            return DriverManager.getConnection(url,userName,password);
        }catch(Exception e){
            e.printStackTrace();
            return null;
        }
    }
    //关闭资源方法
    public void close(ResultSet rs,Statement stmt,Connection conn){
        try{
            if(rs!=null)rs.close();
            if(stmt!=null)stmt.close();
            if(conn!=null)conn.close();
            
        }catch(Exception e){
            e.printStackTrace();
        }
        
        
    }
    public void close(ResultSet rs,Statement stmt){
```

```java
            close(rs,stmt,null);
        }
        public void close(Statement stmt,Connection conn){
            close(null,stmt,conn);
        }
        public void close(Connection conn){
            close(null,null,conn);
        }
        public void close(ResultSet rs){
            close(rs,null,null);
        }
        public void close(Statement stmt){
            close(null,stmt,null);
        }
        //登录方法
        public User login(String name,String password) {
            User user=null;
            Connection conn=null;
            PreparedStatement pstmt=null;
            ResultSet rs=null;
            try{
                conn=getConnection();
                String sql="select * from usertbl where name=? and password=?";
                pstmt=conn.prepareStatement(sql);
                pstmt.setString(1, name);
                pstmt.setString(2, password);
                rs=pstmt.executeQuery();
                if(rs.next()){
                    user=new User();
                    user.setName(rs.getString(1));
                    user.setPassword(rs.getString(2));
                    user.setSex(rs.getString(3));
                    user.setPrivence(rs.getString(4));
                    user.setAuthor(rs.getString(5));
                }
            }catch (SQLException e) {
                e.printStackTrace();
            }finally{
                close(rs, pstmt, conn);
            }
            return user;
        }
}
```

程序 User.java 代码如下：

```java
package com.ch07.model;
public class User {
    private String name;
    private String password;
    private String sex;
    private String privence;
    private String author;
    public User() {
    }
    public User(String name, String password, String sex,
            String privence, String author) {
        this.name=name;
        this.password=password;
        this.sex=sex;
        this.privence=privence;
        this.author=author;
    }
    public String getName() {
        return name;
    }
    public void setName(String name) {
        this.name=name;
    }
    public String getPassword() {
        return password;
    }
    public void setPassword(String password) {
        this.password=password;
    }
    public String getSex() {
        return sex;
    }
    public void setSex(String sex) {
        this.sex=sex;
    }
    public String getPrivence() {
        return privence;
    }
    public void setPrivence(String privence) {
        this.privence=privence;
    }
```

```
    public String getAuthor() {
        return author;
    }
    public void setAuthor(String author) {
        this.author=author;
    }
}
```

程序 DBAccess.java 是一个 JavaBean,实现获得数据库连接、关闭数据库资源和用户登录等数据库操作与业务逻辑处理方法。

程序 User.java 是一个 JavaBean,为实体模型,用来描述用户对象实体。

程序 loginCheck1.jsp 为登录功能流程处理页面。loginCheck1.jsp 首先接收登录页面提交的用户名和密码信息数据,其次调用 DBAccess 对象中的方法来实现数据库访问操作和登录业务逻辑操作,然后根据查询结果跳转到不同的页面。如果存在该用户,将用户对象保存在 Session 对象中,跳转至登录成功页面,否则将登录错误信息 message 对象保存在 Session 对象中,跳转至登录页面供用户重新登录。可以看出程序 loginCheck1.jsp 实现了登录功能的程序流程控制。

例程 7-2 运行的步骤和运行效果与例程 7-1 一样。

从例程 7-2 可以看出,使用改进的 JSP Model 1 进行 Java Web 应用开发,虽然将业务处理代码分离出来,但是程序流程控制代码和数据显示逻辑代码仍然混杂在一起。从应用角度分析,该模型适合简单应用的需要,却不能满足复杂的大型应用程序的实现。使用该模型进行应用开发,应用程序代码扩展和维护的难度依然很大,代码的重用性非常低。

7.1.2 JSP Model 2

JSP Model 2 采用 JSP 页面、Servlet 和 JavaBean 组件分工协作共同完成系统功能的所有任务。其中,JSP 负责数据显示逻辑任务,Servlet 负责程序流程控制逻辑任务,JavaBean 负责处理业务逻辑任务。实际上 JSP Model 2 就是采用 MVC 设计模式思想来设计的。JSP Model 2 如图 7-6 所示。

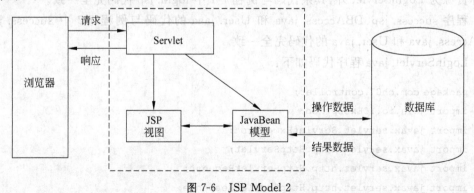

图 7-6　JSP Model 2

7.2 MVC 设计模式

MVC 是 Model View Controller 的缩写,即模型-视图-控制器,是设计创建 Web 应用程序的模式。

模型(Model):实现业务逻辑处理和数据库操作,另外还包括数据模型。

视图(View):实现数据显示逻辑。负责与用户交互,即接收用户数据的输入,显示模型返回的数据。

控制器(Controller):实现程序执行流程的控制,任务的分派。接收用户请求,调用模型完成请求的业务处理功能,并将处理结果返回给视图。

MVC 设计模式中模型、视图和控制器三者之间的关系如图 7-7 所示。

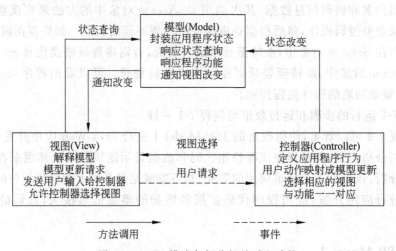

图 7-7 MVC 模式各部分的关系和功能

例程 7-3:利用 MVC 设计模式实现系统登录功能。程序为 login2.jsp、success.jsp、LoginServlet.java、DBAccess.java 和 User.java。

程序 login2.jsp 为登录页面,除了表单属性 action="loginServlet"定义了页面提交目标资源为 LoginServlet 外,其余代码与例程 7-1 中 login.jsp 代码完全一致。

程序 success.jsp、DBAccess.java 和 User.java 的代码与例程 7-2 中 success.jsp、DBAccess.java 和 User.java 的代码完全一致。

LoginServlet.java 程序代码如下:

```
package com.ch07.controller;
import java.io.IOException;
import javax.servlet.ServletException;
import javax.servlet.http.HttpServlet;
import javax.servlet.http.HttpServletRequest;
import javax.servlet.http.HttpServletResponse;
import javax.servlet.http.HttpSession;
```

```java
import com.mvc.model.DBAccess;
import com.ch07.model.User;
public class LoginServlet extends HttpServlet {
    public void doGet(HttpServletRequest request, HttpServletResponse response)
            throws ServletException, IOException {
        doPost(request, response);
    }
    public void doPost(HttpServletRequest request, HttpServletResponse response)
            throws ServletException, IOException {
        HttpSession session=request.getSession();
        request.setCharacterEncoding("UTF-8");
        response.setCharacterEncoding("UTF-8");
        String name=request.getParameter("name");
        String passwd=request.getParameter("password");
        DBAccess dbAccess=new DBAccess();
        User user=dbAccess.login(name,passwd);
        if(user!=null){
            session.setAttribute("user",user);
            response.sendRedirect("success.jsp");
        }else{
            session.setAttribute("message","登录信息有误,请重新登录!!!");
            response.sendRedirect("login2.jsp");
        }
    }
}
```

LoginServlet 在 web.xml 文件中的配置代码如下：

```xml
<servlet>
    <servlet-name>login</servlet-name>
    <servlet-class>com.ch07.controller.LoginServlet</servlet-class>
</servlet>
<servlet-mapping>
    <servlet-name>login</servlet-name>
    <url-pattern>/login/loginServlet</url-pattern>
</servlet-mapping>
```

程序 LoginServlet.java 为登录功能流程控制处理组件。LoginServlet.java 首先接收登录页面提交的用户名和密码信息数据，其次调用 DBAccess 对象中的方法来实现数据库访问操作和登录业务逻辑操作，然后根据查询结果跳转到不同的页面。如果存在该用户，将用户对象保存在 Session 对象中，跳转至登录成功页面，否则将登录错误信息 message 对象保存在 Session 对象中，跳转至登录页面供用户重新登录。可以看出程序 LoginServlet.java 实现了登录功能的程序流程控制。

例程 7-3 运行的步骤和运行效果与例程 7-1 一样。

在例程 7-3 中,程序 DBAccess.java、User.java 是 JavaBean 组件,为 MVC 设计模式中模型部分。DBAccess.java 实现了获得数据库连接、关闭数据库资源和用户登录等数据库操作与业务逻辑处理方法。User.java 实现了描述用户对象实体的实体模型。JSP 页面 login2.jsp 和 success.jsp 为 MVC 设计模式中的视图部分,实现接收用户输入数据和显示结果数据。LoginServlet 为 MVC 设计模式中的控制器部分,实现控制程序执行流程。

从例程 7-3 可以看出,MVC 模式有如下优点:

(1) 将业务处理逻辑、数据显示逻辑和执行流程控制逻辑三者完全分开,使得程序设计的思路更清晰,提高了代码的复用性;

(2) 程序设计采用 MVC 设计思想,适合于多人合作展开并行开发,有效地提高了程序开发效率;

(3) MVC 中模型、视图和控制器三者的责任划分得很清楚,简化了应用程序的测试工作,维护人员很容易了解程序的结构,便于程序维护工作的进行。

当然采用 MVC 设计模式进行 Java Web 应用开发时也有一定的缺点,例如 MVC 设计模式的实施增加了 Web 应用项目的开发难度,同时也提高了对开发人员的技术要求。

7.3 案　　例

本章案例在第 6 章案例的基础上增加实现"基本信息管理"模块中的"课程信息维护"子模块。功能实现主要采用 MVC 开发模式,分别使用 JSP 文件、Servlet 和 JavaBean 分工协作共同完成课程信息维护的增删改查等所有功能。

7.3.1 案例设计

在模型层使用接口、接口实现类以及 JavaBean 等实现业务逻辑处理和对数据库表 course 的操作。

在显示层使用 JSP 技术定义课程信息维护的各个显示页面,如添加信息页面、查看/编辑页面以及显示所有课程信息的页面。

控制层使用 Servlet 实现程序执行流程的控制、任务的分派。每一种操作都对应一个 Servlet,因此针对课程信息的增加、删除、修改、查看等操作都要定义相对应的 Servlet。

本章案例使用的主要文件如表 7-1 所示。

表 7-1 本章案例使用的文件

文　件	所在包/路径	功　能
Course.java	com.imut.javabean	封装课程信息的 JavaBean
ICourse.java	com.imut.javabean	定义操作数据库表增删改查操作方法的接口
CourseImpl.java	com.imut.javabean	具体实现接口定义的增删改查等方法

文　　件	所在包/路径	功　　能
AddCourseServlet.java	com.imut.servlet.base	增加课程信息的 Servlet
CheckCourseServlet.java	com.imut.servlet.base	查询课程信息的 Servlet
DelCourseServlet.java	com.imut.servlet.base	删除课程信息的 Servlet
ListAllCourseServlet.java	com.imut.servlet.base	显示所有课程信息的 Servlet
ShowCourseServlet.java	com.imut.servlet.base	查看/修改课程信息的 Servlet
UpdateCourseServlet.java	com.imut.servlet.base	更新课程信息的 Servlet
addCourse.jsp	/base	增加课程信息的页面
courseList.jsp	/base	显示所有课程信息的页面
courseShow.jsp	/base	查看/修改课程信息的页面

7.3.2　案例演示

在浏览器地址栏中输入"http://localhost:8080/ch07/login.jsp",以管理员用户登录系统,单击"基本数据管理"菜单中的"课程信息维护",显示课程列表页面,效果如图 7-8 所示。单击"添加课程"超链接,出现添加课程页面,效果如图 7-9 所示。输入课程信息,单击"提交"按钮,提示添加课程成功之后跳转到课程列表页面。在课程列表页面中输入查询条件,查询结果将显示在"查询"按钮下方的课程列表处,如果没有输入查询条件则查

图 7-8　课程列表页面

询所有课程信息。根据"课程名称"查询结果的效果图如图 7-10 所示。在课程列表页面中单击"删除"超链接，将会删除该课程信息，删除完成后显示课程列表页面。在课程列表页面中单击"查看/修改"超链接，或者单击课程编号，将显示课程信息查看/修改页面，效果如图 7-11 所示。修改课程信息完成后单击"提交"按钮，提示修改成功后显示课程列表页面。

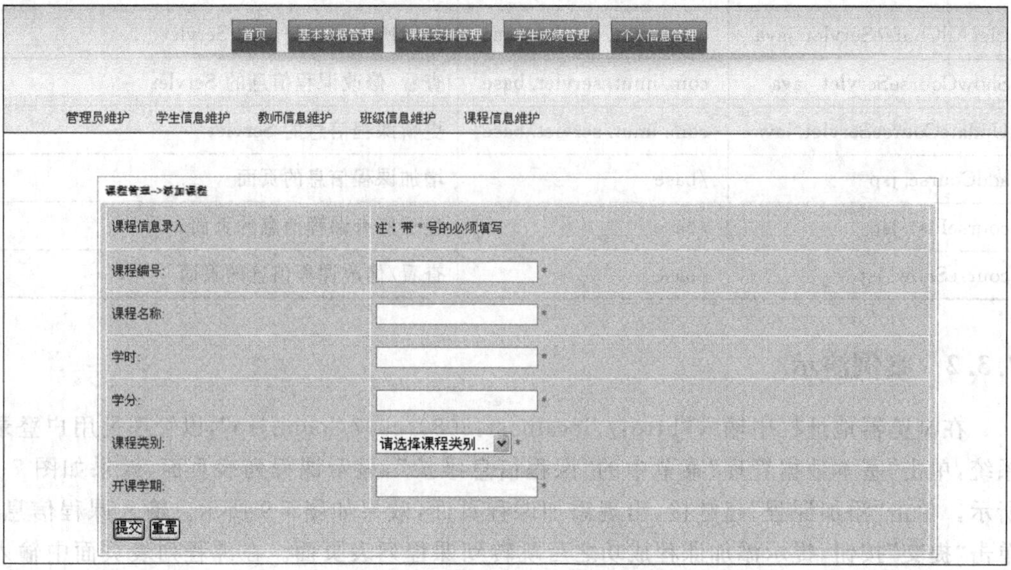

图 7-9 添加课程信息页面

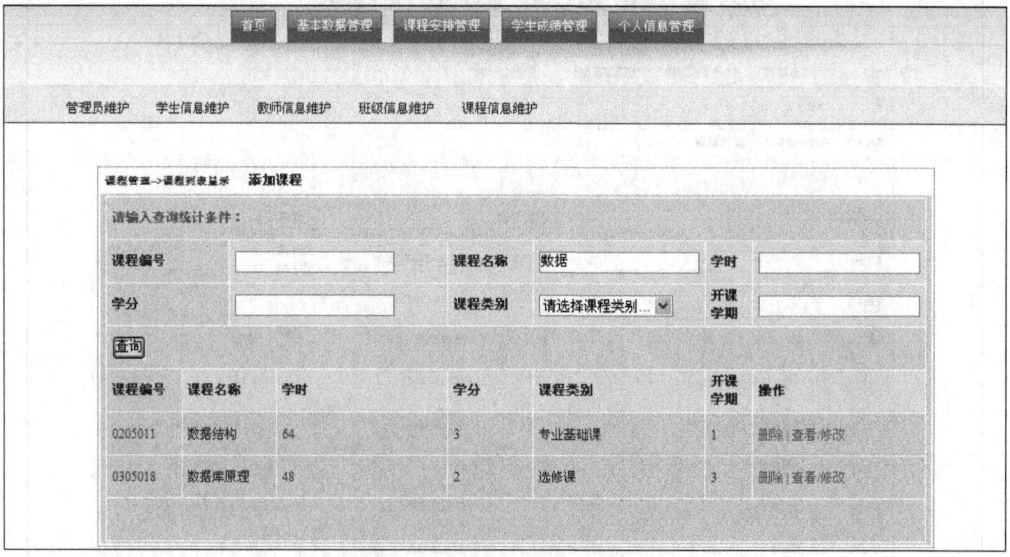

图 7-10 课程查询结果页面

图 7-11 查看/修改课程页面

7.3.3 代码实现

创建工程 ch07，根据案例设计描述需要完成各部分功能的具体实现，本章主要给出课程类 Course.java、接口 ICourse.java、接口实现类 CourseImpl.java、增加课程 Servlet 类 AddCourseServlet.java 以及增加课程页面 addCourse.jsp 的具体代码，其他源文件代码请参考随书电子资源。

程序 Course.java 代码如下：

```java
package com.imut.javabean;
public class Course {
    private String courseNo;
    private String courseName;
    private int studyTime;
    private int grade;
    private int courseType;
    private int term;
    public Course() {
    }
    public Course(String courseNo, String courseName, int studyTime, int grade,
            int courseType, int term) {
        super();
        this.courseNo=courseNo;
        this.courseName=courseName;
        this.studyTime=studyTime;
        this.grade=grade;
        this.courseType=courseType;
```

```java
            this.term=term;
        }
        public String getCourseNo() {
            return courseNo;
        }
        public void setCourseNo(String courseNo) {
            this.courseNo=courseNo;
        }
        public String getCourseName() {
            return courseName;
        }
        public void setCourseName(String courseName) {
            this.courseName=courseName;
        }
        public int getStudyTime() {
            return studyTime;
        }
        public void setStudyTime(int studyTime) {
            this.studyTime=studyTime;
        }
        public int getGrade() {
            return grade;
        }
        public void setGrade(int grade) {
            this.grade=grade;
        }
        public int getCourseType() {
            return courseType;
        }
        public void setCourseType(int courseType) {
            this.courseType=courseType;
        }
        public int getTerm() {
            return term;
        }
        public void setTerm(int term) {
            this.term=term;
        }
}
```

程序 ICourse.java 代码如下：

```java
package com.imut.javabean;
import java.util.List;
import java.util.Map;
```

```java
public interface ICourse {
    //添加课程方法
    public void addCourse(Course course);
    //删除课程方法
    public void delCourse(String courseNo);
    //修改课程信息
    public void updateCourse(Course course);
    //根据课程编号查找课程
    public Course findCourseByCourseNo(String courseNo);
    //列表显示所有课程列表--分页
    public Map findAllCourse(int curPage);
    //列表显示所有课程列表
    public List findAllCourse();
    //多条件查询课程
     public List findAllCourseByMostCon (String courseNo, String courseName,
      Integer studyTime,Integer grade,Integer courseType,Integer term);
}
```

程序 CourseImpl.java 代码如下：

```java
package com.imut.javabean;
import java.sql.Connection;
import java.sql.PreparedStatement;
import java.sql.ResultSet;
import java.sql.SQLException;
import java.sql.Statement;
import java.util.ArrayList;
import java.util.HashMap;
import java.util.List;
import java.util.Map;
import com.imut.commmon.ConnectionFactory;
import com.imut.commmon.Page;
import com.imut.commmon.ResourceClose;
public class CourseImpl implements ICourse{
    //添加课程方法
    public void addCourse(Course course){
        Connection conn=null;
        PreparedStatement pstmt=null;
        ResultSet rs=null;
        try{
            conn=ConnectionFactory.getConnection();
            String sql="insert into Course values(?,?,?,?,?,?)";
            pstmt=conn.prepareStatement(sql);
            pstmt.setString(1, course.getCourseNo());
            pstmt.setString(2, course.getCourseName());
```

```java
            pstmt.setInt(3, course.getStudyTime());
            pstmt.setInt(4, course.getGrade());
            pstmt.setInt(5, course.getCourseType());
            pstmt.setInt(6, course.getTerm());
            pstmt.executeUpdate();
        }catch(Exception e){
            e.printStackTrace();
        }finally{
            ResourceClose.close(rs, pstmt, conn);
        }
    }
    //删除课程方法
    public void delCourse(String courseNo){
        Connection conn=null;
        PreparedStatement pstmt=null;
        ResultSet rs=null;
        try{
            conn=ConnectionFactory.getConnection();
            String sql="delete from Course where courseNo=?";
            pstmt=conn.prepareStatement(sql);
            pstmt.setString(1, courseNo);
            pstmt.executeUpdate();
        }catch (SQLException e) {
            e.printStackTrace();
        }finally{
            ResourceClose.close(rs, pstmt, conn);
        }
    }
    //修改课程信息
    public void updateCourse(Course course){
        Connection conn=null;
        PreparedStatement pstmt=null;
        ResultSet rs=null;
        try{
            conn=ConnectionFactory.getConnection();
            String sql="update Course set courseNo=?,courseName=?,studyTime=?,
            grade=?,courseType=?,term=? where courseNo=?";
            pstmt=conn.prepareStatement(sql);
            pstmt.setString(1, course.getCourseNo());
            pstmt.setString(2, course.getCourseName());
            pstmt.setInt(3, course.getStudyTime());
            pstmt.setInt(4, course.getGrade());
            pstmt.setInt(5, course.getCourseType());
            pstmt.setInt(6, course.getTerm());
```

```java
            pstmt.setString(7, course.getCourseNo());
            pstmt.executeUpdate();
        }catch (SQLException e) {
            e.printStackTrace();
        }finally{
            ResourceClose.close(rs, pstmt, conn);
        }
    }
    //根据课程编号查找课程
    public Course findCourseByCourseNo(String courseNo){
        Course course=null;
        Connection conn=null;
        PreparedStatement pstmt=null;
        ResultSet rs=null;
        try{
            conn=ConnectionFactory.getConnection();
            String sql="select * from Course where courseNo=?";
            pstmt=conn.prepareStatement(sql);
            pstmt.setString(1, courseNo);
            rs=pstmt.executeQuery();
            while(rs.next()){
                course=new Course();
                course.setCourseNo(rs.getString(1));
                course.setCourseName(rs.getString(2));
                course.setStudyTime(rs.getInt(3));
                course.setGrade(rs.getInt(4));
                course.setCourseType(rs.getInt(5));
                course.setTerm(rs.getInt(6));
            }
        }catch (SQLException e) {
            e.printStackTrace();
        }finally{
            ResourceClose.close(rs, pstmt, conn);
        }
        return course;
    }
    //列表显示所有课程列表--分页
    public Map findAllCourse(int curPage){
        Course course=null;
        ArrayList list=new ArrayList();
        Connection conn=null;
        Statement pstmt=null;
        ResultSet rs=null;
        ResultSet r=null;
```

```java
        Map map=null;
        Page pa=null;
        try{
            conn=ConnectionFactory.getConnection();
            String sql="select * from Course order by courseNo";
            pstmt = conn. createStatement (ResultSet. TYPE _ SCROLL _ INSENSITIVE,
            ResultSet.CONCUR_READ_ONLY);
            rs=pstmt.executeQuery(sql);
            pa=new Page();          //声明分页类对象
            pa.setPageSize(5);
            pa.setPageCount(rs);
            pa.setCurPage(curPage);
            r=pa.setRs(rs);
            r.previous();
            for(int i=0;i<pa.getPageSize();i++){
                if(r.next()){
                    course=new Course();
                    course.setCourseNo(r.getString(1));
                    course.setCourseName(r.getString(2));
                    course.setStudyTime(r.getInt(3));
                    course.setGrade(r.getInt(4));
                    course.setCourseType(r.getInt(5));
                    course.setTerm(r.getInt(6));
                    list.add(course);
                }else{
                    break;
                }
            }
            map=new HashMap();
            map.put("list",list);
            map.put("pa",pa);
        }catch (SQLException e) {
            e.printStackTrace();
        }finally{
            ResourceClose.close(rs, pstmt, conn);
            ResourceClose.close(r, null, null);
        }
        return map;
    }
    //列表显示所有课程列表
    public List findAllCourse(){
        Course course=null;
        ArrayList list=new ArrayList();
        Connection conn=null;
```

```java
        Statement pstmt=null;
        ResultSet rs=null;
        try{
            conn=ConnectionFactory.getConnection();
            String sql="select * from Course order by courseNo";
            pstmt=conn.createStatement();
            rs=pstmt.executeQuery(sql);
            while(rs.next()){
                course=new Course();
                course.setCourseNo(rs.getString(1));
                course.setCourseName(rs.getString(2));
                course.setStudyTime(rs.getInt(3));
                course.setGrade(rs.getInt(4));
                course.setCourseType(rs.getInt(5));
                course.setTerm(rs.getInt(6));
                list.add(course);
            }
        }catch (SQLException e) {
            e.printStackTrace();
        }finally{
            ResourceClose.close(rs, pstmt, conn);
        }
        return list;
    }
    //多条件查询课程
    public List findAllCourseByMostCon (String courseNo, String courseName,
Integer studyTime,Integer grade,Integer courseType,Integer term){
        Course course=null;
        ArrayList list=new ArrayList();
        Connection conn=null;
        PreparedStatement pstmt=null;
        ResultSet rs=null;

        //构造多条件查询的 SQL 语句
        String sql="select * from Course where 1=1 ";
        //模糊查询
        if(courseNo!=null&&!courseNo.equals("")){
            sql+=" and courseNo like '%"+courseNo+"%'";
        }
        if(courseName!=null&&!courseName.equals("")){
            sql+=" and courseName like '%"+courseName+"%'";
        }
        if(studyTime!=null&&!studyTime.equals("")){
            sql+=" and studyTime="+studyTime;
```

```java
            }
            if(grade!=null&&!grade.equals("")){
                sql+=" and grade="+grade;
            }
            if(courseType!=null&&!courseType.equals("")){
                sql+=" and courseType="+courseType;
            }
            if(term!=null&&!term.equals("")){
                sql+=" and term="+term;
            }
            sql+=" order by courseNo";
            try{
                conn=ConnectionFactory.getConnection();
                pstmt=conn.prepareStatement(sql);
                rs=pstmt.executeQuery();
                while(rs.next()){
                    course=new Course();
                    course.setCourseNo(rs.getString(1));
                    course.setCourseName(rs.getString(2));
                    course.setStudyTime(rs.getInt(3));
                    course.setGrade(rs.getInt(4));
                    course.setCourseType(rs.getInt(5));
                    course.setTerm(rs.getInt(6));
                    list.add(course);
                }
            }catch (SQLException e) {
                e.printStackTrace();
            }finally{
                ResourceClose.close(rs, pstmt, conn);
            }
            return list;
        }
    }
```

程序 AddCourseServlet.java 代码如下：

```java
package com.imut.servlet.base;
import java.io.IOException;
import javax.servlet.ServletException;
import javax.servlet.http.HttpServlet;
import javax.servlet.http.HttpServletRequest;
import javax.servlet.http.HttpServletResponse;
import javax.servlet.http.HttpSession;
import com.imut.javabean.Course;
import com.imut.javabean.CourseImpl;
```

```java
public class AddCourseServlet extends HttpServlet {
    public void doGet (HttpServletRequest request, HttpServletResponse response)
            throws ServletException, IOException {
        doPost(request, response);
    }
    public void doPost(HttpServletRequest request, HttpServletResponse response)
            throws ServletException, IOException {
        HttpSession session=request.getSession();
        request.setCharacterEncoding("UTF-8");
        response.setCharacterEncoding("UTF-8");
        String courseNo=request.getParameter("courseNo");
        String courseName=request.getParameter("courseName");
        Integer studyTime=new Integer(request.getParameter("studyTime"));
        Integer grade=new Integer(request.getParameter("grade"));
        Integer courseType=new Integer(request.getParameter("courseType"));
        Integer term=new Integer(request.getParameter("term"));
        CourseImpl dbAccess=new CourseImpl();
        if(dbAccess.findCourseByCourseNo(courseNo)!=null){
            session.setAttribute("message","所要添加的课程已存在!");
            request. getRequestDispatcher ( "/base/addCourse. jsp "). forward
                (request, response);
        }else{
            Course course = new Course (courseNo, courseName, studyTime, grade,
                courseType,term);
            dbAccess.addCourse(course);
            session.setAttribute("message","课程信息添加成功!");
            request. getRequestDispatcher ( "/base/listAllCourseServlet ").
                forward(request, response);
        }
    }
}
```

程序 AddCourseServlet.java 在 web.xml 文件中的配置代码如下：

```xml
<servlet>
    <servlet-name>addCourse</servlet-name>
    <servlet-class>com.imut.servlet.base.AddCourseServlet</servlet-class>
</servlet>
<servlet-mapping>
    <servlet-name>addCourse</servlet-name>
    <url-pattern>/base/addCourseServlet</url-pattern>
</servlet-mapping>
```

程序 addCourse.jsp 代码如下：

```jsp
<%@page language="java" contentType="text/html; charset=UTF-8" pageEncoding="UTF-8"%>
<!DOCTYPE html PUBLIC "-//W3C//DTD XHTML 1.0 Transitional//EN" "http://www.w3.org/TR/xhtml1/DTD/xhtml1-transitional.dtd">
<html xmlns="http://www.w3.org/1999/xhtml">
<head>
<title>课程管理</title>
<link rel="stylesheet" type="text/css" id="css" href="../style/main.css" />
<link rel="stylesheet" type="text/css" id="css" href="../style/style1.css" />

<link rel="stylesheet" type="text/css" id="css" href="../style/style.css" />
<style type="text/css">
<!--
table{border-spacing:1px; border:1px solid #A2C0DA;}
td, th{padding:2px 5px;border-collapse:collapse;text-align:left; font-weight:normal;}
thead tr th{height:50px;background:#B0D1FC;border:1px solid white;}
thead tr th.line1{background:#D3E5FD;}
thead tr th.line4{background:#C6C6C6;}
tbody tr td{height:35px;background:#CBE2FB;border:1px solid white; vertical-align:middle;}
tbody tr td.line4{background:#D5D6D8;}
tbody tr th{height:35px;background: #DFEDFF;border:1px solid white; vertical-align:middle;}
tfoot tr td{height:35px;background:#FFFFFF;border:1px solid white; vertical-align:middle;}
-->
</style>
<script type="text/javascript" src="../js/common.js" ></script>
<script type="text/javascript" src="../js/calendar.js" ></script>
<script type="text/javascript" src="../js/ajax.js" ></script>
<script type="text/JavaScript">
    /*判断是否为数字*/
    function isNumber(str) {
        var Letters="1234567890";
        for (var i=0; i <str.length; i=i+1) {
            var CheckChar=str.charAt(i);
            if (Letters.indexOf(CheckChar)==-1) {
                return false;
            }
        }
        return true;
    }
```

```javascript
/*判断是否为Email*/
function isEmail(str) {
    var myReg=/^[-_A-Za-z0-9]+@([_A-Za-z0-9]+\.)+[A-Za-z0-9]{2,3}$/;
    if (myReg.test(str)) {
        return true;
    }
    return false;
}
/*判断是否为空*/
function isEmpty(value) {
    return /^\s*$/.test(value);
}
function check(){
    if(isEmpty(document.myForm.courseNo.value)){
        alert("课程编号不能为空!");
        document.myForm.courseNo.focus();
        return false;
    }
    if(isEmpty(document.myForm.courseName.value)){
        alert("课程名称不能为空!");
        document.myForm.courseName.focus();
        return false;
    }
    if(isEmpty(document.myForm.studyTime.value)){
        alert("课程学时不能为空!");
        document.myForm.studyTime.focus();
        return false;
    }
    if(!isNumber(document.myForm.studyTime.value)){
        alert("课程学时必须是数字!");
        document.myForm.studyTime.focus();
        return false;
    }
    if(isEmpty(document.myForm.grade.value)){
        alert("课程学分不能为空!");
        document.myForm.grade.focus();
        return false;
    }
    if(!isNumber(document.myForm.grade.value)){
        alert("课程学分必须是数字!");
        document.myForm.grade.focus();
        return false;
    }
    if(document.myForm.courseType.value==""){
```

```
            alert("请选择课程类别!");
            document.myForm.courseType.focus();
            return false;
        }
        if(isEmpty(document.myForm.term.value)){
            alert("开课学期不能为空!");
            document.myForm.term.focus();
            return false;
        }
        if(!isNumber(document.myForm.term.value)){
            alert("开课学期必须是数字!");
            document.myForm.term.focus();
            return false;
        }
        return true;
    }
</script>
</head>
<%
    String message= (String)session.getAttribute("message");
    if(message!=null){
%>
        <script type="text/javascript">
            alert('<c:out value="$ {sessionScope.message}"/>');
        </script>
<%
        session.removeAttribute("message");
    }
%>
<body>
<div id="btm">
<div id="main">
    <div id="header">
        <div id="top"></div>
        <div id="logo">
            <h1>课程管理</h1></div>
        <div id="mainnav">
            <ul>
            <li><a href="../index.jsp">首页</a></li>
            <li><a href="listAllTeacherServlet">基本数据管理</a></li>
            <li><a href="#">课程安排管理</a></li>
            <li><a href="#">学生成绩管理</a></li>
            <li><a href="../person/showPersonServlet">个人信息管理</a></li>
```

```html
            </ul>
            <span>
            </span>
        </div>
    </div>
<div id="tabsJ">
    <ul>
        <li><a href="#" title="管理员维护"><span>管理员维护</span></a></li>
        <li><a href="#" title="学生信息维护"><span>学生信息维护</span></a></li>
        <li><a href="listAllTeacherServlet" title="教师信息维护"><span>教师信息维护</span></a></li>
        <li><a href="listAllClassServlet" title="班级信息维护"><span>班级信息维护</span></a></li>
        <li><a href="listAllCourseServlet" title="课程信息维护"><span>课程信息维护</span></a></li>
    </ul>
</div>
    <div id="content" align="center">
        <div id="center">
    <BR /><BR />
    <form name="myForm"    action="addCourseServlet" method="post" >
        <table width="800" align="center" cellpadding="0" cellspacing="0">
            <thead>
                <tr>
                    <td width="70%"><h5>课程管理-->添加课程</h5></td>
                </tr>
            </thead>
            <tr>
                <td colspan="2" width="100%">
                    <table width="100%">
                        <tbody>
                        <tr>
                            <td align="right">课程信息录入</td>
                            <td>注:带<font color="red"> * </font>号的必须填写</td>
                        </tr>
                        <tr>
                            <td align="right">课程编号:</td>
                            <td>< input id=" courseNo" name =" courseNo" />
                            <font color="red"> * </font></td>
                        </tr>
                        <tr>
                            <td align="right">课程名称:</td>
```

```html
                    <td><input id="courseName" name="courseName" />
                    <font color="red">*</font></td>
                </tr>
                <tr>
                    <td align="right">学时:</td>
                    <td><input id="studyTime" name="studyTime" /><font color="red">*</font></td>
                </tr>
                <tr>
                    <td align="right">学分:</td>
                    <td><input id="grade" name="grade" /><font color="red">*</font></td>
                </tr>
                <tr>
                    <td align="right">课程类别:</td>
                    <td>
                    <select id="courseType" name="courseType">
                        <option value="">请选择课程类别...</option>
                        <option value="0">必修课</option>
                        <option value="1">专业必修课</option>
                        <option value="2">专业基础课</option>
                        <option value="3">选修课</option>
                        <option value="4">专业选修课</option>
                    </select>
                    <font color="red">*</font></td>
                </tr>
                <tr>
                    <td align="right">开课学期:</td>
                    <td><input id="term" name="term" /><font color="red">*</font></td>
                </tr>
            </tbody>
                <tr>
                    <td align="center" colspan="2">
                    <input type="submit" value="提交" onclick="return check();"/>
                    <input type="reset" value="重置"/>
                    </td>
                </tr>
            </table>
            </td>
        </tr>
    </table>
</form>
```

```html
            </div>
        </div>
        <div id="footer">
            <div id="copyright">
                <div id="copy">
                    <p align="center">CopyRight&copy;2010</p>
                    <p>内蒙古工业大学信息工程学院软件工程系</p>
                </div>
            </div>
            <div id="bgbottom"></div>
        </div>
    </div>
</div>
</body>
</html>
```

【代码分析】程序 Course.java 封装了课程的信息,并提供了设置和访问属性的方法。

程序 ICourse.java 定义了操作 course 数据表的增删改查等方法,例如,在添加课程信息的方法 public void addCourse(Course course);中,使用 Course 对象作为参数,传递要添加的课程信息。

程序 CourseImpl.java 是接口 ICourse.java 的具体实现类,它实现了接口 ICourse.java 中定义的所有方法。例如,在添加课程信息的方法中首先连接数据库,然后预处理插入记录的 SQL 语句,并通过 setXxx()方法使用参数 Course 对象的属性值为 SQL 语句中的占位符赋值,最后执行 SQL 语句并关闭数据库资源。

程序 AddCourseServlet.java 首先获取从添加课程页面中传递过来的参数,使用 CourseImpl 的 findCourseByCourseNo(String)方法根据课程编号查找课程信息。如果查到的课程信息不为空,说明该课程存在,不能添加,并将提示信息保存到 HttpSession 对象,跳转到添加课程信息页面。如果查到的课程信息为空,说明该课程不存在可以添加,同时将提示信息保存到 HttpSession 对象,跳转到课程列表信息页面。

程序 addCourse.jsp 提供了系统的主菜单以及基本数据管理菜单下的子菜单,通过 form 表单的各组件提供输入课程信息以及提交信息的功能。

习 题

1. 选择题

(1) 下列有关 MVC 体系结构的说法中错误的一项是( )。

    A. 模型是由 Java Web 中的 JavaBean 担当

    B. 视图是由 Java Web 中的 JSP 担当

    C. 控制器是由 Java Web 中的 JavaBean 担当

    D. MVC 体系结构是开发 Java Web 应用程序常用的设计模式

(2) MVC模式中,(　　)负责通知应用程序客户端,应用程序本身有状态改变。
　　A. 模型(Model)　　　　　　　　B. 视图(View)
　　C. 控制器(Controller)　　　　　D. 以上皆不是
(3) 关于JSP Model1模式的缺点,下列的叙述哪一项是不正确的?(　　)
　　A. 应用是基于过程的　　　　　B. 业务逻辑和表示逻辑混合
　　C. 分离了视图层和业务层　　　D. 产生较多的文件
(4) MVC与Model2架构的最大差别在于(　　)。
　　A. Model 2架构的视图是由HTML组成的
　　B. Model2架构中的模型无法通知视图状态已更新
　　C. MVC架构是基于请求/响应模型的
　　D. MVC架构只能用于单机应用程序
(5) 在Java Web中,MVC模式中的控制器角色由(　　)担当。
　　A. JavaBean　　　　　　　　　B. JSP页面
　　C. Servlet　　　　　　　　　　D. HTML页面
(6) 关于MVC的缺点,下列的叙述哪一项是不正确的?(　　)
　　A. 提高了对开发人员的要求
　　B. 代码复用率低
　　C. 增加了文件管理的难度
　　D. 产生较多的文件

2. 填空题

(1) MVC体系结构的设计思想实现了将＿＿＿＿和＿＿＿＿分开,有益于实现多种多样的显示。
(2) 在MVC模式的Web开发中,"视图"、"模型"和"控制器"分别对应着＿＿＿＿、＿＿＿＿和＿＿＿＿,以＿＿＿＿为核心。

3. 简答题

(1) 请简述什么是MVC模式,并说明其优缺点。
(2) 阐述MVC模式的工作原理。

4. 程序设计题

利用MVC模式实现一个猜数游戏,系统产生一个1～100之间的随机数作为被猜的数,用户在页面上输入要猜的数字,经过业务处理后返回是猜大还是猜小,重新再猜,猜对则返回总共猜的次数并且不能再继续猜数。

EL 表达式

表达式语言(Experssion Language,EL)是 JSP 2.0 中引入的新特性,在 JSP 中使用 EL 表达式可以获取并显示页面数据。使用 EL 表达式可以简化对变量和对象的访问,减少 JSP 中的 Java 代码。本章将以理论与实践相结合的方式介绍 EL 表达式语法、EL 表达式访问数据、EL 表达式的内置对象、EL 表达式的运算符以及在 JSP 页面中如何禁用 EL 表达式。

8.1 EL 简介

EL 是一种语法简单、易于学习的语言。只有 JSP 2.0(Servlet 2.4)以上版本支持 EL,因此在使用 EL 以前要先确认服务器是否支持 JSP 2.0。例如,Tomcat 5.0 以上版本才开始支持 JSP 2.0。通过 EL 表达式可以获得 PageContext 的属性值、直接访问 JSP 的内置对象,还可以访问作用域对象、集合对象等。EL 提供了丰富的运算符,可以进行各种运算。

EL 表达式的语法形式:

${ELexpression}

EL 表达式是以"$"符号开始,后面紧跟一对大括号,大括号内部包含合法的表达式。EL 表达式可以直接用在 JSP 页面的静态文本中,也可以作为 JSP 标签的属性值来使用。

8.2 EL 访问数据

通过 EL 表达式可以很方便地访问作用域变量、JavaBean 对象的属性以及集合的元素值。EL 表达式可以通过点号运算符"."和方括号运算符"[]"来访问数据。

点号运算符用来访问 JavaBean 对象的属性值或者是 Map 对象的一个键值。例如,customer 是 CustomerBean 类的一个实例,通过下面的代码访问该实例的 name 属性值:

${customer.name}

方括号运算符可以访问 JavaBean 对象的属性值、数组对象的元素、List 对象的元素以及 Map 对象的一个键值。例如：

```
${customer["name"]}
```

或

```
${customer['name']}
```

8.2.1 访问作用域变量

可以通过${varname}的形式取得作用域变量的值，其中，varname 是要访问作用域变量的名称。

例程 8-1：演示用 EL 表达式访问作用域变量。程序为 ScopeServlet.java 和 getVar.jsp。

程序 ScopeServlet.java 代码如下：

```java
package com.ch08;
import java.io.IOException;
import java.io.PrintWriter;
import javax.servlet.RequestDispatcher;
import javax.servlet.ServletException;
import javax.servlet.http.HttpServlet;
import javax.servlet.http.HttpServletRequest;
import javax.servlet.http.HttpServletResponse;
public class ScopeServlet extends HttpServlet {
    public void doGet(HttpServletRequest request, HttpServletResponse response)
        throws ServletException, IOException {
        this.doPost(request, response);
    }
    public void doPost(HttpServletRequest request, HttpServletResponse response)
        throws ServletException, IOException {
        String attr1="hello world!";
        request.setAttribute("attr1", attr1);
        RequestDispatcher dispatcher;
        dispatcher=request.getRequestDispatcher("/getVar.jsp");
        dispatcher.forward(request, response);
    }
}
```

ScopeServlet.java 在 web.xml 中的配置信息如下：

```xml
<servlet>
    <servlet-name>ScopeServlet</servlet-name>
    <servlet-class>com.ch08.ScopeServlet</servlet-class>
</servlet>
<servlet-mapping>
    <servlet-name>ScopeServlet</servlet-name>
    <url-pattern>/servlet/scopeServlet</url-pattern>
```

```
</servlet-mapping>
```

程序 getVar.jsp 代码如下：

```
<%@page language="java" import="java.util.*" pageEncoding="UTF-8"%>
<!DOCTYPE HTML PUBLIC "-//W3C//DTD HTML 4.01 Transitional//EN">
<html>
    <head>
        <title>获得作用于对象的值</title>
    </head>
    <body>
        ${attr1}
    </body>
</html>
```

程序 ScopeServlet.java 中将变量 attri1 的值保存在 request 的属性范围中，并将请求转向到 getVar.jsp 中显示。

程序 getVar.jsp 的 EL 表达式 ${attr1}的作用是：首先从 page 范围查找属性 attri1，如果找到了则返回并输出当前范围的变量的值；如果没找到，则再从 request 范围查找，如果还没有找到，则再从 session 范围查找，以此类推，如果直到 application 范围还没有找到，则不再查找直接返回并输出空字符串。

在浏览器地址栏中输入"http://localhost:8080/chapter8/servlet/scopeServlet"直接访问 ScopeServlet，服务器将根据 request 对象的 getRequestDispatcher()方法中指定的地址跳转到 getVar.jsp 页面，在该页面执行 EL 表达式 ${attr1}并显示 attri1 变量的值，运行效果如图 8-1 所示。

图 8-1 EL 表达式访问作用域变量

getVar.jsp 页面中 EL 表达式 ${attr1}相当于以下 JSP 脚本代码的作用，运行效果如图 8-1 所示。

```
<%
    String attr1=(String)request.getAttribute("attr1");
    if(attr1!=null){
%>
<%=attr1%>
<%
```

    }
%>
```

## 8.2.2 访问 JavaBean 属性

使用 EL 表达式访问 JavaBean 属性,就可以通过点号运算符来实现,具体使用格式是:作用域对象名.属性名,它相当于调用 JavaBean 对象的 getXxx()方法。

**例程 8-2**：演示用 EL 表达式访问 JavaBean 对象属性。程序为 Book.java、BookServlet.java 和 getBean.jsp。

程序 Book.java 代码如下：

```java
package com.ch08.bean;
public class Book {
    private String isbn;
    private String name;
    public Book() {
    }
    public Book(String isbn, String name) {
        this.isbn=isbn;
        this.name=name;
    }
    public String getIsbn() {
        return isbn;
    }
    public void setIsbn(String isbn) {
        this.isbn=isbn;
    }
    public String getName() {
        return name;
    }
    public void setName(String name) {
        this.name=name;
    }
}
```

程序 BookServlet.java 代码如下：

```java
package com.ch08;
import java.io.IOException;
import java.io.PrintWriter;
import javax.servlet.RequestDispatcher;
import javax.servlet.ServletException;
import javax.servlet.http.HttpServlet;
import javax.servlet.http.HttpServletRequest;
import javax.servlet.http.HttpServletResponse;
```
```

```
import com.ch08.bean.Book;
public class BookServlet extends HttpServlet {
    public void doGet(HttpServletRequest request, HttpServletResponse response)
            throws ServletException, IOException {
        this.doPost(request, response);
    }
    public void doPost(HttpServletRequest request, HttpServletResponse response)
            throws ServletException, IOException {
        Book book=new Book("197231001","web程序设计");
        request.setAttribute("book", book);
        RequestDispatcher dispatcher= request.getRequestDispatcher("/getBean.jsp");
        dispatcher.forward(request, response);
    }
}
```

程序 getBean.jsp 代码如下：

```
<%@page language="java" import="java.util.*" pageEncoding="UTF-8"%>
<!DOCTYPE HTML PUBLIC "-//W3C//DTD HTML 4.01 Transitional//EN">
<html>
    <head>
        <title>EL 表达式访问 JavaBean 属性</title>
    </head>
    <body>
        ISBN:${ book.isbn }<br>
        书名:${ book.name }<br>
    </body>
</html>
```

程序 Book.java 是一个 JavaBean，为实体模型，用来描述图书对象实体。

程序 BookServlet.java 首先使用有参构造方法实例化一个 Book 对象，然后把该对象保存在 request 属性范围中，并跳转到 getBean.jsp 中显示。

程序 getBean.jsp 中使用 EL 表达式从 page 属性范围查找 book 对象的 isbn 属性和 name 属性，如果找到则不再继续查找同时返回并输出该值，否则依次从 request 属性范围、session 属性范围、application 属性范围查找，如果找到则输出，如果没有找到则返回空字符串。

在浏览器地址栏输入"http://localhost:8080/chapter8/servlet/bookServlet"，服务器会跳转到 getBean.jsp 中显示 book 对象属性的值，运行效果如图 8-2 所示。如果 book 对象的属性没有设置值，则会输出一个空字符串，并不会出现异常，显示效果如图 8-3 所示。

以上 getBean.jsp 中的 EL 表达式相当于以下 JSP 代码的作用，运行效果与图 8-2 一致。

图 8-2 EL 访问具有设置值的 JavaBean 属性

图 8-3 EL 访问没有设置值的 JavaBean 属性

```
<%
    Book book=(Book)request.getAttribute("book");
    if(book!=null){
%>
ISBN:<%=book.getIsbn() %><br>
书名:<%=book.getName() %><br>
<%
    }
%>
```

8.2.3 访问集合元素

EL 表达式可以访问数组、List 以及 Map 等集合对象的元素,具体访问形式如表 8-1 所示。

表 8-1 EL 访问集合元素的用法

访问的数据类型	使用示例	说 明	返 回 值	实际调用的方法
数组	${ins[i]}	i 代表数组元素的下标	返回下标为 i 的数组元素的值	ins[i]
	${ins['i']}			
	${ins["i"]}			

续表

访问的数据类型	使用示例	说　　明	返　回　值	实际调用的方法
List	${ins[i]} ${ins['i']} ${ins["i"]}	i 代表 List 元素的下标	返回下标为 i 的 List 集合元素的值	ins.get(i)
Map	${ins.name} ${ins['name']} ${ins["name"]}	name 代表 Map 对象键值	返回对应于键 name 的值	ins.get("name")

例程 8-3：演示通过 EL 表达式访问数组元素的值。程序为 Book.java、ArrayServlet.java 和 getArray.jsp。

程序 Book.java 同例程 8-2 中的 Book.java，不再赘述。

程序 ArrayServlet.java 代码如下：

```
package com.ch08;
import java.io.IOException;
import javax.servlet.RequestDispatcher;
import javax.servlet.ServletException;
import javax.servlet.http.HttpServlet;
import javax.servlet.http.HttpServletRequest;
import javax.servlet.http.HttpServletResponse;
import com.ch08.bean.Book;
public class ArrayServlet extends HttpServlet {
    public void doGet(HttpServletRequest request, HttpServletResponse response)
        throws ServletException, IOException {
            this.doPost(request, response);
        }
    public void doPost(HttpServletRequest request, HttpServletResponse response)
        throws ServletException, IOException {
        Book books[]=new Book[4];
        books[0]=new Book("001","网络编程基础");
        books[1]=new Book("002","Java核心技术");
        books[2]=new Book("003","数据库系统原理");
        books[3]=new Book("004","Oracle数据库应用");
        request.setAttribute("books", books);
        RequestDispatcher dispatcher = request.getRequestDispatcher ( "/getArray.jsp");
        dispatcher.forward(request, response);
    }
}
```

ArrayServle.java 在 web.xml 文件中的配置信息如下：

```xml
<servlet>
    <servlet-name>ArrayServlet</servlet-name>
    <servlet-class>com.ch08.ArrayServlet</servlet-class>
</servlet>
<servlet-mapping>
    <servlet-name>ArrayServlet</servlet-name>
    <url-pattern>/servlet/arrayServlet</url-pattern>
</servlet-mapping>
```

程序 getArray.jsp 代码如下：

```jsp
<%@page language="java" import="java.util.*" pageEncoding="UTF-8"%>
<!DOCTYPE HTML PUBLIC "-//W3C//DTD HTML 4.01 Transitional//EN">
<html>
    <head>
        <title>EL 表达式访问数组元素</title>
    </head>
    <body>
        <center>
        <table border=1>
            <tr>
                <th>ISBN</th>
                <th>书名</th>
            </tr>
            <tr>
                <td>${books[0].isbn }</td>
                <td>${books[0].name }</td>
            </tr>
            <tr>
                <td>${books[1].isbn }</td>
                <td>${books[1].name }</td>
            </tr>
            <tr>
                <td>${books[2].isbn }</td>
                <td>${books[2].name }</td>
            </tr>
            <tr>
                <td>${books[3].isbn }</td>
                <td>${books[3].name }</td>
            </tr>
        </table>
        </center>
    </body>
```

```
</html>
```

程序 ArrayServlet.java 首先实例化具有 4 个元素的一个 Book 类型的对象数组并分别赋值,然后把该对象数组保存到 request 作用域范围并转向到 getArray.jsp 页面。在 getArray.jsp 页面中通过 EL 表达式获取各个数组元素对象的属性。例如,EL 表达式 ${books[0].isbn} 先通过中括号运算符取得 books 数组中第一个元素 books[0],然后通过点号运算符获取 books[0] 的属性 isbn 值,${books[0].name} 是取得第一个元素 books[0] 的 name 属性值,同理通过此种方法可以取得 books 数组的所有元素的属性值。

在浏览器地址栏中输入"http://localhost:8080/chapter8/servlet/arrayServlet",先运行 ArrayServlet,实例化并保存数组对象后跳转到 getArray.jsp 页面中显示 books 数组各元素的属性值,运行效果如图 8-4 所示。

图 8-4 EL 访问数组元素

例程 8-4:演示通过 EL 表达式访问 List 元素的值。程序为 ListServlet.java 和 getList.jsp。

程序 ListServlet.java 代码如下:

```
package com.ch08;
import java.io.IOException;
import java.util.ArrayList;
import java.util.List;
import javax.servlet.RequestDispatcher;
import javax.servlet.ServletException;
import javax.servlet.http.HttpServlet;
import javax.servlet.http.HttpServletRequest;
import javax.servlet.http.HttpServletResponse;
public class ListServlet extends HttpServlet {
    public void doGet(HttpServletRequest request, HttpServletResponse response)
        throws ServletException, IOException {
        this.doPost(request, response);
    }
    public void doPost(HttpServletRequest request, HttpServletResponse response)
        throws ServletException, IOException {
        List department=new ArrayList();
```

```
        department.add("计算机系");
        department.add("软件工程系");
        department.add("电子系");
        request.setAttribute("department", department);
        RequestDispatcher disptcher=request.getRequestDispatcher("/getList.jsp");
        disptcher.forward(request, response);
    }
}
```

ListServlet.java 在 web.xml 文件中的配置信息如下：

```
<servlet>
    <servlet-name>ListServlet</servlet-name>
    <servlet-class>com.ch08.ListServlet</servlet-class>
</servlet>
<servlet-mapping>
    <servlet-name>ListServlet</servlet-name>
    <url-pattern>/servlet/listServlet</url-pattern>
</servlet-mapping>
```

程序 getList.jsp 代码如下：

```
<%@page language="java" import="java.util.*" pageEncoding="UTF-8"%>
<!DOCTYPE HTML PUBLIC "-//W3C//DTD HTML 4.01 Transitional//EN">
    <html>
        <head>
        <title>EL 表达式访问 List 元素</title>
        </head>
        <body>
            <h3>系别</h3>
            ${department[0] }<br>
            ${department[1] }<br>
            ${department[2] }<br>
        </body>
    </html>
```

程序 ListServlet.java 首先实例化一个 List 对象 department 并向其中添加字符串类型的元素，然后将该对象保存到 request 作用范围后跳转到 getList.jsp 页面。getList.jsp 通过 EL 表达式使用中括号运算符获取保存在 request 属性范围中的 department 对象元素值。

在浏览器地址栏中输入 "http://localhost/chapter8/servlet/listServlet" 访问 ListServlet，然后根据跳转语句跳转到 getList 页面，执行 EL 表达式并显示结果，运行效果如图 8-5 所示。

例程 8-5：演示通过 EL 表达式访问 Map 元素的值。程序为 MapServlet.java 和 getMap.jsp。

程序 MapServlet.java 代码如下：

图 8-5 EL 访问 List 集合元素

```
import java.io.IOException;
import java.util.HashMap;
import java.util.Map;
import javax.servlet.RequestDispatcher;
import javax.servlet.ServletException;
import javax.servlet.http.HttpServlet;
import javax.servlet.http.HttpServletRequest;
import javax.servlet.http.HttpServletResponse;
public class MapServlet extends HttpServlet {
    public void doGet(HttpServletRequest request, HttpServletResponse response)
        throws ServletException, IOException {
        this.doPost(request, response);
    }
    public void doPost(HttpServletRequest request, HttpServletResponse response)
        throws ServletException, IOException {
        Map phone=new HashMap();
        phone.put("home", "0471-3277809");
        phone.put("tel", "13333333333");
        phone.put("office", "0471-6666666");
        request.setAttribute("phone", phone);
        RequestDispatcher dispatcher=request.getRequestDispatcher("/getMap.jsp");
        dispatcher.forward(request, response);
    }
}
```

MapServlet.java 在 web.xml 文件中的配置信息如下：

```
<servlet>
    <servlet-name>MapServlet</servlet-name>
    <servlet-class>com.ch08.MapServlet</servlet-class>
</servlet>
<servlet-mapping>
    <servlet-name>MapServlet </servlet-name>
    <url-pattern>/servlet/mapServlet</url-pattern>
</servlet-mapping>
```

程序 getMap.jsp 代码如下：

```
<%@page language="java" import="java.util.*" pageEncoding="UTF-8"%>
<!DOCTYPE HTML PUBLIC "-//W3C//DTD HTML 4.01 Transitional//EN">
<html>
    <head>
        <title>EL 表达式访问 Map 元素</title>
    </head>
    <body>
        <h3>联系方式:通过点号运算符"."访问</h3>
        家庭:$ {phone.home }<br>
        手机:$ {phone.tel }<br>
        办公室:$ {phone.office}<br>
        <h3>联系方式:通过中括号运算符"[]"访问</h3>
        家庭:$ {phone["home"]}<br>
        手机:$ {phone["tel"] }<br>
        办公室:$ {phone["office"]}<br>
    </body>
</html>
```

程序 MapServlet.java 首先实例化一个 Map 对象 phone 并且向其中添加键值对元素，然后将 phone 对象保存到 request 作用范围后跳转到 getMap.jsp 页面进行显示。在 getMap.jsp 页面中通过 EL 表达式采用两种不同的运算符分别访问 Map 对象 phone 中键值对的值。第一种方式是使用点号运算符，例如 ${phone.home} 来访问 phone 中键为"home"所对应的值；第二种方式是使用中括号运算符，例如 ${phone["office"]} 来访问 phone 中键为"office"所对应的值，其中中括号中的内容可以用双引号括起来，也可以用单引号括起来，或者不加引号也可以。

在浏览器地址栏中输入："http://localhost:8080/chapter8/servlet/mapServlet"，首先执行 MapServlet，然后跳转到 getMap.jsp 中执行 EL 表达式的值，将两组不同访问方式的结果显示到页面，运行效果如图 8-6 所示。

图 8-6　EL 访问 Map 集合元素的值

8.3 EL 内置对象

前面曾介绍过 JSP 的内置对象，它们不用显式声明就可以在 JSP 页面中直接使用。同样在 EL 表达式中也定义了一套自己的内置对象，分为 6 大类，共有 11 个，使用它们可以完成 JSP 中的一些常用功能，如表 8-2 所示。

表 8-2 EL 内置对象

内置对象名	类型	说明
pageContext	javax.servlet.jsp.PageContext	当前页面上下文对象
pageScope	java.util.Map	访问 page 属性范围的对象
requestScope	java.util.Map	访问 request 属性范围的对象
sessionScope	java.util.Map	访问 session 属性范围的对象
applicationScope	java.util.Map	访问 application 属性范围的对象
param	java.util.Map	获取其他页面传递过来的参数
paramValues	java.util.Map	获取其他页面传递过来的多值参数
header	java.util.Map	获取头信息
headerValues	java.util.Map	获取头信息的值
cookie	java.util.Map	获取 cookie 的值
initParam	java.util.Map	获取设定的初始化参数的值

1. pageContext 内置对象

pageContext 对象可以取得与用户请求或服务器端相关的信息。它用于访问 JSP 内置对象，如请求、响应、会话、输出、servletContext 等。例如，${pageContext.request}代表页面的请求对象。pageContext 常用的表达式如表 8-3 所示。

表 8-3 常用 pageContext 内置对象的表达式

表达式	作用
${pageContext.request.queryString}	取得请求字符串
${pageContext.request.requestURL}	取得不包括请求字符串的 URL
${pageContext.request.method}	取得 HTTP 方法（GET、POST）
${pageContext.request.contextPath}	取得请求的上下文路径
${pageContext.request.remoteAddr}	取得用户的 IP 地址
${pageContext.session.new}	判断 session 是否是已产生但未使用
${pageContext.session.id}	取得 session 的 id
${pageContext.servletContext.serverInfo}	取得服务器的信息

例程 8-6：演示 pageContext 的使用方法。程序为 pageContext.jsp。

程序 pageContext.jsp 代码如下：

```
<%@page language="java" import="java.util.*" pageEncoding="UTF-8"%>
<!DOCTYPE HTML PUBLIC "-//W3C//DTD HTML 4.01 Transitional//EN">
<html>
    <head>
        <title>EL 内置对象</title>
    </head>
    <body>
        请求方式:${pageContext.request.method}<br>
        响应的编码方式:${pageContext.response.characterEncoding}<br>
        session 的 ID:${pageContext.session.id}<br>
        上下文路径:${pageContext.servletContext.contextPath}<br>
    </body>
</html>
```

在 pageContext.jsp 中通过使用内置对象 pageContext 分别获取请求对象、响应对象、session 对象以及 servletContext 对象各自的属性。运行 pageContext.jsp 页面效果如图 8-7 所示。

图 8-7　pageContext 使用方式

2．作用范围相关的内置对象

与作用范围相关的内置对象共有 4 个，分别是：pageScope、requestScope、sessionScope、applicationScope。它们基本和 JSP 的内置对象 pageContext、request、session 和 application 一样，不过 EL 的这 4 个内置对象只能用来取得作用范围中的值而不能设置值。例如，在 request 中存储一个名称为 username 的对象，那么就可以使用 JSP 的内置对象 request.setAttribute("username");来存储，然后使用 EL 的内置对象 ${requestScope.username} 取出它的值。

例程 8-7：演示 pageScope、requestScope、sessionScope、applicationScope 的使用方法，程序为 scope.jsp。

程序 scope.jsp 代码如下：

```
<%@page language="java" import="java.util.*" pageEncoding="UTF-8"%>
```

```
<!DOCTYPE HTML PUBLIC "-//W3C//DTD HTML 4.01 Transitional//EN">
<html>
    <head>
        <title>EL 内置对象</title>
    </head>
    <body>
        <!--使用 JSP 内置对象设置属性范围的值-->
        <%
            pageContext.setAttribute("bookname","网络编程基础");
            request.setAttribute("bookname", "Java 核心技术");
            session.setAttribute("bookname", "数据库系统原理");
            application.setAttribute("bookname", "Oracle 数据库应用");
        %>
        <!--使用 EL 内置对象获取属性范围的值-->
        获取 page 属性范围的值:$ ={pageScope.bookname }<br>
        获取 request 属性范围的值:${requestScope.bookname }<br>
        获取 session 属性范围的值:${sessionScope.bookname }<br>
        获取 application 属性范围的值:${sessionScope.bookname }<br>
        获取未指定属性范围的值:${bookname }<br>
    </body>
</html>
```

程序 scope.jsp 首先通过 JSP 的内置对象把 bookname 的值分别设置到 4 种不同的作用范围中,然后通过 EL 的 4 种作用范围的内置对象分别将 bookname 中的值取出,最后取出未指定作用范围的 bookname 的值,可以看到未指定的作用范围默认是 page 作用范围。运行效果如图 8-8 所示。

图 8-8 4 种作用范围对象的使用方式

3. 与请求参数相关的内置对象

与请求参数相关的内置对象有两个,分别是 param 和 paramValues。param 获取传递过来的单值参数的值,相当于请求对象的 getParameter()方法的执行效果。paramValues 获取传递过来的多值参数的值,相当于请求对象的 getParameterValues()方法的执行效果。

例程 8-8：演示 param、paramValues 的使用方法。程序为 param.html 和 param.jsp。
程序 param.html 代码如下：

```
<!DOCTYPE html>
<html>
    <head>
        <title>param.html</title>
    </head>
    <body>
        <form action="param.jsp" method="post">
            用户名:<input type="text" name="username"><br>
            密码:<input type="password" name="userpass"><br>
            兴趣:
            <input type="checkbox" name="ins" value="sport">sport
            <input type="checkbox" name="ins" value="reading">reading
            <input type="checkbox" name="ins" value="programming">programming
            <input type="checkbox" name="ins" value="singing">singing <br>
                <input type="submit" value="提交">
                <input type="reset" value="取消">
        </form>
    </body>
</html>
```

程序 getparam.jsp 代码如下：

```
<%@page language="java" import="java.util.*" pageEncoding="UTF-8"%>
<!DOCTYPE HTML PUBLIC "-//W3C//DTD HTML 4.01 Transitional//EN">
<html>
    <head>
        <title>EL 内置对象</title>
    </head>
    <body>
        <!--处理中文参数乱码-->
        <%request.setCharacterEncoding("UTF-8"); %>
        <!--使用 param 获取请求参数-->
        用户名:${param.username }<br>
        密码:${param.userpass}<br>
        <!--使用 paramValues 获取请求参数值-->
        兴趣: ${paramValues.ins[0] }
             ${paramValues.ins[1] }
             ${paramValues.ins[2] }
             ${paramValues.ins[3] }
    </body>
</html>
```

程序 param.html 编写了一个表单，包含两个单值的组件和一个多值的组件，单击

"提交"按钮程序跳转到 getparam.jsp。在 getparam.jsp 中通过 param 内置对象分别获取用户名和密码的值,然后通过 paramValues 内置对象获取兴趣的值。先运行 param.html 程序,在页面中输入用户名、密码并且选择兴趣,效果如图 8-9 所示;之后单击"提交"按钮跳转到 getparam.jsp 中显示表单传递过来的参数值,效果如图 8-10 所示。

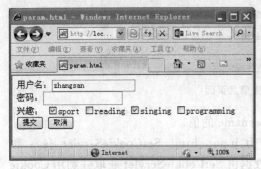
图 8-9 表单页面

图 8-10 获取请求参数页面

4. 与请求头相关的内置对象

与请求头相关的内置对象包括 header 和 headerValues。herder 用来访问 HTTP 请求的一个具体的 Header 值,headerValues 用来访问所有 HTTP 请求的 Header 值。如果头中包含"-",则访问时要使用中括号运算符而不能使用点号运算符。

例程 8-9:演示 header、headerValues 的使用方法。程序为 header.jsp。

程序 header.jsp 代码如下:

```
<%@page language="java" import="java.util.*" pageEncoding="UTF-8"%>
<!DOCTYPE HTML PUBLIC "-//W3C//DTD HTML 4.01 Transitional//EN">
<html>
    <head>
        <title>EL 内置对象</title>
    </head>
    <body>
        主机名:${header.host }<br>
        用户浏览器信息:${header["User-Agent"]}<br>
        浏览器可接受的 MIME 类型:${header.Accept }
    </body>
</html>
```

程序 header.jsp 使用 EL 内置对象 header 获取标头信息,例如,${header.host}是获取标头中主机信息,它相当于 request.getHeader("host");的作用。执行 header.jsp 效果如图 8-11 所示。

5. cookie 内置对象

cookie 相当于 HttpServletRequest.getCookies() 的作用。表达式 ${cookie.

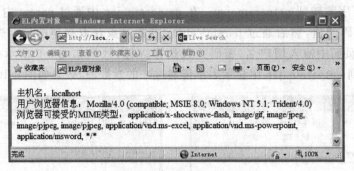

图 8-11 获取头信息页面

username.value}表示返回 cookie 中名称为 username 的值。

例程 8-10：演示 cookie 的使用。程序为 input.html、SetCookieServlet.java 和 getCookie.jsp。例程中 input.html 提交表单数据由 SetCookieServlet 获取并利用 Cookie 保存到客户端，getCookie.jsp 读取 Cookie 数据并显示。

程序 input.html 代码如下：

```html
<!DOCTYPE html>
<html>
    <head>
        <title>input.html</title>
        <meta http-equiv="keywords" content="keyword1,keyword2,keyword3">
        <meta http-equiv="description" content="this is my page">
        <meta http-equiv="content-type" content="text/html; charset=UTF-8">
    </head>
    <body>
        <form action="setCookieServlet" method="post">
        <table>
            <tr>
                <td align="right">用户名：</td>
                <td ><input type="text" name="username"></td>
            </tr>
            <tr>
                <td align="center" colspan="2"><input type="submit" value="提交">
                <input type="reset" value="重置"></td>
            </tr>
        </table>
        </form>
    </body>
</html>
```

程序 SetCookieServlet.java 代码如下：

```java
package com.ch08;
import java.io.IOException;
```

```java
import java.io.PrintWriter;
import java.util.Date;
import javax.servlet.ServletException;
import javax.servlet.http.Cookie;
import javax.servlet.http.HttpServlet;
import javax.servlet.http.HttpServletRequest;
import javax.servlet.http.HttpServletResponse;
public class SetCookieServlet extends HttpServlet {
    public void doGet(HttpServletRequest request, HttpServletResponse response)
        throws ServletException, IOException {
        this.doPost(request, response);
    }
    public void doPost(HttpServletRequest request, HttpServletResponse response)
        throws ServletException, IOException {
        response.setContentType("text/html;charset=UTF-8");
        String username=request.getParameter("username");
        PrintWriter out=response.getWriter();
        String responseContent=null;
        if(username!=null&&!username.equals("")){
            //创建 Cookie 对象
            Cookie c1=new Cookie("username",username);
            Date date=new Date();
            Cookie c2=new Cookie("lastCall",date.toString());
            //设置 Cookie 的存活时间
            c1.setMaxAge(60*60*24*30);
            c2.setMaxAge(60*60*24*30);
            //将 Cookie 发送到客户端保存
            response.addCookie(c1);
            response.addCookie(c2);
            responseContent="本次登录的用户名和时间已成功写入 Cookie.<br><a href
            ='/chapter8/getCookie.jsp'>读取 Cookie 信息</a>";
        }else{
            responseContent="用户名为空,请重新输入.<br><a href='/chapter8/
            input.html'>重新输入</a>";
        }
        out.println ( " <!DOCTYPE HTML PUBLIC \" -//W3C//DTD HTML 4.01
        Transitional//EN\">");
        out.println("<HTML>");
        out.println("<HEAD><TITLE>A Servlet</TITLE></HEAD>");
        out.println("<BODY>");
        out.println("<h2><font color=red>"+responseContent+"</font></h2>");
        out.println("    </BODY>");
        out.println("</HTML>");
```

```
            out.flush();
            out.close();
        }
    }
```

SetCookieServlet 的配置信息如下：

```xml
<servlet>
    <servlet-name>setCookie</servlet-name>
    <servlet-class>com.ch08.SetCookieServlet</servlet-class>
</servlet>
<servlet-mapping>
    <servlet-name>setCookie</servlet-name>
    <url-pattern>/setCookieServlet</url-pattern>
</servlet-mapping>
```

程序 getCookie.jsp 代码如下：

```
<%@page language="java" import="java.util.*" pageEncoding="UTF-8"%>
<!DOCTYPE HTML PUBLIC "-//W3C//DTD HTML 4.01 Transitional//EN">
<html>
    <head>
        <title>My JSP 'getCookie.jsp' starting page</title>
    </head>
    <body>
        <h2>从 Cookie 中读取的用户名和上次登录时间</h2>
        用户名：${cookie.username.value }<br>
        上次登录时间：${cookie.lastCall.value }<br>
    </body>
</html>
```

该例程运行过程与例程 3-1 运行过程及运行效果完全一致。程序 SetCookieServlet.java 类从表单中读取客户端提交的用户名信息，如果用户名为空则向客户端浏览器返回一个带有重新输入超链接的信息页面，让用户重新输入用户名。用户名不为空则创建 Cookie 对象将用户名信息和当前时间保存，同时将 Cookie 对象发送到客户端保存，并向客户端浏览器返回一个带有读取 Cookie 信息（调用 GetGookie.jsp）超链接的信息页面。单击读取 Cookie 信息超链接调用 GetGookie.jsp，GetGookie.jsp 使用 EL 内置对象 cookie 实现了从客户端读取 Cookie 信息功能，并将读取的 Cookie 信息显示在浏览器中。

6. 与初始化参数相关的内置对象

initParam 内置对象用于获取 web.xml 文件中配置的初始化参数，它相当于 ServletContext.getInitParameter(String name) 方法的作用。

例程 8-11：演示 initParam 的使用方法。程序为 initParam.jsp。

首先在 web.xml 文件中配置初始化参数，代码如下：

```
<context-param>
    <param-name>username</param-name>
    <param-value>system</param-value>
</context-param>
<context-param>
    <param-name>password</param-name>
    <param-value>tiger</param-value>
</context-param>
```

程序 initParam.jsp 代码如下：

```
<%@page language="java" import="java.util.*" pageEncoding="UTF-8"%>
<!DOCTYPE HTML PUBLIC "-//W3C//DTD HTML 4.01 Transitional//EN">
<html>
    <head>
        <title>EL内置对象</title>
    </head>
    <body>
        用户名：${initParam.username}<br>
        密　码：${initParam.password}<br>
    </body>
</html>
```

首先在 web.xml 文件中配置了两组初始化参数 username 和 password，在 initParam.jsp 中使用 EL 内置对象 initParam 获取 username 和 password 对应的参数值，运行效果如图 8-12 所示。

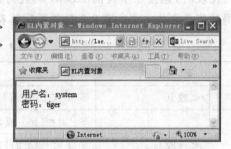

图 8-12　获取初始化配置参数页面

8.4　EL 运算符

EL 作为数据访问语言，它提供了一套自己的运算符，包括算术运算符、关系运算符、逻辑运算符以及 empty 运算符。这些运算符与 Java 语言中的运算符类似，但仍有不同。

8.4.1　算术运算符

算术运算符有加(+)、减(-)、乘(*)、除(/或 div)以及求模(%或 mod)，表 8-4 是算术运算符使用的例子。

表 8-4　算术运算符范例

运算符	表达式	运算结果
+	${2+3}	5
-	${-4-2}	-6
*	${10*2}	20

续表

运算符	表达式	运算结果
/ 或 div	${3/4}或 ${3 div 4}	0.75
/ 或 div	${3/0}或 ${3 div 0}	Infinity
% 或 mod	${10 % 4}或 ${10 mod 4}	2

可以看到,EL 中除法运算符对于除数为 0 的情况也进行了处理,以 Infinity 作为结果而不是抛出异常。

8.4.2 关系运算符

关系运算符有小于(<或 lt)、小于等于(<=或 le)、大于(>或 gt)、大于等于(>=或 ge)、等于(==或 eq)以及不等于(!=或 ne)。关系运算符不仅可以对数字进行比较而且还可以对字符或字符串进行比较,比较结果是布尔类型的值。其中对字符或字符串进行比较时,是按照字符的 ASCII 码进行比较。而==、eq 与!=、ne 也可以用来判断取得的值是否为 null。表 8-5 是关系运算符使用的一些例子。

表 8-5 关系运算符范例

运算符	表达式	运算结果
<或 lt	${1 < 2} 或 ${1 lt 2}	true
<或 lt	${'a'<'b'} 或 ${'a' lt 'b'}	true
<=或 le	${1 <=(4/2)} 或 ${1 le (4/2)}	true
>或 gt	${3.0>4} 或 ${3.0 gt 4}	false
>或 gt	${"hip">"hit"} 或 ${"hip" gt "hit"}	false
>=或 ge	${4>=3} 或 ${4 ge 3}	true
==或 eq	${10 * 10==100} 或 ${10 * 10 eq 100}	true
!=或 ne	${10 * 10!=100} 或 ${10 * 10 eq 100}	false

8.4.3 逻辑运算符

逻辑运算符有逻辑与(&& 或 and)、逻辑或(||或 or)、逻辑非(!或 not)。表 8-6 是逻辑运算符使用的例子。

8.4.4 条件运算符

条件运算符?:是一个三目运算符,使用方式如:exp1? exp2：exp3,其中 exp1 是一个 boolean 型的表达式,如果 exp1 的值为 true,则计算 exp2 的值并且整个条件表达式的值就是 exp2 的值;如果 exp1 的值为 false,则计算 exp3 的值并且整个条件表达式的值就是 exp3 的值。

表 8-6 逻辑运算符范例

运 算 符	表 达 式	运算结果
&& 或 and	${(9.3>=4) && 10 < 63}或 ${(9.3>=4) and 10<63}	true
|| 或 or	${(9.3>=4)||10< 63}或 ${(9.3>=4) or 10<63}	true
! 或 not	${! 4>=9.2}	true

8.4.5 empty 运算符

empty 运算符使用格式为：${empty expression}，它用来判断 expression 的值是否是 null、空字符串、空数组、空 Map 或空集合，若为空则返回 true，否则返回 false。例如，在 JSP 页面中有如下代码段：

```
<%
    pageContext.setAttribute("hello", "Hello World");
%>
${empty hello}
```

则页面运行输出结果是 false，因为判断的 hello 对象的值是"Hello World"，并不为空。

8.5 在页面中禁止使用 EL 表达式

在 JSP 页面中可以禁用 EL 表达式，即将 JSP 页面中形如 ${expr} 的表达式作为模版文本而不作为 EL 处理。

1. 在当前页面中禁止使用 EL 表达式

如果只需要在当前页面中禁用 EL 表达式，则可以通过 JSP 的 page 指令的 isELIgnored 属性设置，格式如下：

```
<%@page isELIgonred=''{true | false}''%>
```

如果属性 isELIgonred 的值为 true，表示禁用 EL，值为 false 表示允许使用 EL，默认值为 false。因此，如果要在某个 JSP 页面中禁用 EL，可以加上下面一行：

```
<%@page isELIgonred=''{true }''%>
```

例程 8-12：演示在当前页面中禁止使用 EL 表达式。程序为 forbidEL1.jsp。

程序 forbidEL1.jsp 代码如下：

```
<%@page language="java" import="java.util.*" pageEncoding="UTF-8"%>
<%@page isELIgnored="true" %>
<!DOCTYPE HTML PUBLIC "-//W3C//DTD HTML 4.01 Transitional//EN">
```

```
<html>
    <head>
        <title>当前页面禁用 EL</title>
    </head>
    <body>
        <%
            pageContext.setAttribute("hello", "Hello World");
        %>
        ${hello}
    </body>
</html>
```

程序 forbidEL1.jsp 中使用 page 指令＜%@ page isELIgnored="true" %＞设置当前页面禁止使用 EL，所以页面中出现的 EL 表达式就不会再起作用而是以文本的形式显示在页面中。运行 forbidEL1.jsp 效果如图 8-13 所示。

图 8-13　当前页面禁用 EL 效果

2. 在多个页面禁用 EL 表达式

在 WEB-INF/web.xml 中配置 jsp-property-ground 元素批量禁用 EL，通过配置可以在多个 JSP 页面中禁止使用 EL。配置代码如下：

```
<? xml version="1.0" encoding="UTF-8"? >
<web-app version="2.5"
    xmlns="http://java.sun.com/xml/ns/javaee"
    xmlns:xsi="http://www.w3.org/2001/XMLSchema-instance"
    xsi:schemaLocation="http://java.sun.com/xml/ns/javaee
    http://java.sun.com/xml/ns/javaee/web-app_2_5.xsd">
    <jsp-config>
        <jsp-property-group>
            <url-pattern>/files/*.jsp</url-pattern>
            <el-ignored>true</el-ignored>
        </jsp-property-group>
    </jsp-config>
</web-app>
```

通过上述配置就会禁止/files 目录中所有 JSP 页面中使用 EL 表达式。

3. 禁止在当前 Web 应用的所有 JSP 页面中使用 EL 表达式

如果要禁止在当前 Web 应用的所有 JSP 页面中使用 EL 表达式，则可以将 web.xml 定义成 2.3 版本。如下所示：

```
<?xml version="1.0" encoding="UTF-8"?>
<!DOCTYPE web-app
    PUBLIC "-//Sun Microsystems, Inc.//DTD Web Application 2.3//EN"
    "http://java.sun.com/dtd/web-app_2_3.dtd">
<web-app>
    ...
</web-app>
```

web.xml 定义了 web-app 的版本号是 2.3，这样所有的 JSP 都无法使用 EL 表达式了，因为 EL 表达式是 2.4 版才开始支持的功能。

8.6 案 例

本章案例在第 7 章案例的基础上增加实现"基本信息管理"模块中的"学生信息维护"子模块功能。在功能实现过程中编写 JSP 页面时主要使用 EL 表达式来减少 JSP 页面中的脚本（＜％ ％＞）或表达式（＜％＝％＞）代码。

8.6.1 案例设计

学生信息维护模块要实现学生信息的增删改查等操作，使用 MVC 模式将页面显示和业务逻辑处理进行分离。为了减少 JSP 页面中的 Java 脚本代码，在添加学生信息、显示学生信息和查看/修改学生信息的页面中使用 EL 表达式进行数据访问。

由于班级表和学生表具有关联关系，因此在删除班级表中的记录时要判断学生表中是否存在关联记录，如果不存在则可以删除，否则禁止删除。本章需要修改删除班级的 Servlet。

本章案例使用的主要文件如表 8-7 所示。

表 8-7 本章案例使用的文件

文 件	所在包/路径	功 能
Studeng.java	com.imut.javabean	封装学生信息的 JavaBean
IStudent.java	com.imut.javabean	定义操作数据库表增删改查操作方法的接口
StudentImpl.java	com.imut.javabean	具体实现接口定义的增删改查等方法
AddStudentServlet	com.imut.servlet.base	增加学生信息的 Servlet

续表

文件	所在包/路径	功能
CheckStudentServlet	com.imut.servlet.base	查询学生信息的 Servlet
DelStudentServlet	com.imut.servlet.base	删除学生信息的 Servlet
ListAllStudentServlet	com.imut.servlet.base	显示所有学生信息的 Servlet
ShowStudentServlet	com.imut.servlet.base	查看/修改学生信息的 Servlet
UpdateStudentServlet	com.imut.servlet.base	更新学生信息的 Servlet
addStudent.jsp	/base	增加学生信息的页面
studentList.jsp	/base	显示所有学生信息的页面
studentShow.jsp	/base	查看/修改学生信息的页面

8.6.2 案例演示

在浏览器地址栏中输入"http://localhost:8080/ch08/login.jsp",以管理员用户登录系统,单击"基本数据管理"菜单中的"学生信息维护",显示学生信息列表页面,效果如图 8-14 所示。单击"添加学生"超链接,出现添加学生信息页面,效果如图 8-15 所示。输入学生信息,单击"提交"按钮,提示添加学生信息成功之后跳转到学生信息列表页面。在学生信息列表页面中输入查询条件,查询结果将显示在"查询"按钮下方的学生信息列表

图 8-14 学生信息列表页面

处，如果没有输入查询条件则查询所有学生信息。根据"姓名"查询结果的效果图如图 8-16 所示。在学生信息列表页面中单击"删除"超链接，将会删除该学生信息，删除完成后显示学生列表页面。在学生列表页面中单击"查看/修改"超链接，或者单击学号，将显示学生信息查看/修改页面，效果如图 8-17 所示。修改学生信息完成后单击"提交"按钮，提示修改成功后显示学生信息列表页面。

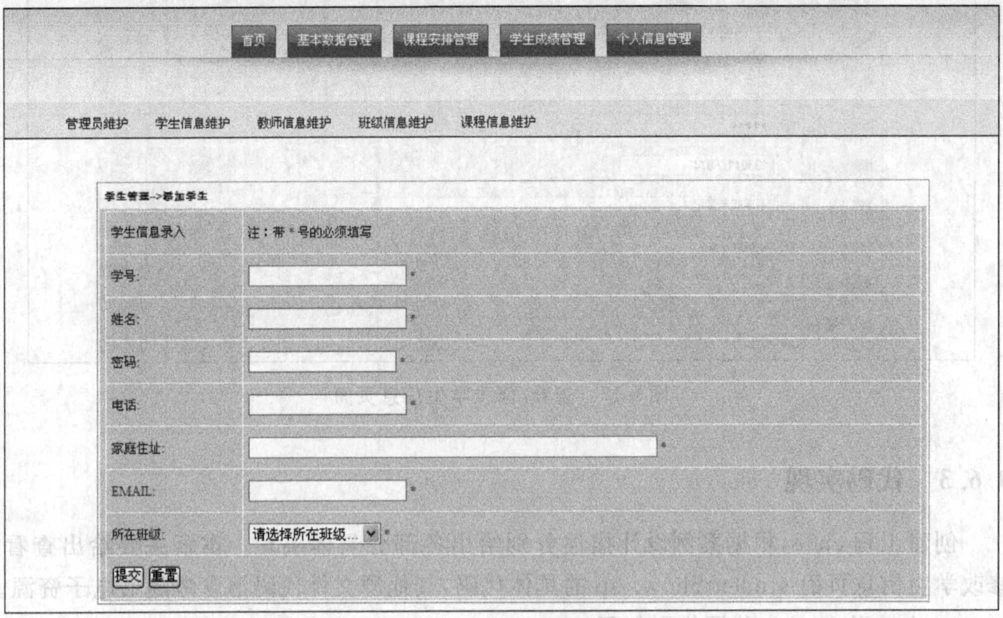

图 8-15 添加学生信息页面

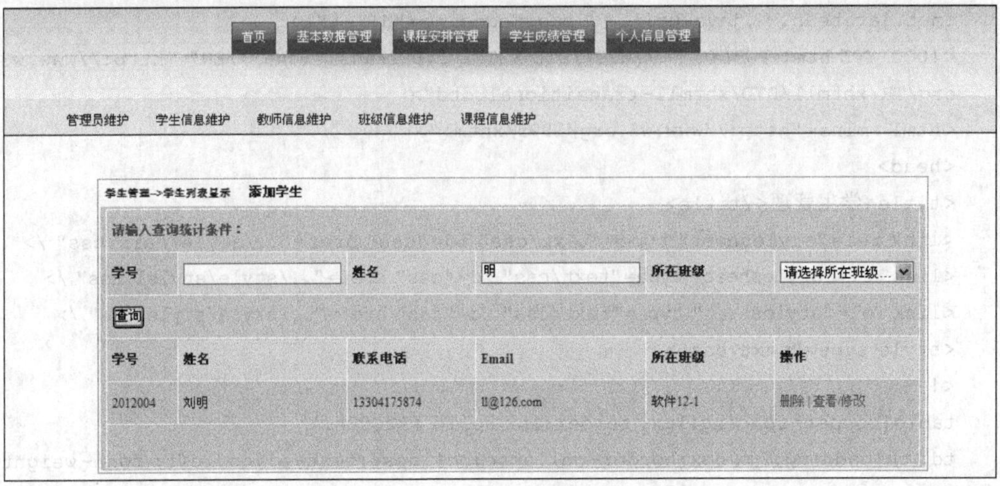

图 8-16 查询学生信息页面

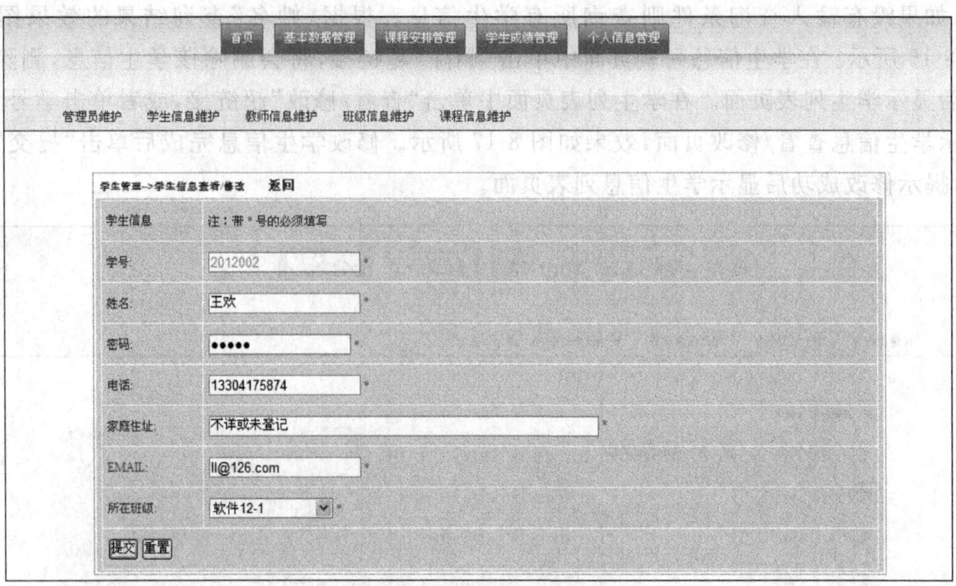

图 8-17 查看/修改学生信息页面

8.6.3 代码实现

创建工程 ch08，根据案例设计描述分别给出各部分具体实现。本章主要给出查看/修改学生信息页面 studentShow.jsp 的具体代码，其他源文件代码请参考随书电子资源。

studentShow.jsp 的源代码如下：

```
<%@page language="java" contentType="text/html; charset=UTF-8" import="com.imut.javabean.*,java.util.*" pageEncoding="UTF-8"%>
<!DOCTYPE html PUBLIC "-//W3C//DTD XHTML 1.0 Transitional//EN" "http://www.w3.org/TR/xhtml1/DTD/xhtml1-transitional.dtd">
<html xmlns="http://www.w3.org/1999/xhtml">
<head>
<title>学生管理</title>
<link rel="stylesheet" type="text/css" id="css" href="../style/main.css" />
<link rel="stylesheet" type="text/css" id="css" href="../style/style1.css" />
<link rel="stylesheet" type="text/css" id="css" href="../style/style.css" />
<style type="text/css">
<!--
table{border-spacing:1px; border:1px solid #A2C0DA;}
td, th{padding:2px 5px;border-collapse:collapse;text-align:left; font-weight:normal;}
thead tr th{height:50px;background:#B0D1FC;border:1px solid white;}
thead tr th.line1{background:#D3E5FD;}
thead tr th.line4{background:#C6C6C6;}
tbody tr td{height:35px;background:#CBE2FB;border:1px solid white; vertical-
```

```
align:middle;}
tbody tr td.line4{background:#D5D6D8;}
tbody tr th{height:35px;background: #DFEDFF;border:1px solid white; vertical-
align:middle;}
tfoot tr td{height:35px;background:#FFFFFF;border:1px solid white; vertical-
align:middle;}
-->
</style>
<script src="js/common.js" type="text/javascript"></script>
<script type="text/javascript" src="../js/common.js" ></script>
<script type="text/javascript" src="../js/calendar.js" ></script>
<script type="text/JavaScript">
    /*判断是否为数字*/
    function isNumber(str) {
        var Letters="1234567890";
        for (var i=0; i<str.length; i=i+1) {
            var CheckChar=str.charAt(i);
            if (Letters.indexOf(CheckChar)==-1) {
                return false;
            }
        }
        return true;
    }
    /*判断是否为Email*/
    function isEmail(str) {
        var myReg=/^[-_A-Za-z0-9]+@([_A-Za-z0-9]+\.)+[A-Za-z0-9]{2,3}$/;
        if (myReg.test(str)) {
            return true;
        }
        return false;
    }
    /*判断是否为空*/
    function isEmpty(value) {
        return /^\s*$/.test(value);
    }
    function check(){
        if(isEmpty(document.myForm.name.value)){
            alert("学生姓名不能为空!");
            document.myForm.name.focus();
            return false;
        }
        if(isEmpty(document.myForm.password.value)){
            alert("密码不能为空!");
            document.myForm.password.focus();
```

```
                return false;
            }
            if(isEmpty(document.myForm.phone.value)){
                alert("电话不能为空!");
                document.myForm.phone.focus();
                return false;
            }
            if(isEmpty(document.myForm.address.value)){
                alert("家庭住址不能为空!");
                document.myForm.address.focus();
                return false;
            }
            if(isEmpty(document.myForm.email.value)){
                alert("email 不能为空!");
                document.myForm.email.focus();
                return false;
            }
            if(!isEmail(document.myForm.email.value)){
                alert("email 格式不正确!");
                document.myForm.email.focus();
                return false;
            }
            if(document.myForm.classNo.value==""){
                alert("请选择学生所在班级!");
                document.myForm.classNo.focus();
                return false;
            }
            return true;
        }
</script>
</head>
<%
    //获得班级列表(在下拉框中使用)
    ClassDBAccess dbAccess=new ClassDBAccess();
    List classtbls=dbAccess.findAllClassTbl();
    request.setAttribute("classtbls",classtbls);
%>
<body>
<div id="btm">
<div id="main">
    <div id="header">
        <div id="top"></div>
        <div id="logo">
            <h1>学生管理</h1></div>
```

```html
                <div id="mainnav">
                    <ul>
                    <li><a href="../index.jsp">首页</a></li>
                    <li><a href="listAllTeacherServlet">基本数据管理</a></li>
                    <li><a href="#">课程安排管理</a></li>
                    <li><a href="#">学生成绩管理</a></li>
                    <li><a href="../person/showPersonServlet">个人信息管理</a></li>
                    </ul>
                    <span>
                    </span>
                </div>
            </div>
            <div id="tabsJ">
                <ul>
                    <li><a href="#" title="管理员维护"><span>管理员维护</span></a></li>
                    <li><a href="listAllStudentServlet" title="学生信息维护"><span>学生信息维护</span></a></li>
                    <li><a href="listAllTeacherServlet" title="教师信息维护"><span>教师信息维护</span></a></li>
                    <li><a href="listAllClassServlet" title="班级信息维护"><span>班级信息维护</span></a></li>
                    <li><a href="listAllCourseServlet" title="课程信息维护"><span>课程信息维护</span></a></li>
                </ul>
            </div>
                <div id="content" align="center">
                    <div id="center">
            <BR /><BR />
            <form action="updateStudentServlet" method="post" name="myForm" >
                <table width="800" align="center" cellpadding="0" cellspacing="0">
                    <thead>
                        <tr>
                            <td width="70%"><h5>学生管理-->学生信息查看/修改  
                                   < a href=" listAllStudentServlet">
                               返回</a></h5></td>
                        </tr>
                    </thead>
                    <tr>
                      <td colspan="2" width="100%">
                        <table width="100%">
                            <tbody>
                            <tr>
                                <td align="right">学生信息</td>
                                <td>注:带<font color="red"> * </font>号的必须填写</
```

```html
                                td>
                        </tr>
                        <tr>
                            <td align="right">学号:</td>
                            <td>< input disabled="disabled" name="studentNo"
                                value="${requestScope.student.studentNo}"/>< font
                                color="red">*</font></td>
                        </tr>
                        <tr>
                            <td align="right">姓名:</td>
                            <td><input id="name" name="name" value=
                                "${requestScope.student.name}"/><font color="red">*
                                </font></td>
                        </tr>
                        <tr>
                            <td align="right">密码:</td>
                            <td><input id="password" type="password" name=
                                "password" value="${requestScope.student.password}"/>
                                <font color="red">*</font></td>
                        </tr>
                        <tr>
                            <td align="right">电话:</td>
                            <td><input id="phone" name="phone" value=
                                "${requestScope.student.phone}" /><font color="red">*
                                </font></td>
                        </tr>
                        <tr>
                            <td align="right">家庭住址:</td>
                            <td><input id="address" name="address" value=
                                "${requestScope.student.address}" size=60/>
                                <font color="red">*</font></td>
                        </tr>
                        <tr>
                            <td align="right">EMAIL:</td>
                            <td><input id="email" name="email" value=
                                "${requestScope.student.email}" /><font color="red">*
                                </font></td>
                        </tr>
                        <tr>
                            <td align="right">所在班级:</td>
                            <td>
                                <select id="classNo" name="classNo">
                                    <option value="">请选择所在班级..</option>
                                    <%
```

```
                List < ClassTbl > list1 = (List) request.
                getAttribute("classtbls");
                Student student = (Student) request.
                getAttribute("student");
                String classNo = student. getClassTbl ( ).
                getClassNo();
                if(list1!=null){
                    Iterator< ClassTbl > it = list1. iterator
                    ();
                      while(it.hasNext()){
                          ClassTbl classtbl=it.next();
                            request. setAttribute ( "classtbl",
                            classtbl);
                            if(classtbl. getClassNo ( ).
                            equals(classNo)){
            %>
                                <option value=" ${classtbl.
                                classNo }" selected =
                                "selected"> ${ classtbl.
                                className }</option>
            <%
                          }else{
            %>
                                <option value=" ${classtbl.classNo
                                }"> ${classtbl.className }</option
                                >
                <%
                          }
                        }
                    }
                %>
                </select>
            <font color="red"> * </font></td>
        </tr>
    </tbody>
    <tr>
        <td align="center" colspan="2">
            <input type="hidden" name="studentNo" value=
            ${ requestScope. student. studentNo } style =
            "width:0;height:0" />
            <input type=" submit" value=" 提交" onclick=
            "return check();"/>
            <input type="reset" value="重置"/>
```

```
                </td>
              </tr>
            </table>
          </form>
        </div>
      </div>
      <div id="footer">
        <div id="copyright">
          <div id="copy">
            <p align="center">CopyRight&copy;2010</p>
            <p>内蒙古工业大学信息工程学院软件工程系</p>
          </div>
        </div>
        <div id="bgbottom"></div>
      </div>
    </div>
  </body>
</html>
```

【代码分析】程序 studentShow.jsp 首先使用脚本获取班级列表,然后作为学生信息中班级信息的下拉列表项使用。在 myForm 表单中使用 EL 表达式为每个组件的 value 属性赋值。在"所在班级"下拉列表中不仅要显示并选中当前学生的所在班级而且还能显示其他班级的信息,这样用户在修改班级信息时才能有其他班级选项。因此从 request 作用范围中分别获取 classtbls 列表对象和 student 对象。迭代访问 classtbls 列表对象,并将每一个迭代项赋值给 ClassTbl 对象 classtbl,同时把 classtbl 保存在 request 作用范围中。比较 classtbl 对象的班级编号和 student 对象的班级编号是否相同,如果相同则使用 EL 表达式将班级名称显示在下拉列表项中并选中,否则只将班级名称显示在下拉列表项中。在"提交"按钮前使用隐藏组件传递当前学生的学号,学号也是以 EL 表达式的形式提供的,这样页面在提交时可以将学号传递到下一个页面使用。

习　题

1. 选择题

(1) 关于 EL 表示式语言,以下哪个说法是错误的?(　　)

　　A. 它和 Java 一样,是一种编程语言

　　B. 它的基本形式为 ${var}

　　C. 只有在 JSP 文件中才能使用 EL 语言,在 Servlet 类的程序代码中不能使用它

D. 它能使JSP文件的代码更加简洁

(2) 表达式${56>12? 56:12}的值是(　　)。

　　A. 56　　　　　　B. 12　　　　　　C. true　　　　　　D. false

(3) 在Web应用程序中有以下的程序代码,执行后转发到JSP页面

```
Map<String,String>map=new HashMap<String,String>();
map.put("key1","string1");
map.put("key2","string2");
request.setAttribute("user",map);
```

以下选项中哪一项可以正确使用EL取得map中的值?(　　)

　　A. ${map.key1}　　　　　　B. ${map["key1"]}
　　C. ${user.key1}　　　　　　D. ${user[key1]}

(4) 在Web应用程序中有以下的程序代码,执行后转发至JSP页面:

```
List<String>namesarr=new ArrayList<String>();
namesarr.add("zhangsan");
request.setAttribute("names",namesarr);
```

以下选项中哪一项可以正确使用EL取得List中的值?(　　)

　　A. ${names.0}　　　　　　B. ${names[0]}
　　C. ${names.[0]}　　　　　　D. ${names["0"]}

(5) 在Web应用范围内存放了一个属性名为"myBean"的CounterBean对象,以下哪个选项不能实现输出myBean的count属性值?(　　)

　　A. ${applicationScope.myBean.count}
　　B. ${myBean.count}
　　C. <%=myBean.count>
　　D. <%CounterBean myBean=(CounterBean)application.getAttribute("myBean");%>
　　　　<%=myBean.getCount()%>

(6) 一个JSP文件中包含如下代码

```
<%int a=0;%>
a=${a}
```

通过浏览器访问这个JSP文件,会出现什么情况?(　　)

　　A. JSP文件输出"a="
　　B. JSP文件输出"a=0"
　　C. JSP文件输出"a=${a}"
　　D. Servlet容器方法编译错误,提示表达式${a}不合法

(7) 以下不是EL内置对象的是(　　)。

　　A. param　　　　B. request　　　　C. pageContext　　　　D. cookie

(8) 以下哪个选项是表达式${3+var}(其中var未定义)的正确输出?(　　)

A. 它将引发异常　　　　　　　　B. 它将给出编译错误
C. 3　　　　　　　　　　　　　D. 它将给出一些无用值

2. 填空题

(1) EL 表达式以_____开头,以_____结束。使用 JSP _____指令能够阻止解析 EL 表达式。

(2) 定义一个 request 范围内的变量 user,其值设为 null,通过 EL 语言中_____运算符判断 user 为空,具体代码为_____。

(3) 在 EL 表达式中可以使用运算符_____和_____来取得对象的属性。

3. 简答题

(1) 什么是 EL 表达式？请描述 EL 表达式的语法。

(2) EL 表达式中提供了哪几个内置对象？分别有什么作用？

(3) 请描述在页面中禁止使用 EL 的基本情况及对应方法。

4. 程序设计题

(1) 在客户端的表单中填写用户注册信息后,创建用户 Bean,在页面上单击"提交"按钮,应用 EL 表达式通过访问 JavaBean 属性的方法显示到页面上。

(2) 编写一个 JSP 程序,显示用户登录信息。要求用 EL 判断在没有输入用户名时,显示用户名为空,否则显示登录用户名;在没有输入密码时,显示密码为空,否则显示登录密码。

(3) 制作一个注册页面,注册项包括用户名、密码、E-mail、地址、性别(采用单选按钮)和业余爱好(采用复选框),提交注册后,用 EL 表达式展示注册信息。

JSP 标签

JSP 标签的引入可以有效地消除 JSP 页面中的 Java 代码,实现 Java 代码的重用。在 JSP 页面编程中,界面显示和程序代码通常是由美工和程序员共同完成的,如果在 JSP 页面中含有大量 Java 代码,一般情况下美工是不懂 Java 语言的,这样就会影响美工的界面设计工作。在大中型项目开发中使用 JSP 标签,有利于团队合作开发,方便后期维护和修改,使得项目更加规范和美观。本章将以理论与实践相结合的方式介绍标签接口与实现类、自定义标签开发、自定义标签应用实例、JSTL 标签库介绍以及常用 JSTL 标签的使用等内容。

9.1 自定义标签

9.1.1 自定义标签简介

为了能够重用 Java 代码,提高 JSP 文件的可维护性,JSP 1.2 规范开始支持在 JSP 文件中使用自定义标签。自定义标签是开发者通过自己开发的 JSP 标签来实现消除 JSP 页面中的 Java 代码。自定义标签具有控制标签体是否执行、控制标签体重复执行、修改标签体内容、控制 JSP 页面是否执行的功能。

9.1.2 标签接口和实现类

开发自定义标签涉及的接口和实现类如图 9-1 所示。

自定义标签的标签处理类必须要实现 Tag 接口,为了降低程序的复杂度,减少开发者的工作量,通常在编写标签处理类时去继承 Tag 接口的默认实现类或者继承 Tag 接口子接口的默认实现类,即继承 TagSupport 类或者 BodyTagSupport 类,在 JSP 2.0 出现之前通常采用上述这种做法。通过实现 Tag、IterationTag 或者 BodyTag 接口开发的标签通常称为传统标签。限于篇幅,有关传统标签的开发本书不再详述。

传统标签使用标签接口 Tag、IterationTag 和 BodyTag 来完成不同的功能,显得过于复杂,不利于标签技术的推广,Sun 公司为降低自定义标签技术的学习难度,在 JSP 2.0 中定义了 SimpleTag 接口,通过调用 SimpleTag 接口来实现标签的功能更为简单、便于编写。实现 SimpleTag 接口的标签通常称为简单标签。SimpleTag 接口有 5 个方法,具体描述如表 9-1 所示。

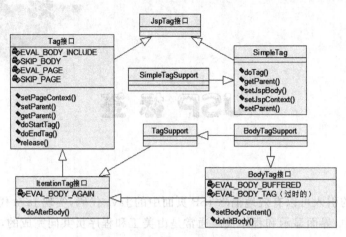

图 9-1 标签接口与实现类类图

表 9-1 SimpleTag 接口方法介绍

方法	说明
public void setJspContext(JspContext pc)	用于把 JSP 页面的 pageContext 对象传递给标签处理器对象
public void setParent(JspTag parent)	用于把父标签处理器对象传递给当前标签处理器对象
public JspTag getParent()	用于获得当前标签的父标签处理器对象
public void setJspBody（JspFragment jspBody）	用于把代表标签体的 JspFragment 对象传递给标签处理器对象
public void doTag()	用于完成所有的标签逻辑，包括输出、迭代、修改标签体内容等。在 doTag 方法中可以抛出 javax.servlet.jsp.SkipPageException 异常，用于通知 Web 服务器不再执行 JSP 页面中位于结束标记后面的内容，这等效于在传统标签的 doEndTag 方法中返回 Tag.SKIP_PAGE 常量的情况

实现 SimpleTag 接口的标签处理器对象方法的执行顺序如下：

（1）Web 服务器调用标签处理器对象的 setJspContext 方法，将代表 JSP 页面的 pageContext 对象传递给标签处理器对象；

（2）在标签具有父标签的情况下，Web 服务器调用标签处理器对象的 setParent 方法，将父标签处理器对象传递给这个标签处理器对象；

（3）如果调用标签时设置了属性，Web 服务器将调用每个属性对应的 setter 方法把属性值传递给标签处理器对象。如果标签的属性值是 EL 表达式或脚本表达式，则 Web 服务器首先计算表达式的值，然后把值传递给标签处理器对象；

（4）如果简单标签含有标签体，Web 服务器将调用 setJspBody 方法把代表标签体的 JspFragment 对象传递进来；

（5）Web 服务器调用标签处理器的 doTag() 方法，开发人员在方法体内通过操作 JspFragment 对象，可以实现是否执行标签体、迭代标签体、修改标签体的目的。

SimpleTag 接口的 setJspBody 方法中涉及 JspFragment 对象,这里对 JspFragment 类做一个简单介绍。javax.servlet.jsp.tagext.JspFragment 类是在 JSP 2.0 中定义的,它的实例对象代表 JSP 页面中的一段符合 JSP 语法规范的 JSP 片段,该片段中不能包含 JSP 脚本元素。

Web 服务器在处理简单标签的标签体时,会把标签体内容用一个 JspFragment 对象表示,并调用标签处理器对象的 setJspBody 方法把 JspFragment 对象传递给标签处理器对象。JspFragment 类中定义了两个方法,介绍如表 9-2 所示。

表 9-2 JspFragment 类的方法介绍

方　　法	说　　明
public abstract JspContext getJspContext()	用于返回代表调用页面的 JspContext 对象
abstract void invoke(java.io.Writer out)	用于执行 JspFragment 对象所代表的 JSP 代码片段。参数 out 用于指定将 JspFragment 对象的执行结果写入到哪个输出流对象中,如果传递给参数 out 的值为 null,则将执行结果写入到 JspContext.getOut() 方法返回的输出流对象中即输出到浏览器

JspFragment.invoke 方法是 JspFragment 最重要的方法,利用这个方法可以控制标签体是否执行、是否输出标签体的内容、是否迭代执行标签体的内容或对标签体的执行结果进行修改后再输出。

(1) 在标签处理器 doTag 方法中如果没有调用 JspFragment.invoke 方法,其结果就相当于忽略标签体内容;

(2) 在标签处理器 doTag 方法中抛出 javax.servlet.jsp.SkipPageException 异常,则不再执行 JSP 页面中位于结束标记后面的内容,这等效于在传统标签的 doEndTag 方法中返回 Tag.SKIP_PAGE 常量的情况;

(3) 在标签处理器 doTag 方法中重复调用 JspFragment.invoke 方法,则标签体内容将会被重复执行;

(4) 若在标签处理器 doTag 方法中修改标签体内容,只需在调用 invoke 方法时指定一个可取出结果数据的输出流对象(例如 StringWriter),让标签体的执行结果输出到该输出流对象中,然后从该输出流对象中取出数据进行修改后再输出到目标设备,即可达到修改标签体的目的。

9.1.3　自定义标签开发

自定义标签的开发包括编写标签处理器类、编写标签描述符文件和在 JSP 中使用标签三步,详细介绍如下。

(1) 编写实现了 SimpleTag 接口的标签处理类,将 JSP 页面中的 Java 代码移植到标签处理类中。在实际编程中,标签处理类通常继承 SimpleTag 接口的默认实现类 SimpleTagSupport,然后去覆盖该类中的 doTag() 方法,实现标签逻辑。

(2) 编写标签描述符(*.tld)文件,在 tld 文件中把标签处理器类描述成一个标签,并把 tld 文件放置在 WebRoot→WEB-INF 下。

任何一个自定义标签都必须在 tld 文件中进行声明,声明之后才能够在 JSP 文件中使用,tld 文件就是一个 XML 文件。tld 文件的根元素是<taglib>,其中包含一个或者多个<tag>标签,该元素用来描述标签处理器类即声明定制标签。<taglib>元素中只有<tlib-version>元素是必需的,其他都是可选的。tld 文件中常用元素介绍如表 9-3 所示。

表 9-3　tld 文件中的元素介绍

元　　素	说　　明
<taglib>	tld 文件的根元素,描述标签库
<tlib-version>	<taglib>的子元素。此标签库的版本,必须有的元素
<short-name>	<taglib>的子元素。当在 JSP 文件中使用标签时,此标签库首选或建议的前缀,可以忽略这个建议
<description>	<taglib>的子元素。此标签库描述信息
<uri>	<taglib>的子元素。指定使用该标签库中标签的 URI,在 JSP 文件中的 taglib 指令元素中 uri 指定
<tag>	<taglib>的子元素。描述一个自定义标签
<name>	<tag>的子元素。标签名称
<tag-class>	<tag>的子元素。Java 标签处理器类的名称。注意这是处理器类的全限定名称,比如 com.xx.tag.TableTag
<body-content>	<tag>的子元素。此标签的主体部分的内容。其值可为 scriptless\tagdependent\empty,默认为 empty。empty:标签主体为空。scriptless:标签主体不为空,并且包含 JSP 的 EL 表达式和动作元素,但不能包含 JSP 的脚本元素。所谓动作元素是指<jsp:include>和<jsp:forward>等以"jsp"为前缀的 JSP 内置标签。所谓脚本元素是指"<%!"和"%>"、"<%"和"%>"和"<%="和"%>"这三种以"<%"开头的 JSP 标记。tagdependent:标签主体不为空,并且标签主体内容由标签处理类来解析和处理。标签主体的所有代码都会原封不动地传给标签处理类,而不是把标签主体的执行结果传给标签处理类。在 JSP 2.0 之前的传统标签开发中,<body-content>还可以有一个可选值 JSP,JSP 代表标签主体不为空,并且包含 JSP 代码。在 JSP 代码中可以包含 EL 表达式、动作元素和脚本元素。<body-content>子元素的 scriptless 可选值与 JSP 可选值的区别在于前者不能包含 JSP 的脚本元素。但是在 JSP2.0 之后,不建议使用 JSP 脚本元素,所以简单标签开发中的<body-content>可选值中不包含 JSP
<attribute>	<tag>的子元素。描述此标签的一个属性
<name>	在 JSP 标签中使用的属性名称
<required>	指定属性是必需的还是可选的,默认为 false,表示属性可选。如果该值为 true,则 JSP 页面必须为该属性提供一个值。可选值为 true、false、yes 或 no
<rtexprvalue>	表示是否可以接受 Java 表达式和 EL 表达式的值。可选值为 true、false、yes 或 no

(3) 在 JSP 页面中通过 taglib 页面指令元素引入标签,然后通过 taglib 指令指定的前缀名和 tld 文件指定的标签名在 JSP 文件中使用标签。

9.1.4 自定义标签应用举例

例程 9-1：演示自定义标签实现控制标签体是否执行。程序为 MySimpleTag1.java 和 myTag1.jsp。

MySimpleTag1 为标签处理器类，其程序代码如下：

```java
package com.ch09;
import java.io.IOException;
import javax.servlet.jsp.JspException;
import javax.servlet.jsp.tagext.JspFragment;
import javax.servlet.jsp.tagext.SimpleTagSupport;
public class MySimpleTag1 extends SimpleTagSupport {
    public void doTag() throws JspException, IOException {
        //获得标签体
        JspFragment jf=this.getJspBody();
        //执行标签体
        jf.invoke(this.getJspContext().getOut());   //也可以使用jf.invoke(null)
    }
}
```

标签处理器类 MySimpleTag1 直接继承了 SimpleTag 接口的默认实现类 SimpleTagSupport，覆盖了 SimpleTagSupport 类的 doTag 方法即编写实现标签功能的代码。在 doTag 方法中首先使用 SimpleTagSupport 类的 getJspBody()方法获得标签体，然后通过 JspFragment 类的 invoke 方法执行标签体。

在 tld 文件中将标签处理器类 MySimpleTag1 描述成一个标签，代码如下：

```xml
<short-name>imut</short-name>
<uri>http://www.imut.edu.cn</uri>
<tag>
    <name>myTag1</name>
    <tag-class>com.ch09.MySimpleTag1</tag-class>
    <body-content>scriptless</body-content>
</tag>
```

上述 tld 文件代码，指定 uri 为"http://www.imut.edu.cn"，标签名称为"myTag1"，标签处理类为"com.ch09.MySimpleTag1"。

JSP 页面 myTag1.jsp 使用自定义标签 myTag1，程序代码如下：

```jsp
<%@page language="java" import="java.util.*" pageEncoding="UTF-8"%>
<%@taglib uri="http://www.imut.edu.cn" prefix="imut"%>
<!DOCTYPE HTML PUBLIC "-//W3C//DTD HTML 4.01 Transitional//EN">
<html>
    <head>
        <title>控制标签体是否执行</title>
    </head>
```

```
        <body>
            <imut:myTag1>
                This is my JSP page.
            </imut:myTag1>
        </body>
</html>
```

页面 myTag1.jsp 中首先通过"<%@ taglib uri="http://www.imut.edu.cn" prefix="imut"%>"将标签库引入 JSP 文件并指定前缀为"imut",然后在页面中使用标签<imut:myTag1>。

通过浏览器访问 myTag1.jsp 页面,运行效果如图 9-2 所示。这是由于在标签处理器类 MySimpleTag1 的 doTag 方法中通过 JspFragment 类的 invoke 方法执行了标签体。如果在 doTag 方法中不去调用 invoke 方法,则将不会执行标签体,运行结果如图 9-3 所示为空白页面。

图 9-2　执行标签体页面运行效果　　图 9-3　不执行标签体页面运行效果

例程 9-2:演示自定义标签实现控制 JSP 页面是否执行。程序为 MySimpleTag2.java 和 myTag2.jsp。

MySimpleTag2 为标签处理器类,其程序代码如下:

```
package com.ch09;
import java.io.IOException;
import javax.servlet.jsp.JspException;
import javax.servlet.jsp.SkipPageException;
import javax.servlet.jsp.tagext.SimpleTagSupport;
public class MySimpleTag2 extends SimpleTagSupport {
    public void doTag() throws JspException, IOException {
        throw new SkipPageException();
    }
}
```

标签处理器类 MySimpleTag2 直接继承了 SimpleTag 接口的默认实现类 SimpleTagSupport,覆盖了 SimpleTagSupport 类的 doTag 方法即编写实现标签功能的代码。在 doTag 方法中抛出了 SkipPageException 异常,则不再执行 JSP 页面中位于结束标记后面的内容。

在 tld 文件中将标签处理器类 MySimpleTag2 描述成一个标签,代码如下:

```xml
    <short-name>imut</short-name>
    <uri>http://www.imut.edu.cn</uri>
    <tag>
        <name>myTag2</name>
        <tag-class>com.ch09.MySimpleTag2</tag-class>
        <body-content>scriptless</body-content>
    </tag>
```

上述 tld 文件代码,指定 uri 为"http://www.imut.edu.cn",标签名称为"myTag2",标签处理类为"com.ch09.MySimpleTag2"。

JSP 页面 myTag2.jsp 使用自定义标签 myTag2,程序代码如下:

```jsp
<%@page language="java" import="java.util.*" pageEncoding="UTF-8"%>
<%@taglib uri="http://www.imut.edu.cn" prefix="imut"%>
<imut:myTag2/>
<!DOCTYPE HTML PUBLIC "-//W3C//DTD HTML 4.01 Transitional//EN">
<html>
    <head>
        <title>控制 JSP 页面是否执行</title>
    </head>
    <body>
        This is my JSP page.
    </body>
</html>
```

页面 myTag2.jsp 中首先通过 taglib 指令元素将标签库引入 JSP 文件并指定前缀为"imut",然后在页面中使用标签<imut:myTag2/>。

通过浏览器访问 myTag2.jsp 页面,运行效果如图 9-4 所示。这是由于在标签处理器类

图 9-4 不执行 JSP 页面代码运行效果

MySimpleTag2 的 doTag 方法中抛出了 SkipPageException 异常,标签<imut:myTag2/>之后的页面代码将不再执行,所以显示为空白页面。

例程 9-3:演示自定义标签实现控制标签体重复执行。程序为 MySimpleTag3.java 和 myTag3.jsp。

MySimpleTag3 为标签处理器类,其程序代码如下:

```java
package com.ch09;
import java.io.IOException;
import javax.servlet.jsp.JspException;
import javax.servlet.jsp.tagext.JspFragment;
import javax.servlet.jsp.tagext.SimpleTagSupport;
public class MySimpleTag3 extends SimpleTagSupport {
    public void doTag() throws JspException, IOException {
        JspFragment jf=this.getJspBody();
```

```
            for(int i=0;i<5;i++){
                jf.invoke(null);
            }
        }
    }
```

标签处理器类 MySimpleTag3 直接继承了 SimpleTag 接口的默认实现类 SimpleTagSupport，覆盖了 SimpleTagSupport 类的 doTag 方法即编写实现标签功能的代码。在 doTag 方法获得标签体之后，循环执行标签体。

在 tld 文件中将标签处理器类 MySimpleTag3 描述成一个标签，代码如下：

```
<short-name>imut</short-name>
<uri>http://www.imut.edu.cn</uri>
<tag>
    <name>myTag3</name>
    <tag-class>com.ch09.MySimpleTag3</tag-class>
    <body-content>scriptless</body-content>
</tag>
```

上述 tld 文件代码，指定 uri 为"http://www.imut.edu.cn"，标签名称为"myTag3"，标签处理类为"com.ch09.MySimpleTag3"。

JSP 页面 myTag3.jsp 使用自定义标签 myTag3，程序代码如下：

```
<%@page language="java" import="java.util.*" pageEncoding="UTF-8"%>
<%@taglib uri="http://www.imut.edu.cn" prefix="imut"%>
<!DOCTYPE HTML PUBLIC "-//W3C//DTD HTML 4.01 Transitional//EN">
<html>
    <head>
        <title>控制标签体重复执行</title>
    </head>
    <body>
        <imut:myTag3>
            This is my JSP page.<br>
        </imut:myTag3>
    </body>
</html>
```

页面 myTag3.jsp 中首先通过 taglib 指令元素将标签库引入 JSP 文件并指定前缀为 "imut"，然后在页面中使用标签<imut:myTag3>，标签体内为文本"This is my JSP page."。

通过浏览器访问 myTag3.jsp 页面，运行效果如图 9-5 所示。标签<imut:myTag3>将标签体文本循环执行了 5 次。

例程 9-4：演示自定义标签实现修改标签体内容。程序为 MySimpleTag4.java 和 myTag4.jsp。

MySimpleTag4 为标签处理器类，其程序代码如下：

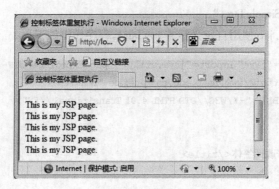

图 9-5 循环执行标签体运行效果

```
package com.ch09;
import java.io.IOException;
import java.io.StringWriter;
import javax.servlet.jsp.JspException;
import javax.servlet.jsp.tagext.JspFragment;
import javax.servlet.jsp.tagext.SimpleTagSupport;
public class MySimpleTag4 extends SimpleTagSupport {
    public void doTag() throws JspException, IOException {
        JspFragment jf=this.getJspBody();
        StringWriter sw=new StringWriter();
        jf.invoke(sw);                              //将标签内容写到 sw
        String str=sw.toString();
        str=str.toUpperCase();                      //将字符串中的字母全变为大写
        this.getJspContext().getOut().write(str);   //将修改后的内容写到浏览器
    }
}
```

标签处理器类 MySimpleTag4 直接继承了 SimpleTag 接口的默认实现类 SimpleTagSupport，覆盖了 SimpleTagSupport 类的 doTag 方法即编写实现标签功能的代码。在 doTag 方法获得标签体之后，调用 invoke 方法时指定 StringWriter 类型对象 sw，该对象是一个可取出结果数据的输出流对象，让标签体的执行结果输出到该输出流对象中，从 sw 中取出标签内容将其中的小写字母转化为大写字母，然后将修改后的结果输出到浏览器。

在 tld 文件中将标签处理器类 MySimpleTag4 描述成一个标签，代码如下：

```
<short-name>imut</short-name>
<uri>http://www.imut.edu.cn</uri>
<tag>
    <name>myTag4</name>
    <tag-class>com.ch09.MySimpleTag4</tag-class>
    <body-content>scriptless</body-content>
</tag>
```

上述 tld 文件代码，指定 uri 为"http://www.imut.edu.cn"，标签名称为"myTag4"，标签处

理类为"com.ch09.MySimpleTag4"。

JSP 页面 myTag4.jsp 使用自定义标签 myTag4，程序代码如下：

```jsp
<%@page language="java" import="java.util.*" pageEncoding="UTF-8"%>
<%@taglib uri="http://www.imut.edu.cn" prefix="imut"%>
<!DOCTYPE HTML PUBLIC "-//W3C//DTD HTML 4.01 Transitional//EN">
<html>
    <head>
        <title>修改标签体</title>
    </head>
    <body>
        <imut:myTag4>
            This is my JSP page.<br>
        </imut:myTag4>
    </body>
</html>
```

页面 myTag4.jsp 中首先通过 taglib 指令元素将标签库引入 JSP 文件并指定前缀为"imut"，然后在页面中使用标签<imut:myTag4>，标签体内为文本"This is my JSP page."。

通过浏览器访问 myTag4.jsp 页面，运行效果如图 9-6 所示。标签<imut:myTag4>将标签体文本字符串中的小写字母转化为大写字母。

图 9-6　修改标签体运行效果

例程 9-5：演示自定义标签实现 JSTL 中的 ForEach 标签功能。程序为 MyForEachTag.java 和 myForEach.jsp。

MyForEachTag 为标签处理器类，其程序代码如下：

```java
package com.ch09;
import java.io.IOException;
import java.lang.reflect.Array;
import java.util.ArrayList;
import java.util.Collection;
import java.util.Iterator;
import java.util.Map;
import javax.servlet.jsp.JspException;
import javax.servlet.jsp.tagext.SimpleTagSupport;
public class MyForEachTag extends SimpleTagSupport {
```

```java
private Object items;
private String var;
private Collection collection;
//属性 items 的 setter 方法
public void setItems(Object items) {
    this.items=items;
    //将 List、Set 转化为 Collection 类型
    if(items instanceof Collection){
        collection=(Collection) items;
    }
    //将 Map 转化为 Collection 类型
    if(items instanceof Map){
        Map map=(Map) items;
        collection=map.entrySet();
    }
    //将数组转化为 Collection 类型
    if(items.getClass().isArray()){
        this.collection=new ArrayList();
        int len=Array.getLength(items);
        for(int i=0;i<len;i++){
            Object obj=Array.get(items, i);
            collection.add(obj);
        }
    }
}
//属性 var 的 setter 方法
public void setVar(String var) {
    this.var=var;
}
public void doTag() throws JspException, IOException {
    Iterator iter=this.collection.iterator();
    //遍历 Collection 中所有元素
    while(iter.hasNext()){
        //获得下一个元素
        Object obj=iter.next();
        //将获得元素放在 pageContext 对象中
        this.getJspContext().setAttribute(var, obj);
        //执行标签体
        this.getJspBody().invoke(null);
    }
}
```

标签处理器类 MyForEachTag 直接继承了 SimpleTag 接口的默认实现类

SimpleTagSupport，覆盖了 SimpleTagSupport 类的 doTag 方法即编写实现标签功能的代码。因为自定义标签具有两个属性，所以在标签处理器中定义两个属性 var 和 items，并提供了对应的 setter 方法。在 setItems 方法中将标签所能遍历的集合、映射和数组均转化为 Collection 类型，这样做的目的是便于在 doTag 方法中进行集合元素的遍历。

在 doTag 方法中采用获得迭代器的方式来遍历 collection 属性，逐一取出其中的元素，将取出的元素放入 pageContext 对象中，然后执行标签体逐一显示元素值。

在 tld 文件中将标签处理器类 MyForEachTag 描述成一个标签，代码如下：

```
<short-name>imut</short-name>
<uri>http://www.imut.edu.cn</uri>
<tag>
    <name>myForEach</name>
    <tag-class>com.ch09.MyForEachTag</tag-class>
    <body-content>scriptless</body-content>
    <attribute>
        <name>items</name>
        <required>true</required>
        <rtexprvalue>true</rtexprvalue>
    </attribute>
    <attribute>
        <name>var</name>
        <required>true</required>
        <rtexprvalue>false</rtexprvalue>
    </attribute>
</tag>
```

上述 tld 文件代码，指定 uri 为"http://www.imut.edu.cn"，标签名称为"myForEach"，标签属性为"items"和"var"，标签处理类为"com.ch09.MyForEachTag"。

JSP 页面 myForEach.jsp 使用自定义标签 myForEach，程序代码如下：

```
<%@page language="java" import="java.util.*" pageEncoding="UTF-8"%>
<%@taglib uri="http://www.imut.edu.cn" prefix="imut"%>
<!DOCTYPE HTML PUBLIC "-//W3C//DTD HTML 4.01 Transitional//EN">
<html>
    <head>
        <title>实现 JSTL 中的 ForEach 标签功能</title>
    </head>
    <body>
        <%
            List list=new ArrayList();
            list.add("Java 核心技术");
            list.add("Java Web 开发技术");
            list.add("JavaEE 开发技术");
```

```
        list.add("SSH框架整合技术");
        request.setAttribute("list", list);
    %>
    <imut:myForEach var="bookName" items="${list}">
        ${bookName}<br>
    </imut:myForEach>
    ----------------------------------------<br>
    <%
        Map map=new HashMap();
        map.put("Java 核心技术", 36.8);
        map.put("Java Web 开发技术", 45.9);
        map.put("JavaEE 开发技术", 78.1);
        map.put("SSH框架整合技术", 99.14);
        request.setAttribute("map", map);
    %>
    <imut:myForEach var="entry" items="${map}">
        ${entry.key}---${entry.value}<br>
    </imut:myForEach>
    ----------------------------------------<br>
    <%
        Integer[] num={1,3,5,7,9};
        request.setAttribute("num", num);
    %>
    <imut:myForEach var="i" items="${num}">
        $ {i}<br>
    </imut:myForEach>
    ----------------------------------------<br>
    <%
        int[] arr={1,4,7,10};
        request.setAttribute("arr", arr);
    %>
    <imut:myForEach var="i" items="${arr}">
        ${i}<br>
    </imut:myForEach>
    </body>
</html>
```

页面myForEach.jsp中通过taglib指令元素将标签库引入JSP文件并指定前缀为"imut"。利用Java脚本分别创建了集合、映射和数组对象并向其中添加了对应的元素，然后在页面中使用标签＜imut:myForEach＞对集合、映射和数组对象进行遍历。

通过浏览器访问myForEach.jsp页面，运行效果如图9-7所示。

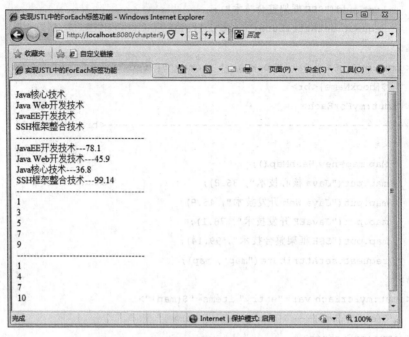

图 9-7　自定义遍历集合元素标签运行效果

9.2　JSTL 标签库

JSP 标准标签库(JSP Standard Tag Library,JSTL)是为简化 JSP 页面和 Web 应用程序功能而开发的标签库。JSTL 只能运行在支持 JSP 1.2 和 Servlet 2.3 规范的容器上,它封装了许多 JSP 应用程序通用的核心功能。

在使用 JSTL 标签之前,首先应该获取 JSTL 包并安装到 Tomcat 服务器中。可以到 http://archive.apache.org/dist/jakarta/taglibs/standard/binaries/下载 JSTL 包。下载的是一个压缩包,将该压缩包解压到一个目录中,然后把 lib 目录中的两个文件 jstl.jar 和 standard.jar 复制到 Web 应用的 WEB-INF/lib 目录中,这样就完成了 JSTL 的安装。或者可以直接在 Tomcat 服务器的示例应用中找到这两个文件,所在目录是:＜CATALINA_HOME＞\webapps\examples\WEB-INF\lib。从该目录中将 jstl.jar 和 standard.jar 复制到 Web 应用的 WEB-INF\lib 目录下即可。如果使用 MyEclipse 开发,在创建工程时如果选择 Java EE 5.0,则自动将 JSTL 的包包含到工程中。创建工程时选择 Java EE 5.0 的效果如图 9-8 所示,JSTL 包所在结构如图 9-9 所示。创建工程时如果选择 J2EE 1.4,则需要手动选择是否需要包含 JSTL,如图 9-10 所示。

JSTL 共提供了 5 种标签库,每种标签库都提供了一组实现特定功能的标签,这些标签库分别如下:

(1) 核心标签库:包括通用处理的标签。

(2) 国际化和格式化库:包括国际化和格式化标签。

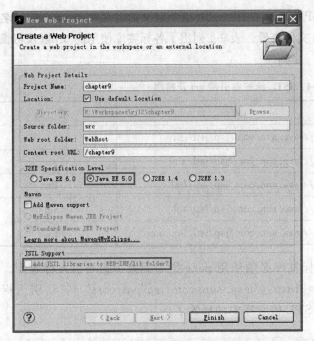

图 9-8 使用 Java EE 5.0 版本的页面　　　　图 9-9 自动包含的 JSTL 包

图 9-10 使用 J2EE 1.4 版本的页面

(3) XML 标签库：包括解析、查询和转换 XML 数据的标签。
(4) SQL 标签库：包括访问关系数据库的标签。
(5) 函数库：包括管理 String 和集合的标签。

使用 JSTL 标签库，必须在 JSP 页面中使用 taglib 指令定义前置名称与 uri。前置名称可以任意指定，而 uri 可以通过 tld 文件查看。表 9-4 列出了各标签库使用的 uri 以及习惯使用的前缀。

表 9-4 JSTL 标签库引用 uri 及前缀

库名称	uri	前缀
核心标签库	http://java.sun.com/jsp/jstl/core	c
XML 标签库	http://java.sun.com/jsp/jstl/xml	x
国际化和格式化库	http://java.sun.com/jsp/jstl/fmt	fmt
SQL 标签库	http://java.sun.com/jsp/jstl/sql	sql
函数库	http://java.sun.com/jsp/jstl/functions	fn

在 JSP 页面中使用 JSTL 的标签库必须使用 page 指令元素 taglib 引入标签库。例如＜%@taglib prefix="c" uri="http://java.sun.com/jsp/jstl/core" %＞，引入标签库，之后就可以通过前缀 c 引用 JSTL 核心库中的标签了。

9.3　JSTL 核心标签库

JSTL 核心标签库标签共有 14 个，从功能上可以分为 4 类，如表 9-5 所示。

表 9-5 JSTL 核心标签

标签类型	标签	说明
表达式标签	＜c:out＞	输出属性内容
	＜c:set＞	设置属性内容
	＜c:remove＞	清除一个作用域变量
	＜c:catch＞	捕获异常
流程控制标签	＜c:if＞	条件判断
	＜c:choose＞	多条件判断，结合＜c:whent＞和＜c:otherwise＞使用
	＜c:when＞	测试一个条件
	＜c:otherwise＞	当 when 条件都为 false 时，执行该标签中的内容
循环标签	＜c:forEach＞	对集合中的每个对象做迭代处理
	＜c:forTokens＞	对给定字符串中的每个子串执行处理
url 相关标签	＜c:url＞	重写 url 并编码
	＜c:import＞	导入网页
	＜c:redirect＞	客户端重定向
	＜c:param＞	传递参数

下面将按照功能分类，分别讲解每个标签的语法和使用方式。

9.3.1 表达式标签

表达式标签包括＜c:out＞、＜c:set＞、＜c:remove＞、＜c:catch＞4 种，下面分别介绍。

1. ＜c:out＞标签

＜c:out＞标签用来输出数据对象的内容。它有以下两种语法格式。

语法一：

```
<c:out value="value" [escapeXML="{true|false}" ] default="defaultValue" />
```

语法二：

```
<c:out value="value" [escapeXML="{true|false}" ]>
    default value
</c:out>
```

其中，value 属性指定要显示的内容，它可以是普通字符串，也可以是 EL 表达式；如果 value 属性的值为 null 则显示 default 属性的内容；escapeXml 表示是否要转换字符，例如将"＞"转换为 &gt。

例程 9-6：演示＜c:out＞标签的使用。程序为 out.jsp。

程序 out.jsp 代码如下：

```
<%@page language="java" import="java.util.*" pageEncoding="UTF-8"%>
<%@taglib uri="http://java.sun.com/jsp/jstl/core" prefix="c" %>
<!DOCTYPE HTML PUBLIC "-//W3C//DTD HTML 4.01 Transitional//EN">
<html>
    <head>
        <title>My JSP 'out.jsp' starting page</title>
    </head>
    <body>
        <%  pageContext.setAttribute("school", "内蒙古工业大学");  %>
        <c:out value="${school }"/><br>
        <c:out value="信息工程学院"/><br>
        <c:out value="${username }" default="没有您要找的名称"/><br>
        <c:out value="${username }">没有您要找的名称</c:out><br>
    </body>
</html>
```

在程序 out.jsp 中首先通过＜taglib＞指令指定自定义标记的 uri 和前缀，之后就可以使用＜c:out＞标签，第一个＜c:out＞标签输出一个作用域中的 school 变量的值，第二个＜c:out＞用来输出一个字符串值，第三个＜c:out＞用来输出作用域中 username 的值，因为作用域中没有该变量，所以输出 default 的值，第四个＜c:out＞中 value 指定的值在作用域中不存在,所以输出标签之间的默认值内容。运行效果如图 9-11 所示。

图 9-11 <c:out>输出结果

2. ＜c:set＞ 标签

＜c:set＞标签用于设置某个作用域变量或者对象(JavaBean 或 Map)的属性值。它有 4 种语法格式,其中语法一和语法二用于设置变量的值,语法三和语法四用于设置对象的属性值,具体格式如下所示。

语法一:

```
< c: set var =" varName" value =" value" [scope =" {page | request | session | application}"]/>
```

语法二:

```
<c:set var="varName" [scope="{page|request|session|application}"]>
    value
</c:set>
```

其中,var 表示要设置内容的变量名,value 表示要设置的内容,scope 表示要设置内容的作用范围。

语法三:

```
<c:set target="target" property="perpertyName" value="value"/>
```

语法四:

```
<c:set target="target" property="perpertyName">
    value
</c:set>
```

其中,target 表示要设置内容属性的对象名,property 表示要设置的属性,其他同语法一中属性相同。

例程 9-7:演示＜c:set＞标签的使用。程序为 SimpleBean.java 和 set.jsp。

程序 SimpleBean.java 代码如下:

```
package com.ch09.bean;
public class SimpleBean {
```

```
    private String info;
    public SimpleBean() {
    }
    public String getInfo() {
        return info;
    }
    public void setInfo(String info) {
        this.info=info;
    }
}
```

程序 set.jsp 代码如下：

```
<%@page language="java" import="java.util.*" pageEncoding="UTF-8"%>
<%@page import="com.ch09.bean.SimpleBean" %>
<%@taglib uri="http://java.sun.com/jsp/jstl/core" prefix="c" %>
<!DOCTYPE HTML PUBLIC "-//W3C//DTD HTML 4.01 Transitional//EN">
<html>
    <head>
        <title>My JSP 'set.jsp' starting page</title>
    </head>
    <body>
        <c:set var="website" value="www.imut.edu.cn" scope="request"/>
        输出变量值:${requestScope.website }<br>
        <%
            SimpleBean bean=new SimpleBean();
            request.setAttribute("simple", bean);
        %>
        <c:set value="我爱内蒙古工业大学" target="${simple }" property="info"/>
        输出 JavaBean 对象属性 info 的值:${requestScope.simple.info }<br>
    </body>
</html>
```

在程序 set.jsp 中通过<taglib>指令指定自定义标记的 uri 和前缀，首先使用<c:set>标签将变量 website 的值设置为 www.imut.edu.cn，同时指定作用域为 request，然后通过 EL 表达式输出变量的值。其次，实例化一个 SimpleBean 对象，并且将其作用范围设置为"request"，使用<c:set>标签将 simple 名代表的对象的属性 info 的值设置为"我爱内蒙古工业大学"，然后使用 EL 表达式输出该属性的值。运行效果如图 9-12 所示。

3. < c:remove> 标签

<c:remove>标签用于从作用域中删除变量。语法格式为：

<c:remove var="varName" [scope="page|request|session|application"]/>

图 9-12　使用<c:set>标签的执行效果

其中,var 代表要删除的变量名,scope 代表要删除变量的作用范围。

例程 9-8：演示<c:remove>标签的使用。程序为 SimpleBean.java 和 remove.jsp。

程序 SimpleBean.java 同例程 9-7 中 SimpleBean.java 相同。

程序 remove.jsp 在例程 9-7 的 set.jsp 的基础上增加一条<c:remove>标签语句,代码如下:

```
<%@page language="java" import="java.util.*" pageEncoding="UTF-8"%>
<%@page import="com.ch09.bean.SimpleBean" %>
<%@taglib uri="http://java.sun.com/jsp/jstl/core" prefix="c" %>
<!DOCTYPE HTML PUBLIC "-//W3C//DTD HTML 4.01 Transitional//EN">
<html>
    <head>
        <title>My JSP 'remove.jsp' starting page</title>
    </head>
    <body>
        <%
            SimpleBean bean=new SimpleBean();
            request.setAttribute("simple", bean);
        %>
        <c:set value="我爱内蒙古工业大学" target="${simple }" property="info"/>
        remove 前,输出 JavaBean 对象属性 info 的值:${requestScope.simple.info }<br>
        <c:remove var="simple"/>
        remove 后,输出 JavaBean 对象属性 info 的值:${requestScope.simple.info }<br>
    </body>
</html>
```

程序 remove.jsp 中为 SimpleBean 的属性 info 设置了值并输出,然后使用<c:remove>标签把 simple 代表的对象删除,之后再输出 info 的值,运行效果如图 9-13 所示。

4. <c:catch> 标签

<c:catch>标签用于处理产生错误的异常情况,并且将信息保存起来。语法格式为:

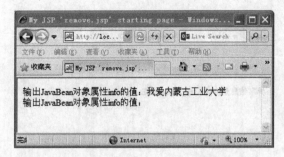

图 9-13 使用＜c:remove＞标签的执行效果

```
<c:catch [var="varName"]>
    可能发生异常的语句
</c:catch>
```

其中，var 用来保存异常信息。

例程 9-9：演示＜c:catch＞标签的使用。程序为 catch.jsp。

程序 catch.jsp 代码如下：

```
<%@page language="java" import="java.util.*" pageEncoding="UTF-8"%>
<%@taglib uri="http://java.sun.com/jsp/jstl/core" prefix="c" %>
<!DOCTYPE HTML PUBLIC "-//W3C//DTD HTML 4.01 Transitional//EN">
<html>
    <head>
        <title>My JSP 'remove.jsp' starting page</title>
    </head>
    <body>
        <c:catch var="error">
            <%int i=5/0; %>
        </c:catch>
        异常信息是:${error }
    </body>
</html>
```

程序 catch.jsp 中通过＜taglib＞指令指定自定义标记的 uri 和前缀，使用＜c:catch＞标签的 var 属性值"error"保存除数为 0 的异常信息，之后输出该异常信息，运行效果如图 9-14 所示。

9.3.2 流程控制标签

流程控制标签包括＜c:if＞、＜c:choose＞、＜c:when＞、＜c:otherwise＞4 种，下面分别介绍。

1.＜c:if＞标签

＜c:if＞标签用于进行条件判断。它有以下两种语法格式。

图 9-14 使用＜c:catch＞标签的执行效果

语法一：

```
<c:if test="testCondition" var="varName" [scope="{page|request|session|application}"]/>
```

语法二：

```
<c:if test="testCondition" var="varName" [scope="{page|request|session|application}"]>
    body content
</c:if>
```

其中，属性 test 指定条件表达式，它可以是 EL 表达式；属性 var 指定用于保存条件表达式结果的变量；scope 用于保存条件表达式结果变量名的作用域。

例程 9-10：演示＜c:if＞标签的使用。程序为 if.jsp。

程序 if.jsp 代码如下：

```
<%@page language="java" import="java.util.*" pageEncoding="UTF-8"%>
<%@taglib uri="http://java.sun.com/jsp/jstl/core" prefix="c" %>
<!DOCTYPE HTML PUBLIC "-//W3C//DTD HTML 4.01 Transitional//EN">
<html>
    <head>
        <title>My JSP 'if.jsp' starting page</title>
    </head>
    <body>
        <c:if test="{2>5}" var="result">
            条件为假时,条件结果的值:${result }<br>
            条件为假时,变量的值:${username }<br>
        </c:if>
        <c:set var="username" value="imut" scope="page"/>
        <c:if test="${username=='imut'}" var="result" scope="page">
            条件为真时,条件结果的值:${result }<br>
            条件为真时,变量的值:${username }<br>
        </c:if>
    </body>
```

</html>

程序 if.jsp 中通过<taglib>指令指定自定义标记的 uri 和前缀,使用<c:if>标签的 test 属性判断条件"2>5"并将判断结果保存在 result 变量中保存,由于结果为 false,则不执行标签体内的语句。再通过<c:set>标签为变量 username 设置值为 imut,设置使用范围为 page,然后使用<c:if>标签的 test 属性判断表达式"username=='imut'"的结果并保存在 result 中,由于结果为 true,所以执行标签体内的语句。效果如图 9-15 所示。

图 9-15 使用<c:if>标签的执行效果

2. <c:choose>标签、<c:when>标签、<c:otherwise>标签

<c:choose>标签需要和<c:when>、<c:otherwise>标签结合使用进行条件判断。

```
<c:choose>
    <c:when test="testCondition">body content</c:when>
    ...
    <c:when test="testCondition">body content</c:when>
    <c:otherwise>
        body content
    </c:otherwise>
<c:choose>
```

<c:choose>标签要作为<c:when>标签和<c:otherwise>标签的父标签使用,<c:choose>根据子标签<c:when>决定要执行哪个内容,如果没有一个条件成立,而存在<c:otherwise>子标签,则执行<c:otherwise>中标签体的内容。<c:otherwise>标签在嵌套中只允许出现一次。

例程 9-11:演示<c:choose>标签、<c:when>标签、<c:otherwise>标签的使用。程序为 choose.jsp。

程序 choose.jsp 代码如下:

```
<%@page language="java" import="java.util.*" pageEncoding="UTF-8"%>
<%@taglib uri="http://java.sun.com/jsp/jstl/core" prefix="c" %>
<!DOCTYPE HTML PUBLIC "-//W3C//DTD HTML 4.01 Transitional//EN">
<html>
```

```
        <head>
            <title>My JSP 'choose.jsp' starting page</title>
        </head>
        <body>
            <c:set var="score" value="80"/>
            <c:choose>
                <c:when test="${score>=90 }">你的成绩为优秀</c:when>
                <c:when test="${score>=75 && score<90 }">你的成绩为良好</c:when>
                <c:when test="${score>=60 && score<75 }">你的成绩为及格</c:when>
                <c:otherwise>你没有通过考试</c:otherwise>
            </c:choose>
        </body>
</html>
```

程序 choose.jsp 首先通过＜c:set＞标签将 score 的值设置为 80，然后在＜c:choose＞中使用＜c:when＞标签进行判断，test 属性的值是一个 EL 条件表达式，如果满足条件则输出＜c:when＞标签体的内容，如果所有＜c:when＞的条件都不满足，则输出＜c:otherwise＞标签体的内容。部署执行该程序，效果如图 9-16 所示。

图 9-16　使用＜c:choose＞、＜c:when＞、＜c:otherwise＞标签的执行效果

9.3.3　循环标签

循环标签包括＜c:forEach＞和＜c:forTokens＞。

1.＜c:forEach＞标签

＜c:forEach＞标签用于循环控制，可以遍历变量也可以遍历集合中的元素。它有以下两种语法格式。

语法一：

```
<c:forEach [var="varName"] [begin="begin" end="end" step="step"] [varStatus="varStatusName"]>
    body content
</c:forEach>
```

语法一类似 Java 语言中的 for 循环，通过 var 指定循环变量、begin 指定变量初值、

end 指定变量结束的值、step 指定每次循环变量增加的步长、varStatus 指定状态。

其中，varStatus 有 4 种状态，分别是：index 表示当前循环的索引值、count 表示已经循环的次数、first 是否为第一个位置、last 是否为最后一个位置。如果指定了 varStatus 属性，值假如为"s"，那么在标签体中就可以通过 s.index 的形式来访问。

语法二：

```
<c:forEach [var="varName"] items="collection" [varStatus="varStatusName"]
[begin="begin" end="end" step="step"]>
    body content
</c:forEach>
```

语法二中的 items 指定要遍历的集合、var 保存集合中的每个元素、begin 指定循环开始的下标、end 指定循环结束的下标、step 指定步长、varStatus 指定状态。

例程 9-12：演示<c:forEach>标签遍历变量。程序为 foreach1.jsp。代码如下：

```
<%@page language="java" import="java.util.*" pageEncoding="UTF-8"%>
<%@taglib uri="http://java.sun.com/jsp/jstl/core" prefix="c"%>
<!DOCTYPE HTML PUBLIC "-//W3C//DTD HTML 4.01 Transitional//EN">
<html>
    <head>
        <title>My JSP 'fortokens.jsp' starting page</title>
    </head>
    <body>
        <table border="1">
            <tr>
                <td>x 的值</td>
                <td>varStatus.index</td>
                <td>varStatus.count</td>
                <td>varStatus.first</td>
                <td>varStatus.last</td>
            </tr>
            <c:forEach var="x" varStatus="status" begin="0" end="9" step="3">
            <tr>
                <td>${x }</td>
                <td>${status.index }</td>
                <td>${status.count }</td>
                <td>${status.first }</td>
                <td>${status.last }</td>
            </tr>
            </c:forEach>
        </table>
    </body>
</html>
```

程序 foreach1.jsp 使用<c:forEach>标签输出变量及各个状态的值。运行效果如

图 9-17 所示。

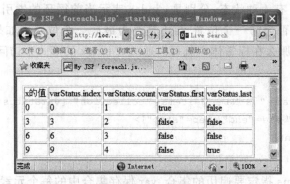

图 9-17 使用＜c:forEach＞输出变量及各状态的值

例程 9-13：演示＜c:forEach＞标签遍历集合元素。程序为 User.java 和 foreach2.jsp。
程序 User.java 代码如下：

```
package com.ch09.bean;
public class User {
    private String username;
    private String userpass;
    public User(){
    }
    public String getUsername() {
        return username;
    }
    public void setUsername(String username) {
        this.username=username;
    }
    public String getUserpass() {
        return userpass;
    }
    public void setUserpass(String userpass) {
        this.userpass=userpass;
    }
}
```

程序 foreach2.jsp 代码如下：

```
<%@page language="java" import="java.util.*" pageEncoding="UTF-8"%>
<%@page import="com.ch09.bean.User" %>
<%@taglib uri="http://java.sun.com/jsp/jstl/core" prefix="c" %>
<!DOCTYPE HTML PUBLIC "-//W3C//DTD HTML 4.01 Transitional//EN">
<html>
    <head>
        <title>My JSP 'foreach2.jsp' starting page</title>
```

```
</head>
<body>
    <%
        List users=new ArrayList();
        for(int i=0;i<3;i++){
            User user=new User();
            user.setUsername("user"+i);
            user.setUserpass("pass"+i);
            users.add(user);
        }
        session.setAttribute("users", users);
    %>
    <c:forEach items="${users }" var="user">
        用户名:${user.username }---密码:${user.userpass }<br>
    </c:forEach>
</body>
</html>
```

程序 User.java 是一个 JavaBean，提供了用户名和密码及其存取方法。程序 foreach2.jsp 通过循环创建三个用户信息放在 List 中，并将该 list 对象存储到 session 作用域中，然后使用＜c:foreach＞循环遍历每个元素。其中，items 指定了要遍历的集合对象为属性范围中的"users"，var 指定了用 user 保存遍历到的每个用户信息，再通过 EL 表达式分别将每个用户的用户名和密码输出。执行程序，运行效果如图 9-18 所示。

图 9-18 使用＜c:forEach＞标签遍历集合元素

2. ＜c:forTokens＞标签

＜c:forTokens＞标签用于浏览字符串中的成员，可以指定一个或者多个分隔符。语法格式如下：

```
<c:forTokens items="stringOfTokens" delims="delimiters"
    [var="varName"] [varStatus="varStatusName"]
    [begin="begin"] [end="end"] [step="step"]>
    body content
</c:forTokens>
```

标签中 items 属性指定要浏览的字符串,它可以是字符串常量也可以是 EL 表达式、delims 指定分割符号,var 定义一个名称,用来保存分割以后的每个子字符串,其他属性与＜c:forEach＞标签中相同。在标签体中可以对分割以后的子字符串进行使用。

例程 9-14:演示＜c:forTokens＞标签的使用。程序为 fortokens.jsp。

程序 fortokens.jsp 代码如下:

```
<%@page language="java" import="java.util.*" pageEncoding="UTF-8"%>
<%@taglib uri="http://java.sun.com/jsp/jstl/core" prefix="c" %>
<!DOCTYPE HTML PUBLIC "-//W3C//DTD HTML 4.01 Transitional//EN">
<html>
    <head>
        <title>My JSP 'fortokens.jsp' starting page</title>
    </head>
    <body>
        <c:set var="user" value="username=zhangsan;userpass=123"/>
        <c:forTokens items="${user}" delims=";" var="infos">
            <c:forTokens items="${infos}" delims="=" var="info">
                ${info }<br>
            </c:forTokens>
        </c:forTokens>
    </body>
</html>
```

程序 fortokens.jsp 首先使用＜c:set＞标签设置了一个字符串变量 user,之后使用＜c:forTokens＞标签对 items 属性指定的字符串进行分割,分隔符是";",分割后的子字符串"username = zhangsan"和"userpass = 123"保存在 var 指定的 infos 中;内层＜c:forTokens＞标签对 items 属性指定的字符串,也即"username = zhangsan"和"userpass=123"分别进行分割,分隔符是"=",此时得到的子字符串保存在 info 中。运行程序,运行效果如图 9-19 所示。

图 9-19 使用＜c:forTokens＞标签分割字符串的执行效果

9.3.4 url 相关标签

JSTL 中包含了 4 个与 URL 相关的标签,分别是＜c:param＞、＜c:import＞标签、＜c:redirect＞标签、＜c:url＞标签。

1. `<c:param>` 标签

`<c:param>`标签用于将参数传递给所包含的文件，主要用在`<c:import>`、`<c:url>`、`<c:redirect>`标签中指定请求参数。它有以下两种语法格式。

语法一：

```
<c:param name="name" value="value"/>
```

语法二：

```
<c:param name="name">
param value
</c:param>
```

其中，name 属性指定参数名，value 属性指定参数值。

2. `<c:import>` 标签

`<c:import>`标签用于将静态或动态文件包含到 JSP 页面中，它与`<jsp:include>`的功能类似。它有以下两种语法格式。

语法一：

```
<c:import url="url" [context="context"] [var="varName"]
    [scope="page|request|session|application"]
    [charEncoding="charEncoding"]>
    body content
</c:import>
```

其中，属性 url 指定要包含资源的 URL，context 指定资源所在的上下文路径，var 定义存储要包含文件内容的变量名，scope 指定 var 的作用范围，charEncoding 指定被包含文件的编码格式。

语法二：

```
<c:import url="url" [context="context"] [varReader="varreaderName"]
    [charEncoding="charEncoding"]>
    body content
</c:import>
```

其中，varReader 指定以 Reader 类型存储被包含内容的名称，其他与语法一相同。

例程 9-15：演示`<c:import>`的使用。程序为 import.jsp，代码如下：

```
<%@page language="java" import="java.util.*" pageEncoding="UTF-8"%>
<%@taglib uri="http://java.sun.com/jsp/jstl/core" prefix="c" %>
<!DOCTYPE HTML PUBLIC "-//W3C//DTD HTML 4.01 Transitional//EN">
<html>
    <head>
        <title>My JSP 'import.jsp' starting page</title>
```

```
        </head>
        <body>
            <c:import url="if.jsp"/>
            <hr>
            <c:import url="http://www.imut.edu.cn" charEncoding="UTF-8"/>
        </body>
    </html>
```

程序 import.jsp 使用＜c:import＞首先引入一个 Web 工程内部资源 if.jsp，在分隔符＜hr＞后使用＜c:import＞引入外部资源 http://www.imut.edu.cn"的内容，同时指定外部资源使用的编码方式，否则会显示乱码。运行程序，执行效果如图 9-20 所示，可以看到，使用＜c:import＞标签只能包含资源的文字内容。

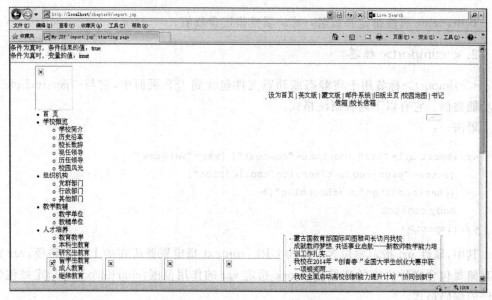

图 9-20　使用＜c:import＞标签的执行效果

3. ＜c:redirect＞ 标签

＜c:redirect＞标签用于将客户端请求从一个 JSP 页面重定向到其他页面，它有以下两种语法格式。

语法一：

```
<c:redirect url="url" [context="context"]/>
```

语法二：

```
<c:redirect url="url" [context="context"]>
    <c:param >子标签
</c:redirect>
```

其中，url 指定重定向的地址，context 指定 url 的上下文。可以使用＜c:param＞为其传递参数。该标签与 HttpServletResponse 的 sendRedirect( )的作用相同。

例程 9-16：演示＜c:redirect＞标签的使用。程序为 redirct1.jsp 和 redirct2.jsp。

程序 redirct1.jsp 代码如下：

```
<%@page language="java" import="java.util.*" pageEncoding="UTF-8"%>
<%@taglib uri="http://java.sun.com/jsp/jstl/core" prefix="c" %>
<!DOCTYPE HTML PUBLIC "-//W3C//DTD HTML 4.01 Transitional//EN">
<html>
    <head>
        <title>My JSP 'redirect1.jsp' starting page</title>
    </head>
    <body>
        <c:redirect url="redirect2.jsp">
            <c:param name="username" value="zhangsan"/>
            <c:param name="age" value="20"/>
        </c:redirect>
    </body>
</html>
```

程序 redirct2.jsp 代码如下：

```
<%@page language="java" import="java.util.*" pageEncoding="UTF-8"%>
<%@taglib uri="http://java.sun.com/jsp/jstl/core" prefix="c" %>
<!DOCTYPE HTML PUBLIC "-//W3C//DTD HTML 4.01 Transitional//EN">
<html>
    <head>
        <title>My JSP 'redirect2.jsp' starting page</title>
    </head>
    <body>
        这是重定向之后的页面 redirect2.jsp<br>
        您的姓名:${param.username }<br>
        您的年龄:${param.age }<br>
    </body>
</html>
```

在程序 redirct1.jsp 中使用＜c:redirect＞标签的属性指定要重定向的页面 URL，同时使用＜c:param＞标签指定了要传递的参数名和值。在 redirct2.jsp 中使用 EL 表达式的内置对象 param 接收参数信息并输出。在浏览器地址中栏输入"http://localhost/chapter9/redirect1.jsp"运行程序，发现地址栏的地址变成了跳转后的页面地址及传递过来的参数：http://localhost/chapter9/redirect2.jsp? username=zhangsan&age=20。运行效果如图 9-21 所示。

4. ＜c:url＞标签

＜c:url＞标签用于生成一个 URL。它有以下两种语法格式。

图 9-21 使用＜c:redirect＞标签的执行效果

语法一：

```
<c:url value="value" [context="context"] [var="varName"]
    [scope="page|request|session|application"]/>
```

语法二：

```
<c:url value="value" [context="context"] [var="varName"]
    [scope="page|request|session|application"]>
    <c:param name="" value="value"/>
</c:url>
```

其中，属性 value 指定一个 URL，当使用相对路径引用外部资源时就用 context 指定上下文，var 指定保存 URL 的名称。可以使用＜c:param＞标签传递属性名和值。

例如，使用＜c:url value="fortokens.jsp" var="path1"/＞的属性 value 构建一个 URL 并保存在 var 指定的变量 path1 中，之后就可以使用 path1 的地址。例如，在＜c:redirect url="${path1}"/＞中使用 path1 中的值作为属性 url 的值，实现重定向的目标地址。

9.4 案　　例

本章案例在第 8 章案例的基础上增加实现"课程安排管理"模块中的全部功能。主要使用 JSTL 标签完成 JSP 页面条件判断及循环结构的功能，进一步减少 JSP 页面中的脚本(＜% %＞)和表达式(＜%= %＞)代码。

9.4.1 案例设计

课程安排管理模块功能为安排课程的任课教师、上课班级及上课地点等。主要实现课程安排的增删改查等操作，使用 MVC 模式将页面显示和业务逻辑处理进行分离。为了减少 JSP 页面中的 Java 脚本代码，使用 JSTL 标签实现条件判断和循环控制。在添加课程安排信息、显示课程安排信息和查看/修改课程安排信息的页面中使用 JSTL 标签进行数据处理。

由于教师表、班级表、课程管理表和课程安排表具有关联关系，因此在删除这些表中的记录时要判断课程安排表中是否存在关联记录，如果不存在则可以删除，否则禁止删除。本章需要修改删除教师的 Servlet、删除班级的 Servlet 以及删除课程信息的 Servlet。本章案例使用的主要文件如表 9-6 所示。

表 9-6　本章案例使用的文件

文　　件	所在包/路径	功　　能
CourseArrange.java	com.imut.javabean	封装课程安排信息的 JavaBean
ICourseArrange.java	com.imut.javabean	定义操作数据库表增删改查操作方法的接口
CourseArrangeImpl	com.imut.javabean	具体实现接口定义的增删改查等方法
AddArrangeServlet.java	com.imut.servlet.arrange	增加课程安排信息的 Servlet
CheckArrangeServlet.java	com.imut.servlet.arrange	查询课程安排信息的 Servlet
DelArrangeServlet.java	com.imut.servlet.arrange	删除课程安排信息的 Servlet
ListAllArrangeServlet.java	com.imut.servlet.arrange	显示所有课程安排信息的 Servlet
ShowArrangeServlet.java	com.imut.servlet.arrange	查看/修改课程安排信息的 Servlet
UpdateArrangeServlet.java	com.imut.servlet.arrange	更新课程安排信息的 Servlet
addArrange.jsp	/arrange	增加课程安排信息的页面
arrangeList.jsp	/arrange	显示所有课程安排信息的页面
arrangeShow.jsp	/arrange	查看/修改课程安排信息的页面

9.4.2　案例演示

在浏览器地址栏中输入"http://localhost:8080/ch09/login.jsp",以管理员用户登录,单击"课程安排管理"菜单,显示课程安排信息列表页面,效果如图 9-22 所示。单击

图 9-22　课程安排列表页面

"添加课程安排"超链接,出现添加课程安排信息页面,效果如图 9-23 所示。输入课程安排信息,单击"提交"按钮,提示添加课程安排信息成功之后跳转到课程安排信息列表页面。在课程安排信息列表页面中输入查询条件,查询结果将显示在"查询"按钮下方的课程安排信息列表处,如果没有输入查询条件则查询所有课程安排信息。根据"上课班级"查询结果的效果图如图 9-24 所示。在课程安排信息列表页面中单击"删除"超链接,将会删除该课程安排信息,删除完成后显示课程安排列表页面。在课程安排列表页面中单击"查看/修改"超链接,或者单击"安排编号",将显示课程安排信息查看/修改页面,效果如图 9-25 所示。修改课程安排信息完成后单击"提交"按钮,提示修改成功后显示课程安排信息列表页面。

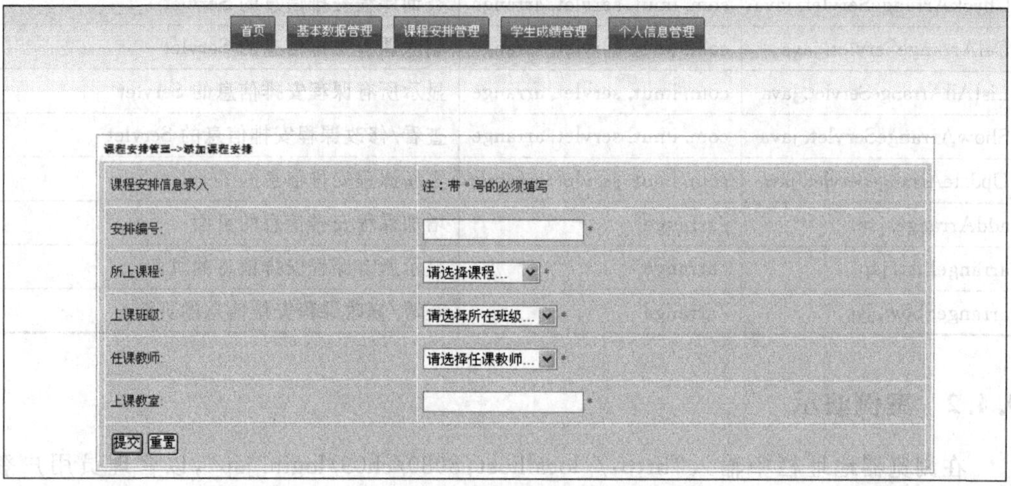

图 9-23 添加课程安排页面

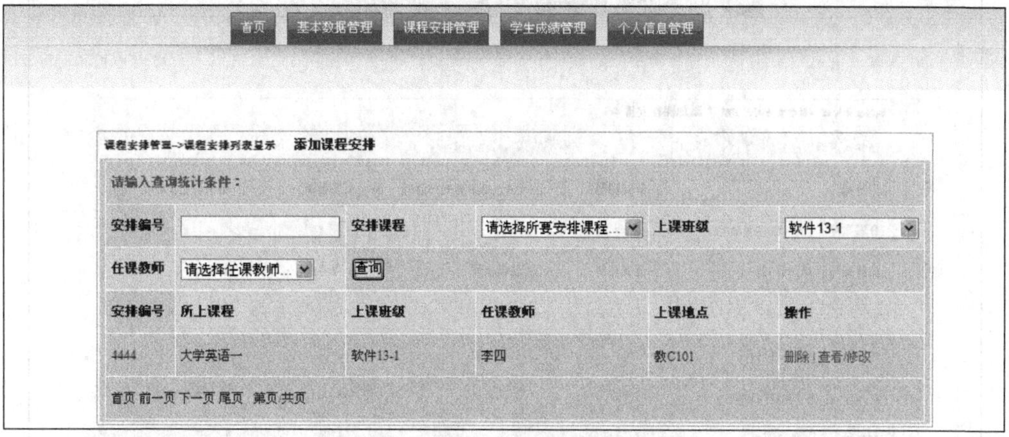

图 9-24 课程安排查询结果页面

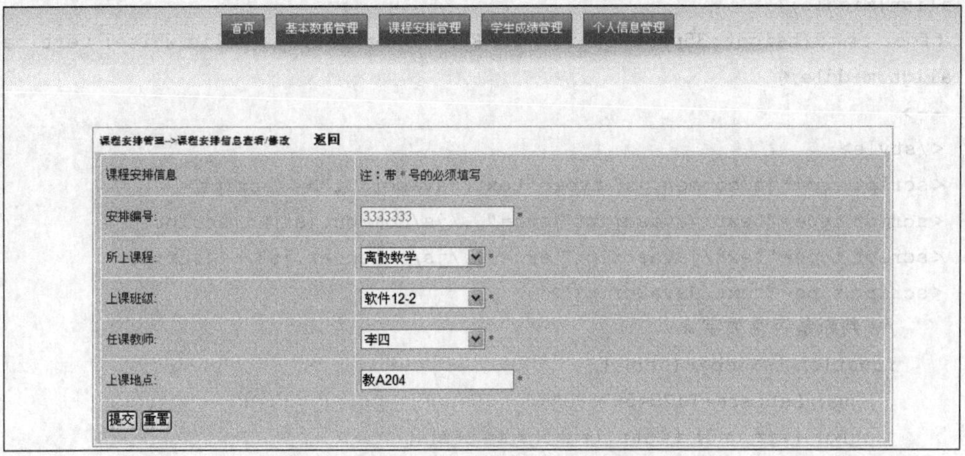

图 9-25　查看/编辑课程安排页面

9.4.3　代码实现

创建工程 ch09，根据案例设计描述分别给出各部分具体实现。本章主要给出 arrangeShow.jsp 的具体代码，其他源文件代码请参考随书电子资源。

程序 arrangeShow.jsp 代码如下：

```
<%@page language="java" contentType="text/html; charset=UTF-8" import="com.imut.javabean.*,java.util.*" pageEncoding="UTF-8"%>
<%@taglib uri="http://java.sun.com/jsp/jstl/core" prefix="c"%>
<!DOCTYPE html PUBLIC "-//W3C//DTD XHTML 1.0 Transitional//EN" "http://www.w3.org/TR/xhtml1/DTD/xhtml1-transitional.dtd">
<html xmlns="http://www.w3.org/1999/xhtml">
<head>
<title>课程安排管理</title>
<link rel="stylesheet" type="text/css" id="css" href="../style/main.css" />
<link rel="stylesheet" type="text/css" id="css" href="../style/style1.css" />
<link rel="stylesheet" type="text/css" id="css" href="../style/style.css" />
<style type="text/css">
<!--
table{border-spacing:1px; border:1px solid #A2C0DA;}
td, th{padding:2px 5px;border-collapse:collapse;text-align:left; font-weight:normal;}
thead tr th{height:50px;background:#B0D1FC;border:1px solid white;}
thead tr th.line1{background:#D3E5FD;}
thead tr th.line4{background:#C6C6C6;}
tbody tr td{height:35px; background:#CBE2FB;border:1px solid white; vertical-align:middle;}
tbody tr td.line4{background:#D5D6D8;}
tbody tr th{height:35px;background: #DFEDFF;border:1px solid white; vertical-
```

```
align:middle;}
tfoot tr td{height:35px;background:#FFFFFF;border:1px solid white; vertical-
align:middle;}
-->
</style>
<script src="js/common.js" type="text/javascript"></script>
<script type="text/javascript" src="../js/common.js" ></script>
<script type="text/javascript" src="../js/calendar.js" ></script>
<script type="text/JavaScript">
    /*判断是否为数字*/
    function isNumber(str) {
        var Letters="1234567890";
        for (var i=0; i<str.length; i=i+1) {
            var CheckChar=str.charAt(i);
            if (Letters.indexOf(CheckChar)==-1) {
                return false;
            }
        }
        return true;
    }
    /*判断是否为Email*/
    function isEmail(str) {
        var myReg=/^[-_A-Za-z0-9]+@([_A-Za-z0-9]+\.)+[A-Za-z0-9]{2,3}$/;
        if (myReg.test(str)) {
            return true;
        }
        return false;
    }
    /*判断是否为空*/
    function isEmpty(value) {
        return /^\s*$/.test(value);
    }
    function check(){
        if(document.myForm.courseNo.value==""){
            alert("请选择所要安排的课程!");
            document.myForm.courseNo.focus();
            return false;
        }
        if(document.myForm.classNo.value==""){
            alert("请选择所要安排的班级!");
            document.myForm.classNo.focus();
            return false;
        }
        if(document.myForm.teacherNo.value==""){
```

```html
                alert("请选择任课教师!");
                document.myForm.teacherNo.focus();
                return false;
            }
            if(isEmpty(document.myForm.studyRoom.value)){
                alert("上课教室不能为空!");
                document.myForm.studyRoom.focus();
                return false;
            }
            return true;
        }
    </script>
</head>
<%
    //获得班级列表(在下拉框中使用)
    ClassDBAccess db1=new ClassDBAccess();
    ICourse db2=new CourseImpl();
    TeacherDBAccess db3=new TeacherDBAccess();
    List classtbls=db1.findAllClassTbl();
    List courses=db2.findAllCourse();
    List teachers=db3.findAllTeacher();
    request.setAttribute("classtbls",classtbls);
    request.setAttribute("courses",courses);
    request.setAttribute("teachers",teachers);
%>
<body>
<div id="btm">
<div id="main">
    <div id="header">
        <div id="top"></div>
        <div id="logo">
            <h1>课程安排管理</h1></div>
        <div id="mainnav">
            <ul>
                <li><a href="../index.jsp">首页</a></li>
                <li><a href="../base/listAllTeacherServlet">基本数据管理</a></li>
                <li><a href="listAllArrangeServlet">课程安排管理</a></li>
                <li><a href="#">学生成绩管理</a></li>
                <li><a href="../person/showPersonServlet">个人信息管理</a></li>
            </ul>
            <span>
            </span>
        </div>
    </div>
```

```html
<div id="content" align="center">
    <div id="center">
<BR /><BR />
    <form action="updateArrangeServlet" method="post" name="myForm">
        <table width="800" align="center" cellpadding="0" cellspacing="0">
            <thead>
                <tr>
                    <td width="70%"><h5>课程安排管理-->课程安排信息查看/修
                    改           <a href=
                    "listAllArrangeServlet">       返回 </a>
                    </h5></td>
                </tr>
            </thead>
            <tr>
                <td colspan="2" width="100%">
                    <table width="100%">
                        <tbody>
                            <tr>
                                <td align="right">课程安排信息</td>
                                <td>注:带<font color="red">*</font>号的必须填
                                写</td>
                            </tr>
                            <tr>
                                <td align="right">安排编号:</td>
                                <td><input disabled="disabled" name="arrangeNo"
                                value=" ${ requestScope. arrange. arrangeNo }"/>
                                <font color="red">*</font></td>
                            </tr>
                            <tr>
                                <td align="right">所上课程:</td>
                                <td>
                                    <select id="courseNo" name="courseNo">
                                        <option value="">请选择所上课程..</option>
                                        <c:forEach items=" ${ requestScope.
                                        courses}" var="course">
                                            <c:choose>
                                                <c:when test=" ${course. courseNo
                                                == requestScope. arrange. course.
                                                courseNo}">
                                                    <option value=" ${ course.
                                                    courseNo }" selected > ${ course.
                                                    courseName}</option>
                                                </c:when>
                                                <c:otherwise>
```

```
                    <option value =" ${ course.
                    courseNo }"  >  ${ course.
                    courseName}</option>
                        </c:otherwise>
                    </c:choose>
                </c:forEach>
            </select>
            <font color="red"> * </font></td>
        </tr>
        <tr>
            <td align="right">上课班级:</td>
            <td>
                <select id="classNo" name="classNo">
                    <option value="">请选择上课班级..</option>
                    <c:forEach  items  ="  ${ requestScope.
                    classtbls}" var="clazz">
                        <c:choose>
                            <c:when test =" ${ clazz. classNo = =
                            requestScope.  arrange.  classTbl.
                            classNo}">
                                <option value="${clazz.classNo}"
                                selected > ${ clazz. className } </
                                option>
                            </c:when>
                            <c:otherwise>
                                <option value="${clazz.classNo}">
                                ${clazz.className}</option>
                            </c:otherwise>
                        </c:choose>
                    </c:forEach>
                </select>
            <font color="red"> * </font></td>
        </tr>
        <tr>
            <td align="right">任课教师:</td>
            <td>
                <select id="teacherNo" name="teacherNo">
                    <option value="">请选择任课教师..</option>
                    <c:forEach items="${requestScope.teachers}"
                    var="teacher">
                        <c:choose>
                            <c:when test =" ${ teacher. teacherNo = =
                            requestScope.arrange.teacher.teacherNo}">
```

```
                            <option value =" ${teacher.
                            teacherNo}" selected > ${teacher.
                            teacherName}</option>
                        </c:when>
                        <c:otherwise>
                            <option value =" ${teacher.
                            teacherNo }" > ${teacher.
                            teacherName}</option>
                        </c:otherwise>
                    </c:choose>
                </c:forEach>
            </select>
            <font color="red">*</font></td>
        </tr>
        <tr>
            <td align="right">上课地点:</td>
            <td><input id="studyRoom" name="studyRoom" value=
            "${requestScope.arrange.studyRoom }"/><font color=
            "red">*</font></td>
        </tr>
        </tbody>
        <tr>
            <td align="center" colspan="2">
                <input type="hidden" name="arrangeNo" value=
                ${requestScope.arrange.arrangeNo } style="width:
                0;height:0" />
                <input type="submit" value="提交" onclick=
                "return check();"/>
                <input type="reset" value="重置"/>
            </td>
        </tr>
        </table>
    </td>
   </tr>
  </table>
 </form>
</div>
</div>
<div id="footer">
    <div id="copyright">
        <div id="copy">
        <p align="center">CopyRight&copy;2010</p>
        <p>内蒙古工业大学信息工程学院软件工程系</p>
        </div>
```

```
            </div>
            <div id="bgbottom"></div>
        </div>
    </div>
  </div>
 </body>
</html>
```

【代码分析】程序 arrangeShow.jsp 使用 myForm 表单的各个组件显示课程安排的值，其中文本框组件直接使用 EL 表达式为 value 属性赋值。对于"所上课程"下拉列表来说，使用 JSTL 中的循环控制标签＜c:foreach＞来迭代处理从 request 作用范围中获取的课程集合对象 courses，然后使用＜c:choose＞和＜c:when＞、＜c:otherwise＞标签进行条件判断，如果某一个课程对象的课程编号和课程安排中对应课程的课程编号相同，则将当前课程和班级选项被选中，同时将其他所有的课程选项和班级选项显示在对应的下拉列表中。"上课班级"以及"任课教师"等下拉列表同样使用 JSTL 标签完成列表项的显示。

习　　题

1. 选择题

(1) JSTL 的以下哪个类别提供执行基本任务的一般标记？（　　）
　　A. 核心操作　　　B. XML 处理操作　　C. 格式化操作　　D. 函数操作
(2) 以下哪个标记提供 c:choose 元素内的备用条件？（　　）
　　A. c:when　　　　B. c:otherwise　　　C. c:param　　　　D. c:catch
(3) 以下哪个 JSTL 标签不能实现 Java 程序中 if、if…else 的功能？（　　）
　　A. ＜c:if＞　　　　　　　　　　　　B. ＜c:else＞
　　C. ＜c:when＞　　　　　　　　　　D. ＜c:otherwise＞
(4) 以下程序段

```
<%
request.setAttribute("user","Tom");
%>
```

以下哪个选项能完成和上述 Java 程序片段同样的功能？（　　）
　　A. ＜c:set var="user" value="Tom"/＞
　　B. ＜c:set var="user" value="Tom" scope="request"/＞
　　C. ＜c:set var="${user}" value="Tom" scope="request"/＞
　　D. ＜c:out var="user" value="Tom" scope="request"/＞
(5) 假定在会话范围内不存在 cart 属性，以下＜c:out＞标签的打印结果是（　　）。

```
<c:out value="${sessionScope.cart.total}">
```

    A. null   B. No ShoppingCart
    C. sessionScope.cart.total   D. 没有任何打印结果

(6) tld 文件的根元素是＜taglib＞,以下哪个元素是＜taglib＞元素中必需的？（ ）
    A. ＜tlib-version＞   B. ＜short-name＞
    C. ＜tag-class＞   D. ＜body-content＞

(7) 开发自定义标签的类和接口包含在哪个包中？（ ）
    A. javax.servlet.jsp.tagext   B. javax.servlet.http
    C. javax.servlet.jsp   D. javax.servlet.Servlet

(8) 在继承 SimpleTagSupport 后,doTag() 的实现如下:

```
public void doTag() throws JspExcepiton{
    try{
        if(test){
            //...
        }
        else{
            throw new
        }catch(java.io.IOException ex){
            Throw new JspException("执行错误",ex);
        }
    }
}
```

如果 test 为 false 时,希望能中断 JSP 后续页面的处理,则空白部分应填入（ ）。
    A. SkipPageException()   B. IOException()
    C. ServletException()   D. Excepion()

(9) 如果 taglib 设定如下:

    ＜%@tablib prefix="x" uri="http://openhome.cc/magic/x"%＞

则下面哪一个会是使用自定义标签的正确方法？（ ）
    A. ＜x:if＞   B. ＜magic:forEach＞
    C. ＜if/＞   D. ＜x:if/＞

2. 填空题

(1) 在 JSTL 中提供了_____、_____、_____和_____4 个条件标签,使用这些条件标签可以处理程序中任何可能发生的事情。

(2) _____标签可以调入站内或其他网站的静态和动态文件到 Web 页面中,而＜jsp:include＞只能导入站内资源。

(3) 任何一个自定义标签都必须在 tld 文件中进行声明,声明之后才能够在 JSP 文件中使用,tld 文件就是一个_____类型文件。tld 文件的根元素是_____,其中包含一个或者多个＜tag＞标签,该元素用来描述标签处理器类即声明定制标签。

(4) 在 tld 文件中＜body-content＞元素的有效值是_____。

(5) web.xml 文件中的以下选段

```
<taglib>
<taglib-uri>/FirstTg.tld</taglib-uri>
<taglib-location>/web-inf/FirstTg.tld</taglib-location>
</taglib>
```

根据以上内容，使用_____指令可以使 JSP 页面能使用此标记。

(6) 假定 hello.jsp 位于 helloapp 应用的 dir1 目录下，运行以下 JSP 代码：

```
<c:url value="hello.jsp" var="myurl">
    <c:param name="username" value="Tom"/>
    <c:param name="age" value="10"/>
</c:url>
<a href="${myurl}">Hello</a>
```

则＜a＞标记生成的代码是_____。

3. 简答题

(1) 请简述 JSTL 与一般 JSP 技术有何差异。
(2) JSTL 标签的分类主要有哪几种？请简单说明。
(3) 在 Tomcat 中安装使用 JSTL 的步骤有哪些？
(4) 创建自定义标签有几种方式？如何使用自定义标签？编写自定义标签的步骤。

4. 程序设计题

(1) 应用＜c:choose＞、＜c:when＞和＜c:otherwise＞标签实现 if…else 功能，即如果 session 变量为空，则显示用户登录表单，要求用户登录；当用户登录后也就是 session 不为空时，将显示当前登录的用户。

(2) 应用 JSTL 标签自定义 application 和 session 的作用域的变量来实现网站计数器。

(3) 创建一个学生 JavaBean，其包含属性姓名、性别和年龄，创建学生实例，放入一个 List 列表中，通过迭代标记和 EL 表达式实现对 List 的遍历，并以表格的形式显示所有学生信息。

(4) 实现一个自定义标签，判断一个 YYYY-MM-DD 格式的日期修改为下面格式输出。

年:XXXX
月:YY
日:DD

中文乱码处理

在 Java Web 应用开发中,经常会遇到中文乱码的问题,所谓中文乱码问题就是本来应该显示中文,但是显示出来的却是乱七八糟的字符或者一大串问号等形式的不能识别的乱码。本章将以理论与实践相结合的方式介绍字符编码、Java Web 应用开发中中文乱码产生的原因、解决方法等内容。

10.1 字符集和字符编码

10.1.1 字符编码目的

计算机是以 0、1 这样的二进制数据进行数据的识别和存储,而人类的语言则多种多样,要想让计算机能够理解众多的人类语言并准确地处理各种字符集文字,就必须将人类语言"翻译"成计算机能够看懂的语言,所以在计算机中需要进行字符编码,以便计算机能够识别和存储各种文字。

10.1.2 字符集与编码分类

字符是各种文字和符号的总称,包括各国家文字、标点符号、图形符号、数字等。字符集是多个字符的集合。

字符集种类较多,每个字符集包含的字符个数不同,常见字符集有:ASCII 字符集、GB2312 字符集、GBK 字符集和 Unicode 字符集等。

1. ASCII 字符集

ASCII(American Standard Code for Information Interchange,美国信息互换标准代码)是基于罗马字母表的一套计算机编码系统,主要用于显示现代英语和其他西欧语言。它是现今最通用的单字节编码系统,并等同于国际标准 ISO 646。ASCII 表示的符号共 128 个,具体包括:回车键、退格、换行键等控制字符;英文大小写字符、阿拉伯数字和西文符号。

2. ISO-8859-1

128 个字符显然是不够用的,于是 ISO 组织在 ASCII 码基础上又制定了一系列标准用来扩展 ASCII 编码,它们是 ISO-8859-1~ISO-8859-15,其中 ISO-8859-1 涵盖了大多数西欧语言字符,所以应用的最广泛。ISO-8859-1 仍然是单字节编码,它总共能表示 256 个字符。

3. GB2312 字符集

GB2312 又称为 GB2312-1980 字符集,全称为《信息交换用汉字编码字符集-基本集》,由原中国国家标准总局发布,1981 年 5 月 1 日实施,是中国国家标准的简体中文字符集。它所收录的汉字已经覆盖 99.75% 的使用频率,基本满足了汉字的计算机处理需要。在中国大陆和新加坡被广泛使用。

4. GBK 字符集

GBK 字符集是 GB2312 的扩展,GBK 1.0 收录了 21 886 个符号,分为汉字区和图形符号区,汉字区包括 21 003 个字符。GBK 字符集主要扩展了繁体中文字的支持。

5. Unicode 字符集

Unicode 字符集编码(Universal Multiple-Octet Coded Character Set)是通用多 8 位编码字符集的简称,是支持世界上超过 650 种语言的国际字符集。Unicode 允许在同一服务器上混合使用不同字符集的不同语言。它是由一个名为 Unicode 学术学会(Unicode Consortium)的机构制定的字符编码系统,支持现今世界各种不同语言的书面文本的交换、处理及显示。该编码于 1990 年开始研发,1994 年正式公布,最新版本是 2014 年 6 月 17 日发布的 Unicode 7.0。Unicode 是一种在计算机上使用的字符编码。它为每种语言中的每个字符设定了统一并且唯一的二进制编码,以满足跨语言、跨平台进行文本转换、处理的要求。

UTF-8 是 Unicode 的其中一个使用方式,UTF 是 Unicode Translation Format,即把 Unicode 转作某种格式的意思。UTF-8 便于不同的计算机之间使用网络传输不同语言和编码的文字,使得双字节的 Unicode 能够在现存的处理单字节的系统上正确传输。UTF-8 使用可变长度字节来储存 Unicode 字符,例如,ASCII 字母继续使用一个字节储存,重音文字、希腊字母或西里尔字母等使用两个字节来储存,而常用的汉字就要使用三个字节,辅助平面字符则使用 4 字节。

GB2312、GBK 和 UTF-8 字符集对中文都能很好地支持,读者在 Java Web 开发中可以选择其中任何一种进行字符集编码,但是 UTF-8 编码是最常用的编码方式,建议读者使用。

10.2 Java Web 中的中文乱码处理

10.2.1 中文乱码产生原因

在 Java Web 应用中,通常都包括客户端页面(HTML 或者 JSP)、Web 服务器(Tomcat)、Web 应用程序(目前为 Servlet 或者 JSP)和数据库等部分。若在各部分不指定编码格式，客户端页面会采用文件默认的字符集（例如GB2312）提交数据，而Web服务器默认采用的是ISO-8859-1的编码方式解析客户端提交的数据。另外，JDBC驱动程序多数也采用ISO-8859-1的编码方式来解析和存储数据，这就会产生中文乱码问题。也就是说客户端页面、Web服务器、Web应用程序和数据库的编码方式不一致，在Web应用程序运行过程中，若输入的中文字符不进行不同的字符集之间的编码转换，就会导致中文乱码问题的频繁出现。

10.2.2 中文乱码问题解决方案

1. 访问HTML页面或者JSP页面出现中文乱码

例程10-1：演示访问HTML页面出现中文乱码的处理。程序为index.html。

程序index.html代码如下：

```
<!DOCTYPE HTML PUBLIC "-//W3C//DTD HTML 4.01 Transitional//EN">
<html>
    <head>
        <title>HTML页面显示乱码</title>
        <meta http-equiv="keywords" content="keyword1,keyword2,keyword3">
        <meta http-equiv="description" content="this is my page">
        <meta http-equiv="content-type" content="text/html; charset=ISO-8859-1">
    </head>
    <body>
        This is my HTML page.<br>
        欢迎您访问该页面！
    </body>
</html>
```

访问index.html运行效果如图10-1所示。

从图10-1可以看出，访问index.html页面出现了中文乱码，其原因是在HTML头文件中＜meta http-equiv="content-type" content="text/html; charset=ISO-8859-1"＞设置的文件编码方式为ISO-8859-1，该编码方式不支持中文，所以出现了中文乱码。解决中文乱码只需要将charset=ISO-8859-1中的编码方式改为GB2312、GBK和UTF-8中任意一种均可，建议使用UTF-8。修改后的index.html页面运行效果如图10-2所示。头文件中设置charset=编码方式，意味着告诉浏览器打开HTML文件的编码方式或者HTML文件的编码方式。在MyEclipse中可以设置HTML文件的默认编码方式，具体操作步骤：进入Window→Preferences→MyEclipse→Files and Editors→HTML，在Encoding下拉框中选择默认编码方式（建议统一使用UTF-8）即可。一般情况下，HTML默认编码方式和页面设置的编码方式应该一致，否则可能产生中文乱码。

例程10-2：演示JSP页面出现中文乱码的处理。程序为index.jsp。

程序index.jsp代码如下：

图 10-1 index.html 运行效果图

图 10-2 修改后 index.html 运行效果

```
<%@page language="java" import="java.util.*" pageEncoding="ISO-8859-1"%>
<html>
    <head>
        <title>JSP 页面显示乱码</title>
    </head>
    <body>
        This is my JSP page.<br>
        欢迎您访问该页面!
    </body>
</html>
```

访问 index.jsp 运行效果如图 10-3 所示。

从图 10-3 可以看出,访问 index.jsp 页面出现了中文乱码,其原因是 JSP 在 page 指令元素＜%@ page language="java" import="java.util.*" pageEncoding="ISO-8859-1"%＞中的页面属性设置的文件编码方式为 ISO-8859-1,该编码方式不支持中文,所以出现了中文乱码。解决中文乱码只需要将 pageEncoding="ISO-8859-1"中的编码方式改为 GB2312、GBK 和 UTF-8 中任意一种均可,建议使用 UTF-8。修改后的 index.jsp 页面运行效果如图 10-4 所示。当 JSP 被翻译成 Servlet 后,pageEncoding="UTF-8"相当于代码 response.setPageEncoding("UTF-8")。在 MyEclipse 中可以设置 JSP 文件的默认编码方式,具体操作步骤:进入 Window→Preferences→MyEclipse→Files and Editors→JSP,在 Encoding 下拉框中选择默认编码方式(建议统一使用 UTF-8)即可。一般情况下,JSP 默认编码方式和页面设置的编码方式应该一致,否则可能产生中文乱码。

图 10-3 index.jsp 运行效果图

图 10-4 修改后 index.jsp 运行效果

JSP 文件的执行是用 Web 服务器调用 JSP 编译器的过程,JSP 编译器第一步检查 JSP 文件中有没有设置文件编码格式,假如 JSP 文件中没有设置 JSP 文件的编码格式,那么 JSP 编译器会调用 JDK 的 JVM(Java 虚拟机)默认的字符编码格式(Web 服务器所在操作系统的默认文件编码)把 JSP 页面转化为临时的 Java 文件,然后再把它编译成 Unicode 格式的 class 中间字节码文件,并保存在临时文件夹中。通过设置 JSP 文件头中页面编码,让 JSP 编译器,选用正确的字符编码格式转化为临时的 Java 文件。例如:<%@ page ContentType=" text/html;charset=gb2312"%> 或 <%@ page ContentType="text/html;charset=gbk"%>或<%@page contentType="text/html;charset=utf-8"%>,使其生成的 class 中间字节码文件的字符串,不再采用系统默认的编码格式,这样就避免了乱码的产生。

2. 访问 Servlet 出现中文乱码

例程 10-3:演示访问 Servlet 时页面出现中文乱码的处理。程序为 HelloServlet.java。

程序 HelloServlet.java 代码如下:

```java
package com.ch10;
import java.io.IOException;
import java.io.PrintWriter;
import javax.servlet.ServletException;
import javax.servlet.http.HttpServlet;
import javax.servlet.http.HttpServletRequest;
import javax.servlet.http.HttpServletResponse;
public class HelloServlet extends HttpServlet {
    public void doGet(HttpServletRequest request, HttpServletResponse response)
        throws ServletException, IOException {
        response.setContentType("text/html;charset=ISO-8859-1");
        PrintWriter out=response.getWriter();
        out.println ( " <! DOCTYPE  HTML  PUBLIC  \" -//W3C//DTD HTML 4.01 Transitional//EN\">");
        out.println("<HTML>");
        out.println("<HEAD><TITLE>A Servlet</TITLE></HEAD>");
        out.println("<BODY>");
        out.println("欢迎访问 Servlet!");
        out.println("</BODY>");
        out.println("</HTML>");
        out.flush();
        out.close();
    }
}
```

访问 HelloServlet 运行效果如图 10-5 所示。

从图 10-5 可以看出,访问 HelloServlet 出现了中文乱码,其原因是 response 对象通过 out.print()或者 out.println()向客户端输出内容之前,使用 response.setContentType("text/html;charset=ISO-8859-1")设置响应编码方式为 ISO-8859-1,该编码方式不支持中文,所以出现了中文乱码。解决中文乱码只需要将 response.setContentType("text/html;charset=ISO-8859-1")中的编码方式改为 GB2312、GBK 和 UTF-8 中任意一种均可,建议使用 UTF-8。修改后的 HelloServlet 运行效果如图 10-6 所示。

图 10-5　HelloServlet 运行效果

图 10-6　修改后 HelloServlet 运行效果

3. GET 方式传递参数出现中文乱码

使用 GET 方式传递参数有两种方法:将<form>表单属性 method 设置为 GET 通过表单传递参数和通过 URL 直接进行参数传递,GET 方式传递参数含有中文,也可能出现中文乱码问题。

例程 10-4:演示 GET 方式传递参数出现中文乱码的处理。程序为 getSubmit.jsp 和 GetServlet.java。

程序 getSubmit.jsp 代码如下:

```
<%@page language="java" import="java.util.*" pageEncoding="UTF-8"%>
<html>
    <head>
        <title>GET方式传递参数</title>
    </head>
    <body>
        <form action="getServlet" method="get">
            用户名:<input type="text" name="name"><br>
            <input type="submit" value="提交">
        </form>
    </body>
</html>
```

程序 GetServlet.java 代码如下:

```
package com.ch10;
import java.io.IOException;
import java.io.PrintWriter;
import javax.servlet.ServletException;
```

```java
import javax.servlet.http.HttpServlet;
import javax.servlet.http.HttpServletRequest;
import javax.servlet.http.HttpServletResponse;
public class GetServlet extends HttpServlet {
    public void doGet(HttpServletRequest request, HttpServletResponse response)
            throws ServletException, IOException {
        response.setContentType("text/html;charset=UTF-8");
        String name=request.getParameter("name");
        PrintWriter out=response.getWriter();
        out.println ( " <! DOCTYPE  HTML  PUBLIC  \" -//W3C//DTD HTML 4. 01 Transitional//EN\">");
        out.println("<HTML>");
        out.println("<HEAD><TITLE>A Servlet</TITLE></HEAD>");
        out.println("<BODY>");
        out.println("name: "+name);
        out.println("</BODY>");
        out.println("</HTML>");
        out.flush();
        out.close();
    }
}
```

运行 getSubmit.jsp 页面,输入用户名"张三"后单击"提交"按钮,运行结果如图 10-7 所示。由于 getSubmit.jsp 页面表单属性设置为 method="get",即表单数据为 GET 方式提交。另一种方式是,在浏览器地址栏中直接输入"http://localhost:8080/chapter10/getServlet? name=张三",在访问 getServlet 的 URL 后直接给定传递参数的值,运行结果如图 10-8 所示。这种方式也为 GET 方式提交参数。

图 10-7 表单 GET 方式传参运行效果　　图 10-8 URL 直接以 GET 方式传参运行效果

从图 10-7 和图 10-8 可以看出,通过 GET 方式传递参数含有中文时出现中文乱码。由于在 GetServlet 中已经通过代码 response.setContentType("text/html;charset=UTF-8")设置了响应的编码方式为 UTF-8,产生中文乱码的原因是从客户端浏览器提交到后台 GetServlet 的过程中出现编码错误导致了出现中文乱码。其根本原因是客户端浏览器页面采用 UTF-8 编码提交数据,而 Web 服务器默认采用的是 ISO-8859-1 的编码方式解析客户端提交的数据,导致出现中文乱码。

上述中文乱码的解决方法有两种,分别如下。

1) 编码转换

在 GetServlet 中读取客户端页面提交的数据时,在程序语句"String name = request.getParameter("name");"之后使用程序语句"name = new String(name.getBytes("ISO-8859-1"),"UTF-8");"做编码变换(将编码方式从 ISO-8859-1 转化为 UTF-8),从而解决中文乱码问题。

2) 修改 Tomcat 中 conf 目录下的 server.xml

在 server.xml 中为<Connector>元素增加属性设置 URIEncoding="UTF-8"。修改后< Connector >元素如下所示:

```
<Connector port="8080" protocol="HTTP/1.1"
    connectionTimeout="20000"
    redirectPort="8443" URIEncoding="UTF-8"/>
```

使用上述两种方式中任意一种方式对例程 10-4 进行修改,均可以解决中文乱码问题,修改后程序的运行效果如图 10-9 所示。

图 10-9　GET 方式传参运行效果

4. POST 方式传递参数出现中文乱码

将<form>表单属性 method 设置为 POST,使用 POST 方式传递含有中文的参数,也可能出现中文乱码问题。

例程 10-5:演示 POST 方式传递参数出现中文乱码的处理。程序为 postSubmit.jsp 和 PostServlet.java。

程序 postSubmit.jsp 代码如下:

```
<%@page language="java" import="java.util.*" pageEncoding="UTF-8"%>
<html>
    <head>
        <title>POST方式传递参数</title>
    </head>
    <body>
        <form action="postServlet" method="post">
            用户名:<input type="text" name="name"><br>
            <input type="submit" value="提交">
        </form>
    </body>
</html>
```

程序 PostServlet.java 代码如下:

```
package com.ch10;
import java.io.IOException;
import java.io.PrintWriter;
```

```java
import javax.servlet.ServletException;
import javax.servlet.http.HttpServlet;
import javax.servlet.http.HttpServletRequest;
import javax.servlet.http.HttpServletResponse;
public class PostServlet extends HttpServlet {
    public void doPost(HttpServletRequest request, HttpServletResponse response)
            throws ServletException, IOException {
        response.setContentType("text/html;charset=UTF-8");
        String name= request.getParameter("name");
        PrintWriter out=response.getWriter();
        out.println ( " <! DOCTYPE  HTML  PUBLIC  \" -//W3C//DTD HTML 4. 01 Transitional//EN\">");
        out.println("<HTML>");
        out.println("<HEAD><TITLE>A Servlet</TITLE></HEAD>");
        out.println("<BODY>");
        out.println("name: "+name);
        out.println("</BODY>");
        out.println("</HTML>");
        out.flush();
        out.close();
    }
}
```

运行 postSubmit.jsp 页面，输入用户名"张三"后单击"提交"按钮，运行结果如图 10-10 所示。由于 postSubmit.jsp 页面表单属性设置为 method="post"，即表单数据为 POST 方式提交。

从图 10-10 可以看出，通过表单 PSOT 方式传递参数含有中文时出现中文乱码。产生中文乱码的根本原因是客户端浏览器页面采用 UTF-8 编码提交数据，而 Web 服务器默认采用的是 ISO-8859-1 的编码方式解析客户端提交的数据，所以导致出现中文乱码。

上述中文乱码的解决方法有两种，分别如下。

1）设置请求对象的编码方式

在 PostServlet 中读取客户端页面提交的数据程序语句"String name＝request.getParameter("name");"之前，加入语句"request.setCharacterEncoding("UTF-8");"来设置请求编码方式，可以解决中文乱码问题。

2）使用编码过滤器

使用编码过滤器的具体实现和配置的详细过程请读者参见例程 5-2，这种方式也可以解决中文乱码问题，并且这种方式为最常用的防止出现中文乱码的解决方案。

使用上述两种方式中任意一种解决方式对例程 10-5 进行修改，均可以解决中文乱码问题，修改后程序的运行效果如图 10-11 所示。

图 10-10　表单 POST 方式传参运行效果

图 10-11　修改后表单 POST 方式传参运行效果

10.3　案　　例

本章案例在第 9 章案例的基础上增加实现"学生成绩管理"中的全部功能,完善"个人信息管理"模块中其他功能,为案例添加字符编码过滤器防止系统运行过程中出现中文乱码问题。案例主要使用 MVC 设计模式、JavaBean 组件和 Servlet 以及 JSP 技术,JSP 中使用 JSTL 标签、EL 表达式以及 JavaScript 技术来完成数据显示功能。

10.3.1　案例设计

学生成绩管理模块功能主要包括成绩录入、成绩修改以及按班级、按课程、按班级和课程进行成绩的综合查询等。对于系统的不同用户对成绩有不同的操作权限,管理员用户可以录入、修改和查询所有班级、所有课程的成绩,任课教师只可以录入、修改和查询本人所带班级、所带课程的成绩,学生用户仅可以查询本人所修课程的成绩。成绩没有删除功能。

在"个人信息管理"模块中增加对学生信息的显示功能以及管理员、教师和学生的信息修改功能。案例实现除了需要编写实现成绩管理的 JavaBean、Servlet 和 JSP,编写修改个人信息的 Servlet,另外还需要修改之前案例中负责控制登录流程的 LoginServlet、负责控制显示个人信息流程的 ShowPersonServlet(在 LoginServlet 和 ShowPersonServlet 中分别增加对学生用户登录判断和显示信息判断)以及相关的 JSP 页面(在 teacherShow.jsp 和 adminShow.jsp 中增加实现单击"提交"按钮执行个人信息修改的 Servlet)等。

编写字符编码过滤器 EncodeFilter.java 并配置过滤器使得访问该系统的所有路径均须通过过滤器 EncodeFilter.java 的过滤,在过滤器中来设置请求和响应的编码一致,此举可以有效避免系统中出现中文乱码问题。

本章案例使用的主要文件如表 10-1 所示。

表 10-1　本章案例使用的文件

文　件	所在包/路径	功　能
EncodeFilter.java	com.imut.filter	字符编码过滤器
Score.java	com.imut.javabean	封装学生成绩信息的 JavaBean
IScore.java	com.imut.javabean	定义操作学生成绩数据库表增删改查操作方法的接口

续表

文　件	所在包/路径	功　能
Score Impl.java	com.imut.javabean	具体实现IScore接口定义的增删改查等方法
SaveGradeServlet.java	com.imut.servlet.grade	保存学生成绩信息的Servlet
CheckGradeServlet.java	com.imut.servlet.grade	查询学生成绩信息的Servlet
InfoUpdateServlet.java	com.imut.servlet.person	修改个人信息的Servlet
StudentScoreServlet.java	com.imut.servlet.person	显示某一学生所有成绩信息的Servlet
ShowPersonServlet.java	com.imut.servlet.person	负责显示个人信息流程控制的Servlet
LoginServlet.java	com.imut.servlet	负责用户登录流程控制的Servlet
submitGrade.jsp	/grade	显示、输入和提交学生成绩的页面
studentScoreList.jsp	/person	显示某一学生所有成绩信息的页面
studentShow.jsp	/person	某一学生个人信息的页面

10.3.2　案例演示

在浏览器地址栏中输入"http://localhost:8080/ch10/login.jsp",以管理员或者教师用户登录,单击"学生成绩管理"菜单,进入学生成绩显示、输入和提交页面,效果如图10-12所示。如果是以管理员身份登录,页面显示全校所有学生的所有课程成绩信息;如果是以

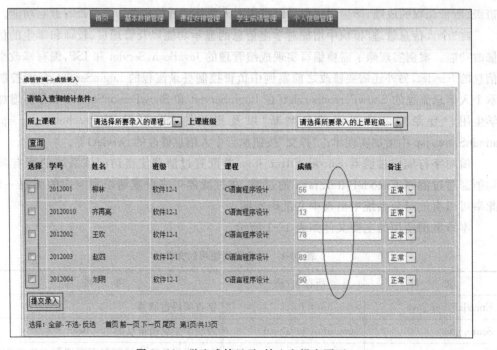

图10-12　学生成绩显示、输入和提交页面

教师身份登录，页面显示登录教师所带课程的学生成绩信息。用户可以通过页面上的"所上课程"和"上课班级"选择对应课程或者班级的学生成绩信息，学生成绩默认是不可操作的，只有选中成绩记录前的复选框，此时该学生成绩记录便可输入或者修改成绩。复选框操作可以通过页面下端左侧的"全部"、"不选"和"反选"链接进行批量操作，"全部"即将当前页面的所有成绩记录选中，"不选"即将所有的成绩记录全部取消选中，"反选"即将所有未选中的记录选中、将选中的取消选中。对于被选中成绩记录的成绩分数和备注信息用户可以进行成绩录入或者修改，然后单击"提交录入"按钮即可。成绩管理还实现了分页显示成绩记录，可以通过页面上的"首页"、"前一页"、"下一页"和"尾页"链接进行操作。

　　以学生用户登录，单击"学生成绩管理"菜单，由于学生不具有学生成绩的输入和修改权限，此时系统弹出提示框，效果如图 10-13 所示。单击"个人信息管理"菜单，进入如图 10-14 所示页面，该页面显示登录学生用户的所有课程成绩列表。在页面中单击"个人信息修改"菜单，运行效果如图 10-15 所示，在该页面上可以对学生信息进行修改，修改完成后，单击"提交"按钮，此时系统弹出学生个人信息修改成功提示框，如图 10-16 所示。

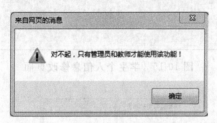

图 10-13　学生用户越权访问提示框

课程	学时	学分	开课学期	任课教师	成绩	备注
高等数学	60	3	1	刘似水	0	
离散数学	48	3	3	刘军	12	正常
Java核心技术	56	3	3	刘军	0	
数据库原理	48	2	3	刘芳	0	
嵌入式技术	32	2	5	刘军	0	
C语言程序设计	40	4	1	王一鸣	56	正常

图 10-14　学生所有课程成绩列表页面

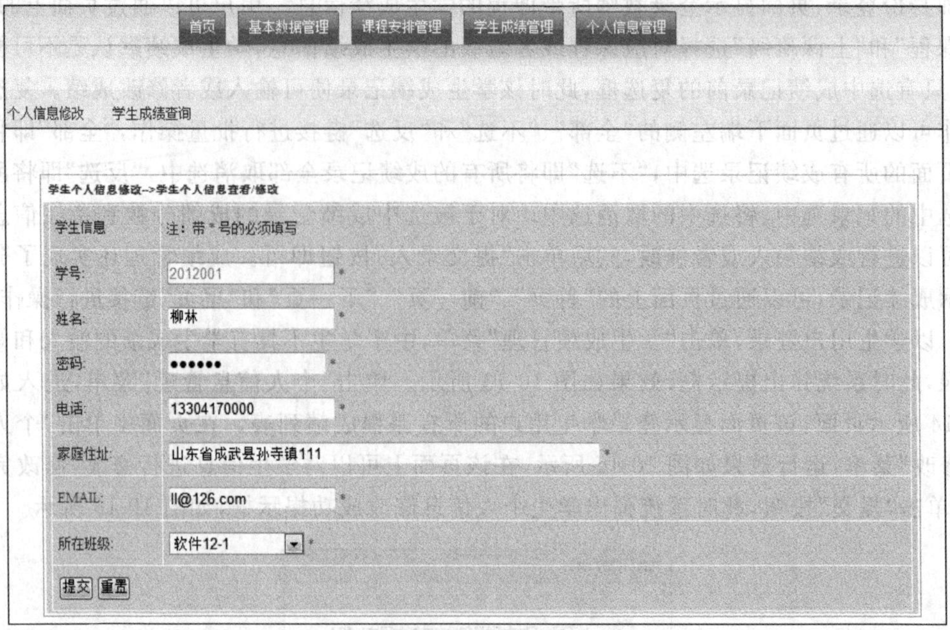

图 10-15　学生个人信息修改页面

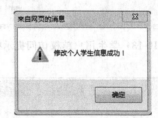

图 10-16　学生个人信息修改成功提示框

10.3.3　代码实现

创建工程 ch10,根据案例设计描述分别给出各部分具体实现。本章主要给出保存学生成绩信息 SaveGradeServlet.java 和显示、查询、输入和提交学生成绩的页面 submitGrade.jsp 的具体代码,其他源文件代码请参考随书电子资源。

程序 SaveGradeServlet.java 代码如下:

```
package com.imut.servlet.grade;
import java.io.IOException;
import java.util.StringTokenizer;
import javax.servlet.ServletException;
import javax.servlet.http.HttpServlet;
import javax.servlet.http.HttpServletRequest;
import javax.servlet.http.HttpServletResponse;
import javax.servlet.http.HttpSession;
```

```java
import com.imut.javabean.IScore;
import com.imut.javabean.Score;
import com.imut.javabean.ScoreImpl;
public class SaveGradeServlet extends HttpServlet {
    public void doGet (HttpServletRequest request, HttpServletResponse response)
          throws ServletException, IOException {
       doPost(request, response);
    }
    public void doPost (HttpServletRequest request, HttpServletResponse response)
          throws ServletException, IOException {
HttpSession session=request.getSession();
                           //获得一个Session对象,用于存放一些提示信息返回到前台
request.setCharacterEncoding("UTF-8");   //设置请求的编码方式
response.setCharacterEncoding("UTF-8");  //设置响应的编码方式
int curPage=1;                    //初始化所要显示的当前页,默认为1
String temp=request.getParameter("curPage");//从前台获得页面数据
if(temp!=null){
    curPage=Integer.parseInt(request.getParameter("curPage"));
}
 //通过request对象从前台表单获得传递到后台的数据
String courseNo=request.getParameter("courseNo");
String classNo=request.getParameter("classNo");
String ids=request.getParameter("ids");
                            //获得提交成绩的记录ID拼接字符串
String scores=request.getParameter("scores");
                            //获得提交的成绩拼接字符串
String remarks=request.getParameter("remarks");
                            //获得提交成绩的备注信息拼接字符串
IScore sdb=new ScoreImpl();
//利用StringTokenizer和"'"将拼接字符串进行分割
StringTokenizer st1=new StringTokenizer(ids,",");
StringTokenizer st2=new StringTokenizer(scores,",");
StringTokenizer st3=new StringTokenizer(remarks,",");
Integer id=null;
Integer grade=null;
String remark=null;
Score score=null;
//通过循环来保存提交的成绩集合
while(st1.hasMoreTokens()&&st2.hasMoreTokens()&&st3.hasMoreTokens()){
    id=Integer.parseInt(st1.nextToken());
    grade=Integer.parseInt(st2.nextToken());
    remark=st3.nextToken();
```

```
            score=sdb.findScoreByID(id);
            score.setScore(grade);
            score.setRemark(remark);
            sdb.updateScore(score);
        }
        //向session对象中存放提示信息
        session.setAttribute("message","学生成绩提交成功!");
        //构造所要跳转路径字符串
        String str="/grade/checkGradeServlet? curPage="+curPage;
        if(courseNo!=null&&!courseNo.equals("")){
            str+="&courseNo="+courseNo;
        }
        if(classNo!=null&&!classNo.equals("")){
            str+="&classNo="+classNo;
        }
        request.getRequestDispatcher(str).forward(request, response);
    }
}
```

程序 submitGrade.jsp 代码如下：

```
<%@page language="java" contentType="text/html; charset=UTF-8" pageEncoding="UTF-8"%>
<%@taglib uri="http://java.sun.com/jsp/jstl/core" prefix="c"%>
<!DOCTYPE html PUBLIC "-//W3C//DTD XHTML 1.0 Transitional//EN" "http://www.w3.org/TR/xhtml1/DTD/xhtml1-transitional.dtd">
<html xmlns="http://www.w3.org/1999/xhtml">
<head>
<title>成绩管理</title>
<link rel="stylesheet" type="text/css" id="css" href="../style/main.css" />
<link rel="stylesheet" type="text/css" id="css" href="../style/style1.css" />
<link rel="stylesheet" type="text/css" id="css" href="../style/style.css" />
<style type="text/css">
<!--
table{border-spacing:1px; border:1px solid #A2C0DA;}
td, th{padding:2px 5px;border-collapse:collapse;text-align:left; font-weight: normal;}
thead tr th{height:50px;background:#B0D1FC;border:1px solid white;}
thead tr th.line1{background:#D3E5FD;}
thead tr th.line4{background:#C6C6C6;}
tbody tr td{height:35px;background:#CBE2FB;border:1px solid white; vertical-align:middle;}
tbody tr td.line4{background:#D5D6D8;}
tbody tr th{height:35px;background: #DFEDFF;border:1px solid white; vertical-align:middle;}
```

```
    tfoot tr td{height:35px;background:#FFFFFF;border:1px solid white; vertical-
align:middle;}
    -->
    </style>
    <script type="text/javascript" src="../js/common.js"></script>
</head>
<c:if test="${!(empty sessionScope.message)}">
    <script type="text/javascript">
        alert('<c:out value="${sessionScope.message}"/>');
    </script>
    <c:remove var="message" scope="session"/>
</c:if>
<script type="text/javascript">
    /*如果单选框被选中使得对应记录的输入控件变为可操作,否则使得对应记录的输入控件
      变为不可操作*/
    function vary(){
        var boxes=document.getElementsByName("ID");
        var score=document.getElementsByName("score");
        var remark=document.getElementsByName("remark")
        for(var i=0;i<boxes.length;i++){
            if(boxes[i].checked==true){
                score[i].disabled=false;
                remark[i].disabled=false;
            }else{
                score[i].disabled=true;
                remark[i].disabled=true;
            }
        }
    }
    /*使得页面所有单选框全部选中*/
    function selAllCheckbox(checkboxName) {
        o=document.getElementsByName(checkboxName);
        for (i=0; i<o.length; i++) {
            o[i].checked=true;
        }
        vary();
    }
    /*使得页面所有单选框不被选中*/
    function unselAllCheckbox(checkboxName) {
        o=document.getElementsByName(checkboxName);
        for (i=0; i<o.length; i++) {
            o[i].checked=false;
        }
        vary();
```

```javascript
    /*使得页面所有单选框被反选,即选中的变为不被选中,不被选中的变为选中 */
    function reAllCheckbox(checkboxName) {
        o=document.getElementsByName(checkboxName);
        for (i=0; i<o.length; i++) {
            if (o[i].checked==false) {
                o[i].checked=true;
            } else {
                o[i].checked=false;
            }
        }
        vary();
    }
</script>
<script type="text/javascript">
    /*保存录入或者修改的成绩*/
    function saveScore(sourceURL,currentPageNum){
        /*通过控件名称来获得列表中所有记录的ID,将其拼接成一个字符串 ids */
        var ids="";
        var boxes=document.getElementsByName('ID');
        for(var i=0;i<boxes.length;i++)
            if(boxes[i].checked==true) ids+=boxes[i].value+",";
        if(ids.length>0){
            ids=ids.substr(0,ids.length-1);
        }
        /*通过控件名称来获得列表中所有记录的录入或者修改的成绩,将其拼接成一个字符
          串 scores */
        var scores="";
        var score=document.getElementsByName('score');
        for(var i=0;i<score.length;i++)
            if(score[i].disabled==false) scores+=score[i].value+",";
        if(scores.length>0){
            scores=scores.substr(0,scores.length-1);
        }
        /*通过控件名称来获得列表中所有记录的成绩备注信息,将其拼接成一个字符串
          remarks */
        var remarks="";
        var remark=document.getElementsByName('remark');
        for(var i=0;i<remark.length;i++)
            if(remark[i].disabled==false) remarks+=remark[i].value+",";
        if(remarks.length>0){
            remarks=remarks.substr(0,remarks.length-1);
        }
        //获得当前页面上班级和课程下拉框的选定值
```

```
            var courseNo=document.getElementById('courseNo').value;
            var classNo=document.getElementById('classNo').value;
            document.location = sourceURL + '? scores = ' + scores + ' &curPage = ' +
            currentPageNum+ ' &ids = ' + ids + ' &remarks = ' + remarks + ' &courseNo = ' +
            courseNo+ '&classNo= '+classNo;
            //执行路径 sourceURL,并传递参数到后台,后台处理需要这些参数。参数包括:ids、
            scores、remarks 三个拼接字符串,courseNo、classNo 当前页面上班级和课程下拉框
            的选定值,当前页面码 curPage。
        }
    </script>
    <body>
    <div id="btm">
    <div id="main">
        <div id="header">
            <div id="top"></div>
            <div id="logo">
                <h1>成绩管理</h1></div>
            <div id="mainnav">
                <ul>
                <li><a href="../index.jsp">首页</a></li>
                <li><a href="../base/listAllTeacherServlet">基本数据管理</a></li>
                <li><a href="../arrange/listAllArrangeServlet">课程安排管理</a></li>
                <li><a href="checkGradeServlet">学生成绩管理</a></li>
                <li><a href="../person/showPersonServlet">个人信息管理</a></li>
                </ul>
                <span>
                </span>
            </div>
        </div>
        <div id="content" align="center">
            <div id="center">
        <BR/><BR/>
            <form method="post" action="checkGradeServlet" >
            <table width="800" align="center" cellpadding="0" cellspacing="0">
                <thead>
                    <tr>
                        <td width="70%"><h5>成绩管理-->成绩录入   
                            </h5></td>
                    </tr>
                </thead>
                <tr>
                    <td colspan="2" width="100%">
                        <table width="100%">
                            <thead>
```

```html
<tr>
    <th align="center" class="line1" scope="col" colspan="12">
        <b><font color="red">请输入查询统计条件:</font></b>
    </th>
</tr>
<tr>
    <th align="center" scope="col" colspan="3">
        <b>所上课程</b>
    </th>
    <th align="center" scope="col" colspan="3">
        <select id="courseNo" name="courseNo">
            <option value="">请选择所要录入的课程…</option>
            <c:forEach items="${requestScope.courses}" var="course">
                <c:choose>
                    <c:when test="${course.courseNo == sessionScope.courseNo}">
                        <option value="${course.courseNo}" selected>${course.courseName}</option>
                    </c:when>
                    <c:otherwise>
                        <option value="${course.courseNo}">${course.courseName}</option>
                    </c:otherwise>
                </c:choose>
            </c:forEach>
        </select>
    </th>
    <th align="center" scope="col" colspan="3">
        <b>上课班级</b>
    </th>
    <th align="center" scope="col" colspan="3">
        <select id="classNo" name="classNo">
            <option value="">请选择所要录入的上课班级…</option>
            <c:forEach items="${requestScope.classTbls}" var="classTbl">
                <c:choose>
                    <c:when test="${classTbl.classNo ==
```

```
                            sessionScope.classNo}">
                        <option value=" ${classTbl.
                        classNo}" selected > ${classTbl.
                        className}</option>
                                                    </c:when>
                        <c:otherwise>
                        <option value=" ${classTbl.
                        classNo}" > ${classTbl.className}
                        </option>
                        </c:otherwise>
                        </c:choose>
                    </c:forEach>
                </select>
            </th>
        </tr>
<tr>
    <th align="center" class="line1" scope="col" colspan="12">
        <input type="submit" value="查询"/>
    </th>
</tr>
<tr>
    <th width="5%" align="center" class="line1" scope="col"
    colspan="1">
        <b>选择</b>
    </th>
    <th width="10%" align="center" class="line1" scope="col"
    colspan="2">
        <b>学号</b>
    </th>
    <th width="10%" align="center" scope="col" colspan="2">
        <b>姓名</b>
    </th>
    <th width="15%" align="center" scope="col" colspan="2">
        <b>班级</b>
    </th>
    <th width="15%" align="center" colspan="2">
        <b>课程</b>
    </th>
    <th width="15%" align="center" colspan="2">
        <b>成绩</b>
    </th>
    <th width="15%" align="center" colspan="1">
        <b>备注</b>
    </th>
```

```
                </tr>
            </thead>
            <tbody>
                <c:forEach items="${requestScope.list}" var="score">
                <tr>
                    <td width="5%" align="center" colspan="1">
                        <input type="checkbox" name="ID" alue="${score.id}"
                        onchange="javascript:vary();" />
                    </td>
                    <td width="10%" align="center" colspan="2">
                        ${score.student.studentNo }
                    </td>
                    <td width="10%" align="center" colspan="2">
                        ${score.student.name }
                    </td>
                    <td width="15%" align="center" colspan="2">
                        ${score.arrange.classTbl.className }
                    </td>
                    <td width="15%" align="center" colspan="2">
                        ${score.arrange.course.courseName }
                    </td>
                    <td width="15%" align="center" colspan="2">
                        <input disabled="disabled" type="text" name="score"
                        value="${score.score }"/>
                    </td>
                    <td width="15%" align="center" colspan="1">
                        <select disabled="disabled" name="remark">
                            <c:if test="${score.remark=='A'}">
                            <option value="A" selected>正常</option>
                            <option value="B">缺考</option>
                            <option value="C">作弊</option>
                            <option value="D">其他</option>
                            </c:if>
                            <c:if test="${score.remark=='B'}">
                            <option value="A">正常</option>
                            <option value="B" selected>缺考</option>
                            <option value="C">作弊</option>
                            <option value="D">其他</option>
                            </c:if>
                            <c:if test="${score.remark=='C'}">
                            <option value="A">正常</option>
                            <option value="B">缺考</option>
                            <option value="C" selected>作弊</option>
                            <option value="D">其他</option>
```

```
                </c:if>
                <c:if test="${score.remark=='D'}">
                            <option value="A">正常</option>
                        <option value="B">缺考</option>
                        <option value="C">作弊</option>
                        <option value="D" selected>其他</option>
                </c:if>
                <c:if test="${score.remark==null}">
                            <option value="A" selected>正常</option>
                        <option value="B">缺考</option>
                        <option value="C">作弊</option>
                        <option value="D">其他</option>
                </c:if>
            </select>
                </td>
                </tr>
            </c:forEach>
                <tr>
                    <td colspan="12" align="right">
<input type="button" value="提交录入" onclick="javascript:saveScore('saveGradeServlet',${requestScope.curPage });"/>
                    </td>
                </tr>
            </tbody>
            <tbody>
                <tr>
                    <td align="center" colspan="12">
                        选择:<a href="#" onclick="javascript:selAllCheckbox('ID');">全部</a>-
                            <a href="#" onclick="javascript:unselAllCheckbox('ID');">不选</a>-
                            <a href="#" onclick="javascript:reAllCheckbox('ID');">反选</a>    
    <c:if test="${requestScope.pageCount!=0&&requestScope.curPage!=1}">
            <a href='<c:url value="checkGradeServlet?curPage=1&classNo=${sessionScope.classNo }&courseNo=${sessionScope.courseNo }"/>'>首页</a>
                    <a href='<c:url value="checkGradeServlet?curPage=${requestScope. curPage - 1 } &classNo = ${ sessionScope. classNo } &courseNo =${sessionScope.courseNo }"/>'>前一页</a>
        </c:if>
<c:if test="${requestScope.pageCount==0||requestScope.curPage==1}">首页 前一页</c:if>
                            <c:if test="${requestScope.pageCount!=0&&requestScope.curPage!=requestScope.pageCount}"><a href='<c:url
```

```
                        value =" checkGradeServlet? curPage = ${ requestScope.
                        curPage+1}&classNo=${sessionScope.classNo }&courseNo=
                        ${sessionScope.courseNo }"/>'>下一页</a><a href='<c:url
                        value =" checkGradeServlet? curPage = ${ requestScope.
                        pageCount}&classNo=${sessionScope.classNo }&courseNo=
                        ${sessionScope.courseNo }"/>'>尾页</a>
            </c:if>
            < c: if test =" ${ requestScope. pageCount = = 0 | | requestScope. curPage = =
        requestScope.pageCount}">下一页 尾页</c:if> 第${requestScope.curPage}页/
        共${requestScope.pageCount}页  </td></tr>
                    </tbody>
                </table>
            </td>
                </tr>
                </table>
            </form>
        </div>
    </div>
    <div id="footer">
        <div id="copyright">
            <div id="copy">
            <p align="center">CopyRight&copy;2010</p>
            <p>内蒙古工业大学信息工程学院软件工程系</p>
                </div>
        </div>
        <div id="bgbottom"></div>
    </div>
</div>
</div>
</body>
</html>
```

【代码分析】程序 SaveGradeServlet.java 从前台页面获得当前页码、班级编号、课程编号和一组成绩的 ID、分数、备注信息合并串,分别做相应的处理后进行数据的保存或者修改,最后跳转到成绩录入、提交和显示页面。程序 submitGrade.jsp 使用 JSTL 中的循环控制标签<c:foreach>来迭代处理从 request 作用范围中获取的学生成绩集合对象 list,并通过实现分页显示。页面还通过使用 JSTL 中的循环控制标签<c:foreach>来迭代处理从 request 作用范围中获取的课程集合对象 courses 和班级集合对象 classTbls,然后使用<c:choose>和<c:when>、<c:otherwise>标签进行条件判断,使得当前课程和班级选项被选中,同时将其他所有的课程选项和班级选项显示在对应的下拉列表中。另外,使用 JavaScript 技术实现了函数 vary()、selAllCheckbox(checkboxName)、unselAllCheckbox(checkboxName)和 reAllCheckbox(checkboxName),函数 vary()实现了如果单选框被选中使得对应记录的输入控件变为可操作,否则使得对应记录的输入控件变为不可操作;函数 selAllCheckbox(checkboxName)实现了使页面所有单选框全部选中;函数 unselAllCheckbox(checkboxName)实现了使页面所有单选框不被选中;函数 reAllCheckbox(checkboxName)实现了使得页面所有单选框被反选,即选中的变为不被选中,不被选中的变为选中。

习 题

1. 选择题

(1) UTF-8 编码方式存储字符使用的字节数为( )。
    A. 1 字节   B. 2 字节   C. 3 字节   D. 变长字节

(2) 以下哪一种编码不支持中文字符?( )
    A. ISO-8859-1   B. GB2312   C. GBK   D. UTF-8

(3) 当利用 request 的方法获取 Form 中元素时,默认情况下字符编码是( )。
    A. ISO-8859-1   B. GB2312   C. GBK   D. UTF-8

2. 填空题

(1) JSP 中的内置对象 response 中的_____方法,能够用来解决中文乱码问题。

(2) 阅读以下 JSP 页面代码:

```
<%@page language="java" import="java.util.*" pageEncoding="ISO-8859-1"%>
<html>
    <head>
        <title>JSP 页面显示乱码</title>
    </head>
    <body>
        This is my JSP page.<br>
        欢迎您访问该页面!
    </body>
</html>
```

以上代码运行后页面出现了中文乱码,解决中文乱码只需将_____修改为_____。

3. 简答题

(1) 简述在 Java Web 开发中中文乱码产生的原因。
(2) 简述在页面中出现中文乱码的解决方法。

Java Web 中的异常处理

在 Java Web 应用开发中,程序出现一些异常是不可避免的,对于这些异常如果程序员不采取任何处理措施,那么异常信息将直接打印到客户端浏览器页面。用户多为非专业人员,这些信息可能会让他们感觉莫名其妙,这是一种极不友好的做法,所以进行异常处理是 Java Web 程序开发者必须考虑的问题。本章将以理论与实践相结合的方式介绍 Java Web 程序中的异常、异常处理一般准则、异常处理,以及利用 Web 服务器对异常的处理等内容。

11.1 Java Web 程序异常处理

11.1.1 Java Web 异常概述

一般而言,对于 Java Web 应用来讲,其执行流程为:用户通过客户端页面向 Web 服务器发送请求之后,首先是到控制器(通常为 MVC 的 Controller 即 Servlet),在控制层会调用业务逻辑层 Service 的相应方法,Service 层会调用数据操作层 Dao 相应方法获取数据,最后操作结果数据会汇总到控制器,然后通过控制器控制转发到指定的结果显示页面到客户端浏览器。

在 Java Web 应用执行过程中,在控制层、业务逻辑层 Service、数据操作层 Dao 这三层其实都有可能发生异常,例如,Dao 层可能会产生 SQLException,Service 层可能会产生 NullPointException,控制层可能会产生 IOException,一旦发生异常但程序员未做处理,那么该层将不会再继续向下执行,而是将异常向上抛出即抛出异常给调用自己的方法,如果 Dao 层、Service 层、控制层都没有进行异常处理,异常信息会抛到服务器,然后服务器会把异常信息直接打印到页面,其实这种错误信息对于用户来说是毫无意义的,因为作为非专业人员他们通常看不懂这是什么意思。

Java 中的异常处理有两种方法:一是通过关键字 throws 直接抛出异常,将异常抛给方法的调用者;另一种是使用 try…catch…finally 块来捕获处理异常,但是在 catch 块中不做任何操作,或者仅调用方法 printStackTrace()把异常信息打印到控制台。第一种方法最后造就了上述的结果将异常信息直接打印到页面上,而第二种方法尽管页面没有报错,但也不执行用户的请求,其实这就是出现了程序 Bug。

11.1.2　Java Web异常处理一般准则

通过11.1.1节可以看出，对于Java Web应用程序中出现的异常采用Java常用的异常处理方式是行不通的。那是不是对于Java Web应用程序中控制层、业务逻辑层Service、数据操作层Dao都需要进行异常处理呢？

1. Dao层异常处理

Dao层是用来实现数据库操作的模块，就目前读者所掌握的知识而言，只能用JDBC编程来实现数据库操作，不可能不做异常处理，因为在JDBC编程中SQLException异常一定会出现。SQLException异常为检查型异常即非运行时异常，要求程序员必须进行处理，否则程序编译无法通过。此时程序员不能简单地使用try…catch块对异常进行处理，这样所能做的也只是将异常信息打印到控制台，上层模块将无法获得该异常相关情况。在Dao层最好的处理办法是将异常以另外一种非检查的方式向上抛出。具体做法就是自定义一个异常类使其继承运行时异常类型RuntimeException，在catch块中抛出自定义异常类型对象。这样做的好处如下。

（1）Dao层的接口不被异常"污染"。根据Java语法规则，如果Dao层方法抛出了SQLException，那么在Dao层接口中也要抛出SQLException。随着学习的深入，Dao层的数据库操作换作采用Hibernate框架（实现数据持久化的Java开源框架）实现，那么Dao层接口将不再抛出SQLException异常，这样Dao层接口抛出异常就是对接口的"污染"（Dao层方法改变对应的接口方法也要改变，违背了模块独立性原则）。

（2）Dao层产生的异常向上传递给更高层处理，异常的错误原因不会丢失，便于排查错误或进行捕获处理。

2. Service层异常处理

Service层实现程序的业务逻辑，通常依赖Dao层，Service层的一个方法有可能调用一个或者多个Dao层方法。该层的异常处理通常根据实际情况对Dao层抛出的异常可以不处理或者进行异常处理，并将本层产生的异常以检查的方式向上抛出。具体做法就是如果对Dao层抛出的异常进行处理，在catch块中将异常继续向上抛出。并自定义一个异常类继承Exception，如果本层产生异常将向上抛出该异常类的对象。

3. 控制层异常处理

控制层实现程序的执行流程控制，通常依赖Service层，控制层的一个方法有可能调用一个或者多个Service层方法。该层中在控制转发之前一般要将Service层抛出的异常进行处理，根据异常的不同，转向不同的异常页面，然后在异常页面上以友好的错误提示信息告诉用户系统程序出现异常状况。

11.1.3　Java Web异常处理实例

例程11-1：通过MVC设计模式实现登录功能演示Java Web中的程序异常处理。

程序由数据操作层Dao(IUserDao.java、UserDaoImpl.java)、业务逻辑处理层Service(IUserService.java、UserServiceImpl.java、User.java)、控制层(LoginServlet.java)、自定义异常类(DaoException.java、ServiceException.java)和表现层(login.jsp、index.jsp、daoException.jsp、serviceException.jsp)组成。

自定义异常类DaoException.java代码如下。

```java
package com.ch11.common;
public class DaoException extends RuntimeException {
    public DaoException() {
        super();
    }
    public DaoException(String message, Throwable cause,
            boolean enableSuppression, boolean writableStackTrace) {
        super(message, cause, enableSuppression, writableStackTrace);
    }
    public DaoException(String message, Throwable cause) {
        super(message, cause);
    }
    public DaoException(String message) {
        super(message);
    }
    public DaoException(Throwable cause) {
        super(cause);
    }
}
```

自定义异常类ServiceException.java代码如下。

```java
package com.ch11.common;
public class ServiceException extends Exception{
    public ServiceException() {
        super();
    }
    public ServiceException(String message, Throwable cause,
            boolean enableSuppression, boolean writableStackTrace) {
        super(message, cause, enableSuppression, writableStackTrace);
    }
    public ServiceException(String message, Throwable cause) {
        super(message, cause);
    }
    public ServiceException(String message) {
        super(message);
    }
    public ServiceException(Throwable cause) {
        super(cause);
```

        }
    }

数据操作层 Dao(IUserDao.java、UserDaoImpl.java)程序代码如下。

接口 IUserDao.java：

```java
package com.ch11.dao;
import com.ch11.service.User;
public interface IUserDao {
    //用户登录
    public User login(String name,String password);
}
```

接口 IUserDao.java 实现类 UserDaoImpl.java：

```java
package com.ch11.dao;
import java.sql.Connection;
import java.sql.DriverManager;
import java.sql.PreparedStatement;
import java.sql.ResultSet;
import java.sql.SQLException;
import com.ch11.common.DaoException;
import com.ch11.service.User;
public class UserDaoImpl implements IUserDao {
    private final static String driver="oracle.jdbc.driver.OracleDriver";
    private final static String url="jdbc:oracle:thin:@127.0.0.1:1521:XE";
    private final static String userName="webbook";
    private final static String pwd="webbook";
    //用户登录
    public User login(String name, String password) {
        Connection conn=null;
        PreparedStatement pstm=null;
        ResultSet rs=null;
        User user=null;
        try {
            Class.forName(driver);
            conn=DriverManager.getConnection(url,userName,pwd);
            String sql="select * from usertbl where name=? and password=?";
            pstm=conn.prepareStatement(sql);
            pstm.setString(1, name);
            pstm.setString(2, password);
            rs=pstm.executeQuery();
            if(rs.next()){
                user=new User();
                user.setName(rs.getString(1));
                user.setPassword(rs.getString(2));
                user.setSex(rs.getString(3));
```

```
                    user.setPrivence(rs.getString(4));
                    user.setAuthor(rs.getString(5));
                }
            } catch (ClassNotFoundException e) {
                e.printStackTrace();
                throw new DaoException("数据库操作异常,请稍后重试!");
            } catch (SQLException e) {
                e.printStackTrace();
                throw new DaoException("数据库操作异常,请稍后重试!");
            } finally{
                try {
                    if(rs!=null)rs.close();
                    if(pstm!=null)pstm.close();
                    if(conn!=null)conn.close();
                } catch (SQLException e) {
                    e.printStackTrace();
                    throw new DaoException("数据库操作异常,请稍后重试!");
                }
            }
            return user;
        }
}
```

业务逻辑处理层 Service(IUserService.java、UserServiceImpl.java、User.java)程序代码如下。

接口 IUserService.java：

```
package com.ch11.service;
import com.ch11.common.ServiceException;
public interface IUserService {
    //用户登录
    public User login(String name,String password) throws ServiceException;
}
```

接口 IUserService.java 实现类 UserServiceImpl.java：

```
package com.ch11.service;
import com.ch11.common.DaoException;
import com.ch11.common.ServiceException;
import com.ch11.dao.IUserDao;
import com.ch11.dao.UserDaoImpl;
public class UserServiceImpl implements IUserService{
    private IUserDao iud=new UserDaoImpl();
    public User login(String name, String password) throws ServiceException {
        User user=null;
        try{
            user=iud.login(name, password);
```

```
            if(user==null){
                throw new ServiceException("用户不存在,请重试!");
            }
        }catch(DaoException e){
            e.printStackTrace();
            throw new DaoException(e.getMessage(), e);
        }
        return user;
    }
}
```

User.java 程序代码与例程 7-2 中的 User.java 代码一致。
控制层 LoginServlet.java 程序代码如下：

```
package com.ch11.controller;
import java.io.IOException;
import javax.servlet.ServletException;
import javax.servlet.http.HttpServlet;
import javax.servlet.http.HttpServletRequest;
import javax.servlet.http.HttpServletResponse;
import com.ch11.common.DaoException;
import com.ch11.common.ServiceException;
import com.ch11.service.IUserService;
import com.ch11.service.User;
import com.ch11.service.UserServiceImpl;
public class LoginServlet extends HttpServlet {
    private IUserService ius=new UserServiceImpl();
    protected void doGet(HttpServletRequest req, HttpServletResponse resp)
            throws ServletException, IOException {
        doPost(req, resp);
    }
    protected void doPost(HttpServletRequest req, HttpServletResponse resp)
            throws ServletException, IOException {
        req.setCharacterEncoding("UTF-8");
        resp.setContentType("text/html;charset=UTF-8");
        String name=req.getParameter("name");
        String password=req.getParameter("password");
        User user=null;
        try {
            user=ius.login(name, password);
            req.setAttribute("name", user.getName());
            req.getRequestDispatcher("index.jsp").forward(req, resp);
        } catch (ServiceException e) {
            e.printStackTrace();
            req.setAttribute("se", e);
```

```
            req.getRequestDispatcher("serviceException.jsp").forward(req,
resp);
        } catch (DaoException e) {
            e.printStackTrace();
            req.setAttribute("de", e);
            req.getRequestDispatcher("daoException.jsp").forward(req, resp);
        }
    }
}
```

表现层(login.jsp、index.jsp、daoException.jsp、serviceException.jsp)程序代码如下：
登录页面 login.jsp：

```
<%@page language="java" import="java.util.*" pageEncoding="UTF-8"%>
<html>
    <head>
        <title>登录页面</title>
    </head>
    <body>
        <form action="loginServlet" method="psot">
            用户名:<input type="text" name="name"><br>
            密码:<input type="password" name="password"><br>
            <input type="submit" value="登录">
        </form>
    </body>
</html>
```

主页面 index.jsp：

```
<%@page language="java" import="java.util.*" pageEncoding="UTF-8"%>
<html>
    <head>
        <title>系统主页面</title>
    </head>
    <body>
        恭喜您<%=request.getAttribute("name") %>,登录成功,欢迎使用 XX 系统!<br>
    </body>
</html>
```

数据操作层异常信息显示页面 daoException.jsp：

```
<%@page language="java" import="java.util.*" pageEncoding="UTF-8"
isErrorPage="true"%>
<html>
    <head>
        <title>数据操作层异常页面</title>
    </head>
```

```
        <body>
                <%=((Exception)request.getAttribute("de")).getMessage() %><br>
        </body>
</html>
```

业务逻辑操作层异常信息显示页面serviceException.jsp：

```
<%@page language="java" import="java.util.*" pageEncoding="UTF-8"
isErrorPage="true"%>
<html>
    <head>
            <title>业务层异常页面</title>
    </head>
    <body>
        <%=((Exception)request.getAttribute("se")).getMessage() %><br>
    </body>
</html>
```

LoginServlet的配置代码如下：

```
<servlet>
<servlet-name>login</servlet-name>
<servlet-class>com.ch11.controller.LoginServlet</servlet-class>
    </servlet>
    <servlet-mapping>
<servlet-name>login</servlet-name>
<url-pattern>/loginServlet</url-pattern>
    </servlet-mapping>
```

例程11-1中自定义异常类DaoException和ServiceException分别为Dao层和Service层抛出的异常类型。DaoException继承了RuntimeException，为运行时异常即非检查型异常，对于这种异常在Service层编译器不强制要求程序员进行异常处理，程序员可以根据实际需要选择处理或者不处理。ServiceException继承了Exception，为检查型异常，对于这种异常在控制层编译器强制要求程序员必须进行异常处理。自定义异常类DaoException和ServiceException实现比较简单，定义几个与父类构造方法对应的构造方法，在构造方法中直接调用父类的构造方法。

数据操作层Dao的IUserDao.java和UserDaoImpl.java分别实现了登录操作的接口和接口实现类，在接口实现类UserDaoImpl的登录方法中对于产生的数据库相关操作DaoException异常采用向上抛出异常的方式抛给Service层。

业务逻辑处理层Service的IUserService.java和UserServiceImpl.java分别实现了登录业务的接口和接口实现类，在接口实现类UserServiceImpl的登录方法中对于产生的ServiceException异常采用向上抛出异常的方式抛给控制层。

控制层LoginServlet.java中调用Service中的登录方法，并对Service抛出的异常进行处理，根据是否产生异常与产生异常类型的不同结果转向登录成功页面与不同异常结果页面。在转向结果页面之前,将用户信息或者产生异常信息保存在 request 对象中,便于页面显示对应信息。程序不发生异常转向 index.jsp 页面,产生 DaoException.java 或者 ServiceException.java 类型异常将转向 daoException.jsp 页面或者 serviceException.jsp 页面。

部署项目启动 Web 服务器,首先访问登录页面 login.jsp,如图 11-1 所示。输入正确的用户名"zhangsan"和密码"zs123",单击"登录"按钮,程序运行至登录成功页面,如图 11-2 所示。如果在登录页面输入不正确的用户名或者密码,程序运行至业务层出现异常页面,如图 11-3 所示。如果程序运行过程中,出现了数据库关闭,SQL 语句拼写错误等数据库问题时,登录操作后程序运行至数据库操作出现异常页面,如图 11-4 所示。

图 11-1 登录页面

图 11-2 登录成功页面

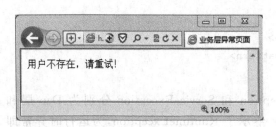
图 11-3 业务逻辑层出现异常页面

图 11-4 数据操作层出现异常页面

11.2 Web 服务器中处理异常

11.1 节内容介绍了在 Java Web 应用程序中通过编程进行异常处理,在程序中对产生的异常不做处理,异常就会抛给 Web 服务器,异常信息直接打印到页面上。实际上应用程序中不进行异常处理,在 Web 服务器中也可以对异常进行处理,这样也可以避免将异常信息内容直接显示给访问者。

Web 服务器对异常的处理是通过 web.xml 配置 <error-page> 元素来实现的,通过这种方式可以支持两种类型的异常处理:根据 HTTP 请求返回的状态码和 Java 异常类型对异常进行拦截处理。

11.2.1 HTTP 状态码拦截

例程 11-2：演示根据 HTTP 请求返回的状态码进行异常拦截处理。程序为 error_404.jsp、error_500.jsp 和 typeTransfer.jsp。

程序 error_404.jsp 代码如下：

```
<%@page language="java" import="java.util.*"
pageEncoding="UTF-8"%>
<!DOCTYPE HTML PUBLIC "-//W3C//DTD HTML 4.01 Transitional//EN">
<html>
    <head>
        <title>404 错误</title>
    </head>
    <body>
        <h1>系统找不到您所访问的资源,请重试!</h1>
    </body>
</html>
```

程序 error_500.jsp 代码如下：

```
<%@page language="java" isErrorPage="true" import="java.util.*"
pageEncoding="UTF-8"%>
<!DOCTYPE HTML PUBLIC "-//W3C//DTD HTML 4.01 Transitional//EN">
<html>
    <head>
        <title>500 错误</title>
    </head>
    <body>
        <h1>服务器出现异常,原因:<%=exception.getMessage() %></h1>
    </body>
</html>
```

程序 typeTransfer.jsp 代码如下：

```
<%@page language="java" import="java.util.*" pageEncoding="UTF-8"%>
<!DOCTYPE HTML PUBLIC "-//W3C//DTD HTML 4.01 Transitional//EN">
<html>
    <head>
        <title>类型转换</title>
    </head>
    <body>
        <% Integer.parseInt("aaa"); %>
    </body>
</html>
```

配置文件 web.xml 中对错误页面配置代码如下：

```
<error-page>
    <error-code>404</error-code>
    <location>/error_404.jsp</location>
</error-page>
<error-page>
    <error-code>500</error-code>
    <location>/error_500.jsp</location>
</error-page>
```

配置信息中＜error-page＞元素用来对错误页面进行配置。＜error-code＞元素信息表示错误代码，404 表示找不到用户所访问的资源错误，500 表示程序运行过程中发生服务器错误，＜location＞元素信息表示当发生对应代码的异常错误时将执行的页面。当程序运行过程中，发生找不到用户所访问的资源错误或者服务器错误，程序将自动跳转到 error_404.jsp 或者 error_500.jsp 页面。error_404.jsp 和 error_500.jsp 页面就是用来显示异常信息给用户。typeTransfer.jsp 页面中 Java 脚本要把字符串"aaa"转化为一个整数，运行时将发生服务器错误（类型转换异常）。

重新部署工程项目并启动 Web 服务器，当在浏览器地址栏中输入"http://localhost:8080/chapter11/xxxx"，"xxxx"为应用 chapter11 中不存在的资源，程序运行结果如图 11-5 所示。当用户访问 typeTransfer.jsp 页面时，程序运行结果如图 11-6 所示。有一点读者需要注意，运行该例程时需要对浏览器进行设置："工具"→"Internet 选项"→"高级"，取消"显示友好 http 错误信息"选项。

图 11-5　404 错误异常页面

图 11-6　500 错误异常页面

11.2.2　Java 异常类型拦截

例程 11-3：演示根据 Java 异常类型进行异常拦截处理。程序为 error_exception.jsp 和 NullPointExceptionTest.jsp。

程序 NullPointExceptionTest.jsp 代码如下：

```
<%@page language="java" import="java.util.*" pageEncoding="UTF-8"%>
<html>
    <head>
        <title>NullPointExceptionTest 页面测试</title>
    </head>
    <%
```

```
        String str=null;
    %>
    <body>
            <%=str.toString() %><br>
    </body>
</html>
```

程序 error_exception.jsp 代码如下:

```
<%@page language="java" isErrorPage="true" import="java.util.*"
pageEncoding="UTF-8"%>
<!DOCTYPE HTML PUBLIC "-//W3C//DTD HTML 4.01 Transitional//EN">
<html>
    <head>
        <title>Exception 页面</title>
    </head>
    <body>
        <h1>程序出现异常,原因:<%=exception.getMessage() %></h1>
    </body>
</html>
```

配置文件 web.xml 中对异常页面配置代码如下:

```
<error-page>
    <exception-type>java.lang.Exception</exception-type>
    <location>/error_exception.jsp</location>
</error-page>
```

配置信息表明当程序运行过程中产生 Exception 类型异常时,程序跳转到 error_exception.jsp 页面。程序 NullPointExceptionTest.jsp 中字符串对象 str 为 null,此时通过 str 调用其方法将产生空指针异常(NullPointerException),异常类 Exception 为异常类 NullPointerException 的父类,所以当访问 NullPointExceptionTest.jsp 页面时将直接跳转到 error_exception.jsp 页面。程序 error_exception.jsp 中调用内置对象 exception 显示异常信息。

重新部署工程项目并启动 Web 服务器,访问 NullPointExceptionTest.jsp 页面,程序运行结果如图 11-7 所示。

图 11-7 空指针异常页面

11.3 案　　例

本章案例在第 10 章案例的基础上完善"管理员管理"模块中的其他功能，改变综合案例实现的分层模式并为案例增加异常处理，通过本章案例的实现结合之前各章案例将形成一个完整的学生成绩管理系统。案例主要使用 MVC 设计模式、Dao 模式、JavaBean 组件和 Servlet 以及 JSP 技术，JSP 中使用 JSTL 标签、EL 表达式以及 JavaScript 技术来完成数据显示功能。

11.3.1 案例设计

整合之前各章案例，统一采用 MVC 设计模式、Dao 模式、JavaBean 组件和 Servlet 以及 JSP 技术将系统程序框架分为数据库层、Dao 层（数据操作层）、Service 层（业务逻辑操作层）、控制层（Servlet）和视图层（JSP）。Dao 层包括 Dao 接口和 Dao 接口实现类，Service 层包括 Service 接口和 Service 接口实现类。

管理员管理模块功能主要包括管理员信息的增删改查，实现该模块功能需要定义 Dao 层中的接口 AdminIDao.java 和实现类 AdminDaoImpl.java，Service 层接口 AdminIService.java 和实现类 AdminServiceImpl.java，负责添加、删除、修改、查询和显示管理员流程控制的 AddAdminServlet.java、DelAdminServlet.java、UpdateAdminServlet.java、CheckAdminServlet.java 和 ShowAdminServlet.java，以及用来添加管理员、列表显示管理员记录、显示某一管理员详细信息的 JSP 页面 addAdmin.jsp、adminList.jsp 和 adminShow.jsp。除了上述管理员管理模块功能的源代码文件，由于要修改程序分层结构和添加异常处理功能，另外还要修改整个案例的源代码文件，限于篇幅其他源文件代码请参考随书电子资源。

为了给系统用户提供友好的用户界面，在综合案例中需要添加程序异常处理。首先自定义两个异常类 DaoException.java 和 ServiceException.java。Dao 层异常处理办法是将异常以另外一种非检查的方式向上抛出。Service 层的异常处理通常根据实际情况对 Dao 层抛出的异常可以不处理或者进行异常处理，并将本层产生的异常以检查的方式向上抛出。对于整个案例均按照这种异常处理原则进行异常处理。

本章案例使用的主要文件如表 11-1 所示。

表 11-1 本章案例使用的文件

文　件	所在包/路径	功　能
DaoException.java	com.imut.common	案例中所有的 Dao 层数据库操作实现类中的方法抛出的异常类型，该类继承一个非检查型异常类 RuntimeException
ServiceException.java	com.imut.common	案例中所有的 Service 层业务逻辑操作实现类中的方法抛出的异常类型，该类继承一个检查型异常类 Exception
daoException.jsp	/	显示 Dao 层产生的异常信息页面

续表

文件	所在包/路径	功能
serviceException.jsp	/	显示 Service 层产生的异常信息页面
AdminIDao.java	com.imut.javabean.dao	该接口中定义管理员数据库操作的方法
AdminDaoImpl.java	com.imut.javabean.dao	具体实现 AdminIDao 接口定义的增删改查等方法
AdminIService.java	com.imut.javabean.service	该接口中定义管理员业务逻辑操作方法
AdminServiceImpl.java	com.imut.javabean.service	具体实现 AdminIService 接口定义的增删改查等方法
AddAdminServlet.java	com.imut.servlet.base	负责控制添加管理员操作流程的 Servlet
DelAdminServlet.java	com.imut.servlet.base	负责控制删除管理员操作流程的 Servlet
UpdateAdminServlet.java	com.imut.servlet.base	负责控制修改管理员操作流程的 Servlet
CheckAdminServlet.java	com.imut.servlet.base	负责控制列表显示/查询管理员操作流程的 Servlet
ShowAdminServlet.java	com.imut.servlet.base	负责控制显示管理员个人详细信息操作流程的 Servlet
addAdmin.jsp	/base	添加管理员页面
adminList.jsp	/base	管理员信息查询列表显示页面
adminShow.jsp	/base	管理员个人信息显示页面

11.3.2 案例演示

在浏览器地址栏中输入"http://localhost:8080/studentManager/login.jsp",以管理员用户登录,单击"基本数据管理"菜单,进入管理员信息列表显示面,运行效果如图 11-8 所示。在该页面上可以通过单击每条管理员信息记录的"删除"链接实现管理员信息的删除;可以通过页面上的"首页"、"前一页"、"下一页"和"尾页"链接进行分页显示操作;单击"添加管理员"链接,进入如图 11-9 所示的管理员添加页面,在页面上输入管理员相关信息之后单击"提交"按钮即完成管理员信息的添加;单击每条管理员信息记录的"查看/修改"链接,进入如图 11-10 所示的管理员信息查看和修改页面,如果只是查看,查看完成之后单击"返回"链接即返回管理员列表显示页面,如果要修改管理员信息,在页面上直接输入要修改的信息,之后单击"提交"按钮即完成管理员信息的修改。

11.3.3 代码实现

创建工程 studentManager,根据案例设计描述分别给出各部分的具体实现。本章主要给出 Dao 层操作管理员数据库表的接口 AdminIDao.java 和实现类 AdminDaoImpl.java,Service 层管理员业务逻辑操作的接口 AdminIService.java 和实现类 AdminServiceImpl.java,以及负责控制添加管理员信息执行流程的 AddAdminServlet.java,其他源文件代码请参考随书电子资源。

图 11-8 管理员信息列表显示/查询结果显示页面

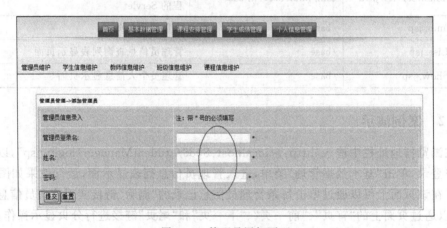

图 11-9 管理员添加页面

图 11-10 管理员信息修改页面

程序 AdminIDao.java 代码如下：

```java
package com.imut.javabean.dao;
import java.util.List;
import java.util.Map;
import com.imut.javabean.entity.Admin;
public interface AdminIDao {
    //添加管理员方法
    public void addAdmin(Admin admin);
    //删除管理员方法
    public void delAdmin(String loginName);
    //修改管理员信息
    public void updateAdmin(Admin admin);
    //根据登录名查找管理员
    public Admin findAdminByLoginName(String loginName);
    //列表显示所有管理员列表
    public Map findAllAdmin(int curPage);
    //多条件查询管理员
     public List findAllAdminByMostCon (String loginName, String name, String
     password);
    //多条件查询管理员
    public Map findAllAdminByMostCon (String loginName, String name, String
    password,int curPage);
    //用户登录验证方法
    public Admin login(String loginName,String password);
}
```

程序 AdminDaoImpl.java 代码如下：

```java
package com.imut.javabean.dao;
import java.sql.Connection;
import java.sql.PreparedStatement;
import java.sql.ResultSet;
import java.sql.SQLException;
import java.sql.Statement;
import java.util.ArrayList;
import java.util.HashMap;
import java.util.List;
import java.util.Map;
import com.imut.commmon.ConnectionFactory;
import com.imut.commmon.DaoException;
import com.imut.commmon.Page;
import com.imut.commmon.ResourceClose;
import com.imut.javabean.entity.Admin;
public class AdminDaoImpl implements AdminIDao {
```

```java
//添加管理员方法
public void addAdmin(Admin admin){
    Connection conn=null;
    PreparedStatement pstmt=null;
    ResultSet rs=null;
    try{
        conn=ConnectionFactory.getConnection();
        String sql="insert into admin values(?,?,?)";
        pstmt=conn.prepareStatement(sql);
        pstmt.setString(1, admin.getLoginName());
        pstmt.setString(2, admin.getName());
        pstmt.setString(3, admin.getPassword());
        pstmt.executeUpdate();
    }catch(Exception e){
        e.printStackTrace();
        throw new DaoException("数据库操作异常,请稍后重试!");
    }finally{
        ResourceClose.close(rs, pstmt, conn);
    }
}

//删除管理员方法
public void delAdmin(String loginName){
    Connection conn=null;
    PreparedStatement pstmt=null;
    ResultSet rs=null;
    try{
        conn=ConnectionFactory.getConnection();
        String sql="delete from admin where loginName=?";
        pstmt=conn.prepareStatement(sql);
        pstmt.setString(1, loginName);
        pstmt.executeUpdate();
    }catch (SQLException e) {
        e.printStackTrace();
        throw new DaoException("数据库操作异常,请稍后重试!");
    }finally{
        ResourceClose.close(rs, pstmt, conn);
    }
}

//修改管理员信息
public void updateAdmin(Admin admin){
    Connection conn=null;
    PreparedStatement pstmt=null;
    ResultSet rs=null;
    try{
```

```java
            conn=ConnectionFactory.getConnection();
            String sql="update admin set loginName=?,name=?,password=? where
            loginName=?";
            pstmt=conn.prepareStatement(sql);
            pstmt.setString(1, admin.getLoginName());
            pstmt.setString(2, admin.getName());
            pstmt.setString(3, admin.getPassword());
            pstmt.setString(4, admin.getLoginName());
            pstmt.executeUpdate();
        }catch (SQLException e) {
            e.printStackTrace();
            throw new DaoException("数据库操作异常,请稍后重试!");
        }finally{
            ResourceClose.close(rs, pstmt, conn);
        }
    }
    //根据登录名查找管理员
    public Admin findAdminByLoginName(String loginName){
        Admin admin=null;
        Connection conn=null;
        PreparedStatement pstmt=null;
        ResultSet rs=null;
        try{
            conn=ConnectionFactory.getConnection();
            String sql="select * from admin where loginName=?";
            pstmt=conn.prepareStatement(sql);
            pstmt.setString(1, loginName);
            rs=pstmt.executeQuery();
            while(rs.next()){
                admin=new Admin();
                admin.setLoginName(rs.getString(1));
                admin.setName(rs.getString(2));
                admin.setPassword(rs.getString(3));
            }
        }catch (SQLException e) {
            e.printStackTrace();
            throw new DaoException("数据库操作异常,请稍后重试!");
        }finally{
            ResourceClose.close(rs, pstmt, conn);
        }
        return admin;
    }
    //列表显示所有管理员列表
    public Map findAllAdmin(int curPage){
```

```java
Admin admin=null;
ArrayList list=new ArrayList();
Connection conn=null;
Statement pstmt=null;
ResultSet rs=null;
ResultSet r=null;
Map map=null;
Page pa=null;
try{
    conn=ConnectionFactory.getConnection();
    String sql="select * from admin order by loginName";
    pstmt= conn. createStatement (ResultSet. TYPE _ SCROLL _ INSENSITIVE,
    ResultSet.CONCUR_READ_ONLY);
    rs=pstmt.executeQuery(sql);
    pa=new Page();                      //声明分页类对象
    pa.setPageSize(5);
    pa.setPageCount(rs);
    pa.setCurPage(curPage);
    r=pa.setRs(rs);
    r.previous();
    for(int i=0;i<pa.getPageSize();i++){
        if(r.next()){
            admin=new Admin();
            admin.setLoginName(r.getString(1));
            admin.setName(r.getString(2));
            admin.setPassword(r.getString(3));
            list.add(admin);
        }else{
            break;
        }
    }
    map=new HashMap();
    map.put("list",list);
    map.put("pa",pa);
}catch (SQLException e) {
    e.printStackTrace();
    throw new DaoException("数据库操作异常,请稍后重试!");
}finally{
    ResourceClose.close(rs, pstmt, conn);
    ResourceClose.close(r, null, null);
}
return map;
}
//多条件查询管理员
```

```java
public List findAllAdminByMostCon (String loginName, String name, String password){
    Admin admin=null;
    ArrayList list=new ArrayList();
    Connection conn=null;
    PreparedStatement pstmt=null;
    ResultSet rs=null;
    //构造多条件查询的 SQL 语句
    String sql="select * from admin where 1=1 ";
    //模糊查询
    if(loginName!=null&&!loginName.equals("")){
        sql+=" and loginName like '%"+loginName+"%'";
    }
    if(name!=null&&!name.equals("")){
        sql+=" and name like '%"+name+"%'";
    }
    if(password!=null&&!password.equals("")){
        sql+=" and password like '%"+password+"%'";
    }
    sql+=" order by loginName";
    try{
        conn=ConnectionFactory.getConnection();
        pstmt=conn.prepareStatement(sql);
        rs=pstmt.executeQuery();
        while(rs.next()){
            admin=new Admin();
            admin.setLoginName(rs.getString(1));
            admin.setName(rs.getString(2));
            admin.setPassword(rs.getString(3));
            list.add(admin);
        }
    }catch (SQLException e) {
        e.printStackTrace();
        throw new DaoException("数据库操作异常,请稍后重试!");
    }finally{
        ResourceClose.close(rs, pstmt, conn);
    }
    return list;
}
//多条件查询管理员
public Map findAllAdminByMostCon (String loginName, String name, String password,int curPage){
    Admin admin=null;
    ArrayList list=new ArrayList();
```

```java
Connection conn=null;
Statement pstmt=null;
ResultSet rs=null;
ResultSet r=null;
Map map=null;
Page pa=null;
//构造多条件查询的 SQL 语句
String sql="select * from admin where 1=1 ";
//模糊查询
if(loginName!=null&&!loginName.equals("")){
    sql+=" and loginName like '%"+loginName+"%'";
}
if(name!=null&&!name.equals("")){
    sql+=" and name like '%"+name+"%'";
}
if(password!=null&&!password.equals("")){
    sql+=" and password like '%"+password+"%'";
}
sql+=" order by loginName";
try{
    conn=ConnectionFactory.getConnection();
    pstmt= conn. createStatement (ResultSet. TYPE_SCROLL_INSENSITIVE,
    ResultSet.CONCUR_READ_ONLY);
    rs=pstmt.executeQuery(sql);
    pa=new Page();                   //声明分页类对象
    pa.setPageSize(5);
    pa.setPageCount(rs);
    pa.setCurPage(curPage);
    r=pa.setRs(rs);
    r.previous();
    for(int i=0;i<pa.getPageSize();i++){
        if(rs.next()){
            admin=new Admin();
            admin.setLoginName(rs.getString(1));
            admin.setName(rs.getString(2));
            admin.setPassword(rs.getString(3));
            list.add(admin);
        }else{
            break;
        }
    }
    map=new HashMap();
    map.put("list",list);
    map.put("pa",pa);
```

```java
        }catch (SQLException e) {
            e.printStackTrace();
            throw new DaoException("数据库操作异常,请稍后重试!");
        }finally{
            ResourceClose.close(rs, pstmt, conn);
            ResourceClose.close(r, null, null);
        }
        return map;
    }
    //用户登录验证方法
    public Admin login(String loginName,String password){
        Admin admin=null;
        Connection conn=null;
        PreparedStatement pstmt=null;
        ResultSet rs=null;
        try{
            conn=ConnectionFactory.getConnection();
            String sql="select * from admin where loginName=? and password=?";
            pstmt=conn.prepareStatement(sql);
            pstmt.setString(1, loginName);
            pstmt.setString(2, password);
            rs=pstmt.executeQuery();
            while(rs.next()){
                admin=new Admin();
                admin.setLoginName(rs.getString(1));
                admin.setName(rs.getString(2));
                admin.setPassword(rs.getString(3));
            }
        }catch (SQLException e) {
            e.printStackTrace();
            throw new DaoException("数据库操作异常,请稍后重试!");
        }finally{
            ResourceClose.close(rs, pstmt, conn);
        }
        return admin;
    }
}
```

程序 AdminIService.java 代码如下：

```java
package com.imut.javabean.service;
import java.util.Map;
import com.imut.commmon.ServiceException;
import com.imut.javabean.entity.Admin;
public interface AdminIService {
```

```java
//添加管理员方法
public void addAdmin(Admin admin) throws ServiceException;
//删除管理员方法
public void delAdmin(String loginName) throws ServiceException;
//修改管理员信息
public void updateAdmin(Admin admin) throws ServiceException;
//根据登录名查找管理员
public Admin findAdminByLoginName(String loginName) throws ServiceException;
//列表显示所有管理员列表
public Map findAllAdmin(int curPage) throws ServiceException;
//多条件查询管理员
public Map findAllAdminByMostCon (String loginName, String name, String password,int curPage) throws ServiceException;
//用户登录验证方法
public Admin login(String loginName,String password) throws ServiceException;
}
```

程序 AdminServiceImpl.java 代码如下：

```java
package com.imut.javabean.service;
import java.util.Map;
import com.imut.commmon.DaoException;
import com.imut.commmon.ServiceException;
import com.imut.javabean.dao.AdminDaoImpl;
import com.imut.javabean.dao.AdminIDao;
import com.imut.javabean.entity.Admin;
public class AdminServiceImpl implements AdminIService {
    private AdminIDao adminDao=new AdminDaoImpl();
    //添加管理员方法
    public void addAdmin(Admin admin) throws ServiceException{
        try{
            adminDao.addAdmin(admin);
        }catch(DaoException e){
            e.printStackTrace();
            throw new DaoException(e.getMessage(),e);
        }
    }
    //删除管理员方法
    public void delAdmin(String loginName) throws ServiceException{
        try{
            adminDao.delAdmin(loginName);
        }catch (DaoException e) {
            e.printStackTrace();
            throw new DaoException(e.getMessage(),e);
        }
```

```java
}
//修改管理员信息
public void updateAdmin(Admin admin) throws ServiceException{
    try{
        adminDao.updateAdmin(admin);
    }catch (DaoException e) {
        e.printStackTrace();
        throw new DaoException(e.getMessage(),e);
    }
}
//根据登录名查找管理员
public Admin findAdminByLoginName(String loginName) throws ServiceException{
    Admin admin=null;
    try{
        admin=adminDao.findAdminByLoginName(loginName);
    }catch (DaoException e) {
        e.printStackTrace();
        throw new DaoException(e.getMessage(),e);
    }
    return admin;
}
//列表显示所有管理员列表
public Map findAllAdmin(int curPage) throws ServiceException{
    Map map=null;
    try{
        map=adminDao.findAllAdmin(curPage);
    }catch (DaoException e) {
        e.printStackTrace();
        throw new DaoException(e.getMessage(),e);
    }
    return map;
}
//多条件查询管理员--带分页功能
public Map findAllAdminByMostCon (String loginName, String name, String
password,int curPage) throws ServiceException{
    Map map=null;
    try{
        map = adminDao. findAllAdminByMostCon (loginName, name, password,
        curPage);
    }catch (DaoException e) {
        e.printStackTrace();
        throw new DaoException(e.getMessage(),e);
    }
    return map;
```

```java
        }
    //用户登录验证方法
    public Admin login(String loginName,String password) throws ServiceException{
        Admin admin=null;
        try{
            admin=adminDao.login(loginName, password);
        }catch (DaoException e) {
            e.printStackTrace();
            throw new DaoException(e.getMessage(),e);
        }
        return admin;
    }
}
```

程序 AddAdminServlet.java 代码如下:

```java
package com.imut.servlet.base;
import java.io.IOException;
import javax.servlet.ServletException;
import javax.servlet.http.HttpServlet;
import javax.servlet.http.HttpServletRequest;
import javax.servlet.http.HttpServletResponse;
import javax.servlet.http.HttpSession;
import com.imut.commmon.DaoException;
import com.imut.commmon.ServiceException;
import com.imut.javabean.entity.Admin;
import com.imut.javabean.service.AdminIService;
import com.imut.javabean.service.AdminServiceImpl;
public class AddAdminServlet extends HttpServlet {
    public void doGet(HttpServletRequest request, HttpServletResponse response)
            throws ServletException, IOException {
        doPost(request, response);
    }
    public void doPost(HttpServletRequest request, HttpServletResponse response)
            throws ServletException, IOException {
        //声明 Service 对象
        AdminIService service=new AdminServiceImpl();
        HttpSession session=request.getSession();
        request.setCharacterEncoding("UTF-8");
        response.setCharacterEncoding("UTF-8");
        String loginName=request.getParameter("loginName");
        String name=request.getParameter("name");
        String password=request.getParameter("password");
        try{
            if(service.findAdminByLoginName(loginName)!=null){
                session.setAttribute("message","用户登录名已存在!");
```

```
                request.getRequestDispatcher("/base/addAdmin.jsp").forward
                    (request,response);
            }else{
                Admin admin=new Admin(loginName,name,password);
                service.addAdmin(admin);
                session.setAttribute("message","管理员信息添加成功!");
                request.getRequestDispatcher("/base/listAllAdminServlet").
                    forward(request,response);
            }
        }catch (ServiceException e) {
            e.printStackTrace();
            request.setAttribute("se", e);
            request.getRequestDispatcher("/serviceException.jsp").forward
                (request,response);
        } catch (DaoException e) {
            e.printStackTrace();
            request.setAttribute("de", e);
            request.getRequestDispatcher("/daoException.jsp").forward
                (request,response);
        }
    }
}
```

【代码分析】Dao 层负责数据操作的 AdminIDao.java 和 AdminDaoImpl.java 分别实现了管理员相关操作的接口和接口实现类,在接口实现类 AdminDaoImpl 的所有方法中对于产生的数据库相关操作 DaoException 异常采用向上抛出异常的方式抛给 Service 层。

Service 层负责业务逻辑处理的 AdminIService.java 和 AdminServiceImpl.java 分别实现了管理员业务逻辑的接口和接口实现类,在接口实现类 AdminServiceImpl.java 的所有方法中对于产生的 ServiceException 异常采用向上抛出异常的方式抛给控制层。

控制层的 AddAdminServlet.java 中调用 Service 中的添加管理员方法 addAdmin (admin),并对 Service 抛出的异常进行处理,根据是否产生异常与产生异常类型的不同结果转向登录成功页面与不同异常结果页面。在转向结果页面之前,将用户信息或者产生异常信息保存在 request 对象中,便于页面显示对应信息。程序不发生异常转向 index.jsp 页面,产生 DaoException.java 或者 ServiceException.java 类型异常将转向 daoException.jsp 页面或者 serviceException.jsp 页面。

习　题

1. 选择题

通过配置 web.xml 实现异常处理,配置的是以下哪个元素?(　　)
A. errorpage　　　　B. errormessage　　　　C. error-page　　　　D. error

2. 填空题

(1) 在 Web 服务器对异常处理是通过 web.xml 配置_____元素来实现的,通过这种方法可以实现两个类型的异常处理,它们分别是_____和_____。

(2) Java 提供了几个异常处理特性,以_____,_____和_____关键字的形式内建于语言自身之中。Java 编程语言也允许用户创建新的异常,并通过使用_____和_____关键字抛出它们。

(3) 若程序运行过程中产生空指针异常(NullPointerException)时,页面会直接跳转到 error.jsp 页面,通过配置文件 web.xml 对异常实现异常类型拦截,请完成以下配置文件内容。

```
<error-page>
    <exception-type>_____</exception-type>
    <location>_____</location>
</error-page>
```

3. 简答题

(1) 简述 Java Web 中处理异常的两种基本方法。

(2) 请描述通过 HTTP 状态码拦截实现异常处理方法。

综合案例使用说明

A.1 数据库安装和导入

首先安装 Oracle 数据库管理系统,然后创建数据库,创建数据库用户,综合案例数据库服务器默认为本机即"localhost"或者"127.0.0.1",数据库名"XE",用户名"webbook",密码"webbook"。如果读者采用不同的数据库服务器、数据库名称和数据库用户,请读者务必修改综合案例工程的源文件"src/com/imut/common/ConnectionFactory.java"中 url、userName 和 password 的属性值,将 url 中"localhost"和"XE"改为所用数据库服务器和数据库名称,userName 和 password 改为数据库用户名和密码。最后将电子资源中数据库脚本文件类下的数据库脚本 student.dmp 导入到数据库即可。

A.2 开发工具的安装和案例工程导入

选用 MyEclipse 10 作为本次案例的开发工具。它的安装较为简单,直接解压 myeclipse 10.zip,双击可执行文件 myeclipse-10.7.1-offline-installer-windows.exe,按照提示,单击 Next 按钮进行安装即可。

MyEclipse10 安装完成后,将项目工程导入到 MyEclipse 中。

(1) 将 Java 工程压缩文件解压到硬盘上,例如 D:\Workspace;

(2) 在 MyEclipse 的 package Explorer 中单击右键,并选择 Impor 命令,进入 Import 对话框;

(3) 在 Import 对话框中选择 Existing projects into Workspace,然后单击 Next 按钮;

(4) 在 Import 对话框中,单击 Browse…按钮,选择所要导入的 Web Project(是一个文件夹,但必须是 Web 工程,否则无法导入),此时 Finish 按钮是不可操作的;

(5) 选中所要导入的 Web Project,同时选中 Copy projects into Workspace,单击 Finish 按钮即可。教材案例分为章节例程和章节综合案例,源代码以 MyEclipse 项目工程的形式分别放置在电子资源中的"程序源代码"文件夹中的子文件夹"章节例程"和"章节综合案例",读者可以直接导入使用,具体参见"程序源代码"文件夹下的"使用说明.txt"。教材最终的综合案例项目"大学生成绩管理系统"工程为电子资源中的 Java Web

工程"程序源代码\章节综合案例\studentManager"。

A.3 综合案例运行

综合案例学生成绩管理系统采用 B/S 架构设计,若运行系统首先应将系统工程部署在 Web 服务器 Tomcat 上,再启动 Tomcat,然后读者便可以通过浏览器访问学生成绩管理系统。Tomcat 服务器为电子资源中的压缩文件 apache-tomcat-7.0.55.rar,解压后通过配置即可使用。

注意事项:

1) 案例中 Web 服务器 Tomcat 所在机器默认为本机即"localhost"或者"127.0.0.1",Tomcat 默认端口号为"8080"。

2) 初始化登录用户:管理员类型用户用户名为"zhangsan",密码为"zs123";教师类型用户用户名为"2143122",密码为"123";学生类型用户用户名为"2012001",密码为"123456"。

参考文献

[1] 夏帮贵. Java Web 开发完全掌握[M]. 北京：中国铁道出版社，2011.
[2] 李兴华，王月清. Java Web 开发实战经典基础篇[M]. 北京：清华大学出版社，2013.
[3] 林信良. JSP&Servlet 学习笔记[M]. 2 版. 北京：清华大学出版社，2013.
[4] Budi Kurniawan. Servlet 和 JSP 学习指南[M]. 崔毅，等译. 北京：机械工业出版社，2013.
[5] 孙卫琴. Tomcat 与 Java Web 开发技术详解[M]. 2 版. 北京：电子工业出版社，2009.
[6] 颜志军. JSP 与 Servlet 程序设计实践教程[M]. 北京：清华大学出版社，2012.
[7] 谢孟军. Go Web 编程[M]. 北京：电子工业出版社，2013.
[8] 朱雪琴，常建功. 亮剑 Java Web 项目开发案例导航[M]. 北京：电子工业出版社，2012.
[9] 刘勇军，韩最蛟. Java Web 核心编程技术[M]. 北京：电子工业出版社，2014.
[10] 明日科技. Java Web 从入门到精通[M]. 北京：清华大学出版社，2014.
[11] 软件开发技术联盟. Java Web 开发实战[M]. 北京：清华大学出版社，2013.
[12] 沈泽刚，秦玉平. Java Web 编程技术[M]. 北京：清华大学出版社，2010.
[13] 厉小军. Web 编程技术[M]. 北京：机械工业出版社，2009.
[14] （美）达科特. Web 编程入门经典——HTML、XHTML 和 CSS[M]. 2 版. 杜静，等译. 北京：清华大学出版社，2010.
[15] 王国辉等. Java Web 入门经典[M]. 北京：机械工业出版社，2013.
[16] 杨树林，胡洁萍. Java Web 应用技术与案例教程[M]. 北京：人民邮电出版社，2011.